T0335571

THE ENDOCANNABINOID SYSTEM

THE ENDOCANNABINOID SYSTEM

Genetics, Biochemistry, Brain Disorders, and Therapy

Edited by

ERIC MURILLO-RODRÍGUEZ, PhD

Universidad Anáhuac Mayab, Mexico

ACADEMIC PRESS

An imprint of Elsevier

Academic Press is an imprint of Elsevier
125 London Wall, London EC2Y 5AS, United Kingdom
525 B Street, Suite 1800, San Diego, CA 92101-4495, United States
50 Hampshire Street, 5th Floor, Cambridge, MA 02139, United States
The Boulevard, Langford Lane, Kidlington, Oxford OX5 1GB, United Kingdom

Library of Congress Cataloging-in-Publication Data
A catalog record for this book is available from the Library of Congress

British Library Cataloguing-in-Publication Data
A catalogue record for this book is available from the British Library

ISBN: 978-0-12-809666-6

For information on all Academic Press publications visit our website at
https://www.elsevier.com/books-and-journals

**Working together
to grow libraries in
developing countries**

www.elsevier.com • www.bookaid.org

Publisher: Mara Conner
Acquisition Editor: Natalie Farra
Editorial Project Manager: Kathy Padilla
Production Project Manager: Julia Haynes
Designer: Matt Limbert

Typeset by TNQ Books and Journals

CONTENTS

LIST OF CONTRIBUTORS

John C. Ashton
University of Otago, Dunedin, New Zealand

Balapal S. Basavarajappa
New York State Psychiatric Institute, New York, NY, United States; Columbia University, New York, NY, United States; Nathan Kline Institute for Psychiatric Research, Orangeburg, NY, United States

Natalia Battista
University of Teramo, Teramo, Italy

Megan J. Dowie
University of Auckland, Auckland, New Zealand

Andrea Giuffrida
University of Texas Health Science Center, San Antonio, TX, United States

Michelle Glass
University of Auckland, Auckland, New Zealand

Bernard Le Foll
Centre for Addiction and Mental Health (CAMH), Toronto, ON, Canada; University of Toronto, Toronto, ON, Canada

F. Markus Leweke
Heidelberg University, Mannheim, Germany

Mauro Maccarrone
Campus Bio-Medico University of Rome, Rome, Italy; European Center for Brain Research/IRCCS Santa Lucia Foundation, Rome, Italy

Alex Martinez
University of Texas Health Science Center, San Antonio, TX, United States

Cathrin Rohleder
Heidelberg University, Mannheim, Germany

Jose M. Trigo
Centre for Addiction and Mental Health (CAMH), Toronto, ON, Canada

PREFACE

In January 2017, the National Academy of Sciences, Engineering and Medicine (NASEM) of the United States published a new report on the health effects of cannabis (http://nationalacademies.org/hmd/reports/2017/health-effects-of-cannabis-and-cannabinoids.aspx). This 500-page document provides an overview of the research of the past 15 years on the health impacts of cannabis and cannabis-derived products, covering a vast amount of scientific material—more than 10,000 published articles—in a rigorous and balanced manner. The report discusses the health risks associated with cannabis use as well as its potential therapeutic effect, and draws conclusions on almost 100 highly relevant topics, ranging from lung cancer and diabetes to schizophrenia and addiction. The sheer size and span of this document bears testimony to the growing importance of its subject matter. For many years now, cannabis has been the most popular illicit drug in the United States and Europe. A recent nationwide survey estimates that 22.2 million Americans aged 12 years and older consumed cannabis in the past 30 days, in the vast majority of cases, for recreational purposes. Although still relatively limited, medical use will undoubtedly expand in the near future as acceptance and accessibility of cannabis as a medicine increase. Basic and clinical research has grown in parallel with public and medical interest, such that more than 24,000 scientific articles have been published on cannabis and the cannabinoids since publication of the previous NASEM report in 1999.

Far from writing the last word on the health impact of cannabinoids, the new NASEM review emphasizes the need to better understand how cannabis and its chemical constituents interact with the human body, and to consolidate new data gathered through animal and human research into solid scientific knowledge. The present book, edited by Dr. Eric Murillo-Rodríguez (at the University Anàhuac Mayab, Merida, Mexico), aims to address those very same needs. A distinguished group of authors, each an expert in their discipline, tackle a broad set of topics centered around the endogenous cannabinoid complex, the signaling system that mediates the action of Δ^9-tetrahydrocannabinol in cannabis.

The book starts with an overview of the biochemical mechanisms through which the endocannabinoid transmitters—anandamide and 2-arachidonoylglycerol (2-AG)—are produced and eliminated. Drs. Natalia Battista (University of Teramo, Italy) and Mauro Maccarrone (University of

Rome, Italy), outline the distinct enzyme pathways involved in the formation and degradation of these lipid-derived substances and highlight key knowledge gaps that remain to be filled to arrive at a full molecular understanding of endocannabinoid signaling. In the following chapter, Dr. Balapal Basavarajappa (New York State Psychiatric Institute and Columbia University, New York, USA) provides an introduction to CB_1 and CB_2 cannabinoid receptors, the G-protein-coupled receptors targeted by exogenous and endogenous cannabinoids. The chapter also describes the neuroanatomical localization of these receptors, along with the intracellular signaling cascades recruited by them.

After these two introductory chapters, the book develops into a series of small monographs, each centered on the roles played by endocannabinoid signals in human pathological conditions. Drs. Andrea Giuffrida and Alex Martinez (both at the University of Texas Health Science Center, San Antonio, USA) summarize current views about the contribution of anandamide and 2-AG to Parkinson's disease. This balanced résumé of current knowledge is particularly welcome at a time when anecdotal evidence spread through social media has fueled public interest (and speculation) on the possible therapeutic utility of cannabis in this neurodegenerative disorder (https://www.youtube.com/watch?v=zNT8Zo_sfwo).

In the next chapter, Drs. Markus Leweke and Cathrin Rohleder (both at the Central Institute on Mental Health, Heidelberg, Germany) review the important and complex issue of the link between cannabis use and schizophrenia. The NASEM report devotes a substantial amount of space to this issue, which remains hard to disentangle after 30 years of intense epidemiological research. The thoughtful perspective offered by Drs. Leweke and Rohleder is that the pathophysiology of schizophrenia may be underpinned by a deficiency in anandamide-mediated signaling at CB_1 receptors, which implies that agents that heighten the actions of anandamide might be protective in this disorder. In Chapter 5, Dr. Michelle Glass and her collaborators (at the Universities of Otago and Auckland, New Zealand) take a broader view of the pathological roles of the endocannabinoid system and discuss multiple possible implications of this signaling complex in cognitive and motor dysfunction as well as in mood disorders. Lastly, Dr. Bernard Le Foll (University of Toronto, Canada) closes this interesting aperçu of endocannabinoid physiopathology with a thought-provoking section on the role of endocannabinoid signals in drug addiction. Summarizing the available preclinical evidence, Dr. Le Foll concludes that CB_1 receptors serve an important modulatory function in drug-seeking behavior, and that elevating anandamide levels in the brain (for example, by preventing its enzyme-mediated degradation) may attenuate drug seeking.

In sum, this new book offers an interesting bird's-eye view of the many possible roles played by the endocannabinoid signaling system in brain psychopathology. Read alongside the 2007 NASEM report, it will represent a useful teaching and learning tool for students and scientists who enter this exciting and important field of investigation.

Daniele Piomelli
Departments of Anatomy and Neurobiology
Pharmacology and Biological Chemistry
University of California, Irvine

ACKNOWLEDGMENTS

I would like to express my gratitude to my family, with special love to my sister Linda and beautiful niece Shaula.

Thanks to my friends, colleagues, and students for sharing my happiness when starting this project.

An extra thanks to collaborators for your trust in the project, as well as for your magnificent contributions to this book! Mauro, Andrea, and Daniele grazie per il vostro sostegno! Markus dank für ihre unterstützung!

Thanks to Natalie Farra, Kathy Padilla, and all Elsevier staff for believing in this project as well as for your support and patience. I would like to offer special thanks to the many people who provided support and assistance in editing, proofreading, and design.

Last but not the least: I beg forgiveness of all those who have been with me over the course of the years and whose names I have failed to mention.

Eric Murillo-Rodríguez

INTRODUCTION

The endocannabinoid system includes endogenous cannabinoidlike compounds such as anandamide; metabotropic receptors named CB_1 and CB_2; enzymes like fatty acid amide hydrolase, which synthesizes anandamide; as well as the anandamide membrane transporter.

Significant advances have been achieved by studying the physiological properties of the endocannabinoid system. The spectrum of evidence of the role of the endocannabinoid system includes neurobiological areas from gene expression to putative modulation of health issues.

At present, observations confirm that the endocannabinoid system displays critical functions by controlling a wide spectrum of physiological conditions. This book highlights the current evidence of the synthesis and degradation of endocannabinoids, as well as the role of the endocannabinoid system in neurodegenerative illness such as Parkinson disease. Moreover, the relevant function of the endocannabinoid system in mental disorders, including schizophrenia, and its critical role in human brain functions such as memory, motor behavior, and mood pathologies is highlighted and remarks on the current experimental evidence regarding the link between the endocannabinoid system and addiction are provided.

A number of world-leading scientists who are experts in the endocannabinoid area agreed to contribute to the book by providing current evidence of the relevance of the endocannabinoid system in multiple neurobiological functions. Thus this book deals with various aspects of the endocannabinoid system, from phenomena to molecular processes. I am sincerely grateful to all the contributors for their scientific contribution.

Eric Murillo-Rodríguez, PhD
Laboratorio de Neurociencias Moleculares e Integrativas
Escuela de Medicina, División Ciencias de la Salud
Universidad Anáhuac Mayab
Mérida, Yucatán, Mexico

CHAPTER 1

Basic Mechanisms of Synthesis and Hydrolysis of Major Endocannabinoids

Natalia Battista[1], Mauro Maccarrone[2,3]
[1]University of Teramo, Teramo, Italy; [2]Campus Bio-Medico University of Rome, Rome, Italy; [3]European Center for Brain Research/IRCCS Santa Lucia Foundation, Rome, Italy

ENDOCANNABINOIDS AND THEIR ANALOGS

The isolation of the psychoactive ingredient of cannabis (*Cannabis sativa*), Δ^9-tetrahydrocannabinol, back in 1964 led almost 30 years later to the discovery of the endogenous lipophilic molecules, collectively termed endocannabinoids (eCBs), able to activate the same two G-protein-coupled receptors, type 1 (CB_1) and type 2 (CB_2) cannabinoid receptors (Maccarrone et al., 2015).

On the basis of their main role in controlling biological processes both in human health and disease, *N*-arachidonoylethanolamine (anandamide) (AEA) and 2-arachidonoylglycerol (2-AG) are presently recognized as the two main members of this family (Fig. 1.1). Other important ω-6 (n-6) fatty acid compounds with cannabimimetic properties, such as *N*-arachidonoyldopamine (NADA), 2-arachidonoylglycerylether (noladin ether), and *O*-arachidonoylethanolamine (virodhamine), have also been listed among these bioactive lipids, although their pharmacology and biological relevance remain to be clarified (Fezza et al., 2014). In addition, in the past 5 years it has been demonstrated that eCBs are closely interconnected to docosahexaenoic acid (DHA) throughout the eicosanoid pathway. Indeed, *N*-docosahexaenoylethanolamine, a derivative of DHA that is thought to be produced by the same pathway that generates AEA, acts as an eCB not only in its ability to activate CB receptors, but also in its susceptibility to be metabolized by the catabolic enzyme fatty acid amide hydrolase (FAAH) (Brown et al., 2010; Cascio, 2013). eCB congeners [i.e., *N*-palmitoylethanolamine (PEA), *N*-oleoylethanolamine (OEA), and *N*-stearoylethanolamine] that exert their effects independently

The Endocannabinoid System
ISBN 978-0-12-809666-6
http://dx.doi.org/10.1016/B978-0-12-809666-6.00001-0

1

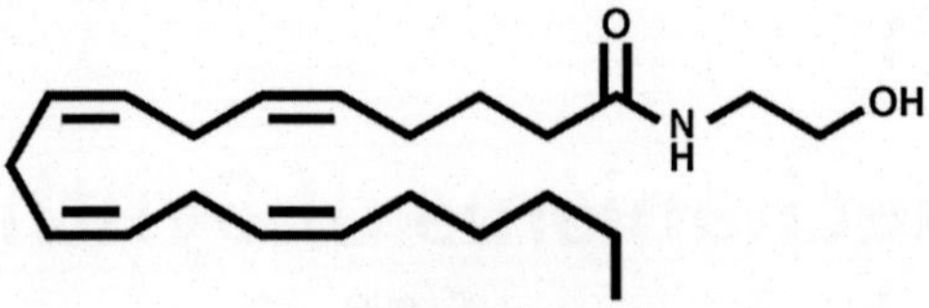

Anandamide

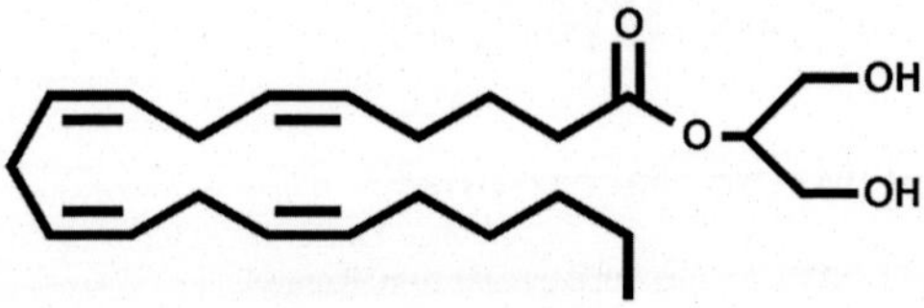

2-arachidonoylglycerol

Figure 1.1 Chemical structures of AEA and 2-AG.

of CB receptors are commonly classified as "endocannabinoid-like" compounds (Ben-Shabat et al., 1998; Costa, Comelli, Bettoni, Colleoni, & Giagnoni, 2008; Fezza et al., 2014; Ho, Barrett, & Randall, 2008). Indeed, these molecules might significantly contribute to an entourage effect that prevents true eCBs from being degraded by specific metabolic enzymes, or from allosterically modulating receptor binding (Fezza et al., 2014).

The knowledge of metabolic pathways that regulate the endogenous tone of eCBs, as well as the discovery of the proteins that bind them, shed light on the critical functions of these molecules at central (Maccarrone, Guzmán, Mackie, Doherty, & Harkany, 2014) and peripheral levels (Maccarrone et al., 2015). The most relevant proteins identified to date in controlling eCB tone will be described in the following sections.

N-ARACHIDONOYLETHANOLAMINE METABOLISM: SYNTHESIS

The classical dogma that eCBs are synthesized and released "on demand" via hydrolysis of cell membrane phospholipid precursors has been revisited on the basis of unexpected evidence for intracellular reservoirs and transporters of eCBs. These new entities have been shown to drive intracellular trafficking of eCBs, adding a new dimension to the regulation of their biological activity (Maccarrone, Dainese, & Oddi, 2010). However, several routes have been proposed to explain the metabolic pathways of AEA biosynthesis (Ueda, Tsuboi, & Uyama, 2013).

The possibility to synthesize AEA in vitro by a simple condensation of arachidonic acid (AA) and ethanolamine catalyzed by a reverse FAAH or an AEA hydrolase "working in reverse" was not feasible in vivo (Kurahashi, Ueda, Suzuki, Suzuki, & Yamamoto, 1997; Ueda, Kurahashi, Yamamoto, Yamamoto, & Tokunaga, 1996). Indeed, the substrate concentrations required to form AEA are much higher than those actually detected in cells, and, nowadays, it is accepted that the generation of AEA in cells and tissues occurs mainly as a result of the hydrolysis of a minor membrane phospholipid, N-arachidonoylphosphatidylethanolamine (NArPE) (Di Marzo et al., 1994). Although an ever-growing number of enzymes have been ascribed to AEA biosynthesis (Cascio & Marini, 2015), the orchestrated and sequential action of N-acyltransferase (NAT) and N-acyl-phosphatidylethanolamine (NAPE)-specific phospholipase D (NAPE-PLD) is believed to be the most relevant pathway to generate AEA. Indeed, NAT acts by transferring AA from the sn-1 position of 1,2-sn-di-arachidonoylphosphatidylcholine (PC) to phosphatidylethanolamine (PE), thus generating NArPE.

The latter compound is next cleaved to yield AEA and phosphatidic acid (PA) by NAPE-PLD. The enzyme is encoded by the $Nape$-pld gene on chromosome 5, position 9.97 cM (21662901–21701396 bp, —strand) (Zimmer, 2015). The transcript has five exons. Exons 2–5 contain the open reading frame of the 396-amino acid NAPE-PLD protein. This 46-kDA protein belongs to the zinc metallohydrolase family of the metallo-β-lactamase fold, is characterized by a highly conserved residues involved in the binding of substrates, is stimulated by divalent cations such as Mg^{2+} and Ca^{2+}, and is involved in the formation of other cannabinoid-receptor-inactive, N-acylethanolamines (NAE) (Liu, Tonai, & Ueda, 2002; Okamoto, Morishita, Tsuboi, Tonai, & Ueda, 2004; Petersen & Hansen, 1999; Ueda, Liu, & Yamanaka, 2001). Thanks to the generation of NAPE-PLD-deficient mice, alternative pathways to form NAE in brain have been suggested (Leung et al., 2006; Simon & Cravatt, 2010). Simon and Cravatt (2006, 2008) proposed an NAPE-PLD-independent route, which includes a double-O-deacylation of NAPE via N-acyl-lyso-PE and further hydrolysis of the resultant glycerophospho-N-acylethanolamine to NAE. This multistep pathway, previously hypothesized by other groups (Natarajan, Schmid, Reddy, & Schmid, 1984; Sun et al., 2004), requires the involvement of a fluorophosphonate-sensitive serine hydrolase and a metal-dependent phosphodiesterase, identified, respectively, as α/β-hydrolase 4 (Abh4) (Simon & Cravatt, 2006) and glycerophosphodiesterase (GDE) 1 (Simon & Cravatt, 2008). Notably, as a lysophospholipase substrate, Abh4 prefers lyso-NAPE to other lysophospholipids such as lyso-PE, lyso-PC, and lysophosphatidylserine (LPS) (Simon & Cravatt, 2006).

On the other hand, the integral membrane GDE1 possesses Mg^{2+}-dependent phosphodiesterase activity selectively hydrolyzing glycero-phosphoinositol (Simon & Cravatt, 2008). Interestingly, double-knockout mice of NAPE-PLD and GDE1 did not exhibit a notable decrease in NAE levels, suggesting the existence of other biochemical routes (Simon & Cravatt, 2010).

Two additional pathways for AEA biosynthesis include (1) sequential hydrolysis of NArPE to phospho-AEA by a phospholipase C (PLC), followed by a dephosphorylation reaction catalyzed by a nonreceptor protein tyrosine phosphatase, identified as PTPN22, to produce AEA (Liu et al., 2006); and (2) hydrolysis of NArPE to N-arachidonoyl-lyso-PE (lyso-NArPE), which in turn is converted to AEA by a lyso-PLD (Sun et al., 2004).

Incidentally, it should be mentioned that although the production of NAPE by N-acylation of PE is thought to be the rate-limiting step of NAE biosynthesis, the ability to generate the AEA precursor should be equally considered. Several evidence reported that this step might be ascribed also to a calcium-independent NAT, designated as rat LRAT-like protein (RLP-1) or more commonly as iNAT, that removes a fatty acyl group from both the sn-1 and sn-2 positions (where AA is most abundant) of PC, which is the acyl donor, producing the aforementioned substrate (Jin et al., 2007, 2009).

In the past years, it has been proposed that also N-acylated plasmalogen-type ethanolamine phospholipid (N-acyl-plasmenylethanolamine, pNAPE) might be converted to NAE through NAPE-PLD-dependent (Petersen, Pedersen, Pickering, Begtrup, & Hansen, 2009; Schmid, Reddy, Natarajan, & Schmid, 1983) and independent pathways (Tsuboi et al., 2011). The latter requires a preliminary deacetylation reaction, probably catalyzed by esterases including phospholipase A_2 ($sPLA_2$)e Abh4, to form a lyso-pNAPE, which is then hydrolyzed to generate NAEs. The identity of the enzyme involved in the last step has to be still ascertained, although the lyso-PLD activity in the brain might be in part attributed to GDE1 (Simon & Cravatt, 2008; Tsuboi et al., 2011). Based on the sequence similarity and the tissue distribution between GDE1 and GDE4, it has been lately suggested that GDE4 might also be responsible for NAE-generating lyso-PLD activities with preferred substrate specificity for N-palmitoyl-lysoplasmenylethanolamine (Tsuboi et al., 2015).

It has been reported that five members of the HRAS-like suppressor (HRASLS) family, that are better known as tumor suppressors, possess phospholipid-metabolizing activities including NAPE-forming NAT activity, and it has been proposed to call them HRASLS1–5 PLA/acyltransferase (PLA/AT)-1–5. Among the five members, PLA/AT-1 shows high NAT activity and is mainly expressed in human and murine testis, skeletal muscle, brain, and heart, suggesting this enzyme as mammalian source of NAPE substrates (Uyama et al., 2012, 2013).

N-ARACHIDONOYLETHANOLAMINE METABOLISM: DEGRADATION

The enzymatic breaking down of the amide bond of AEA is mainly ascribed to the action of an FAAH (Cravatt et al., 1996). FAAH is the best characterized enzyme involved in the degradation of AEA and its cloning, crystal structure, kinetic properties, and distribution in the body have been the subject of many interesting reviews (Cravatt & Lichtman, 2003; Fezza, De Simone, Amadio, & Maccarrone, 2008; Maccarrone, 2006; McKinney & Cravatt, 2005).

The *Faah* gene is located on chromosome 4 at position 53,08 cM (115967145–116017926 bp, − strand), has at least 15 exons and produces a transcript of approximately 3.8 kb (Zimmer, 2015). FAAH contains 597 amino acids, corresponding to a molecular weight of ~60 kDa; the optimal pH working value is between 8 and 10, and its distribution in the mammalian brain as well as in peripheral tissues (except for the heart or lung) (Giang & Cravatt, 1997), is overlapping with CB_1 distribution in the same organs (Fezza et al., 2008). This enzyme belongs to a protein family called "amidase signature (AS)," appears as a homodimer bound to membrane lipids via α-helices 18 and 19, exists as a dimer in solution (McKinney & Cravatt, 2005), and is characterized by an unusual serine-serine-lysine (S241-S217-K142) catalytic triad and several channels that allow this enzyme to integrate into cell membranes and establish direct access to the bilayer from its active site (Bracey, Hanson, Masuda, Stevens, & Cravatt, 2002). Circular dichroism and other spectroscopic techniques unveiled the structural features that determine FAAH activity and its membrane-binding properties (Mei et al., 2007). In this context, it has been reported that the structure, subcellular localization, and activity of FAAH are modulated by membrane lipids (Dainese et al., 2014). Indeed, FAAH membrane-binding affinity and enzymatic activity have been shown experimentally to be higher within membranes containing cholesterol and anandamide.

Additionally, the colocalization of cholesterol, AEA, and FAAH in a cell line of mouse neuroblastoma (SH-SY5Y) suggested a mechanism through which cholesterol increases the substrate accessibility of FAAH (Dainese et al., 2014). FAAH displays broad substrate selectivity and hydrolyzes also a variety of fatty acid amides and a large number of monounsaturated and saturated compounds (Fegley et al., 2005; Ho & Hillard, 2005; Lo Verme et al. 2005; Ueda, Yamanaka, & Yamamoto, 2001), such as NADA, PEA (De Petrocellis, Davis, & Di Marzo, 2001; Lo Verme et al., 2005; Ueda, Yamanaka et al., 2001), OEA, N-arachidonoylserine, and N-arachidonoylglycine, the latter two also being able to act as FAAH inhibitors (Bradshaw & Walker, 2005; Ho & Hillard, 2005; Sheskin, Hanus, Slager, Vogelm, & Mechoulamm, 1997). Interestingly, genetic manipulation of *Faah* gene allowed to generate not only FAAH knockout mice but also FAAH knockdown mice, where AEA hydrolase is deleted from peripheral tissues only (Cravatt et al., 2004). Studies performed on these mice allelic variants highlighted a different susceptibility to drug/alcohol abuse (Blednov, Cravatt, Boehm, Walker, & Harris, 2007; Chiang, Gerber, Sipe, & Cravatt, 2004; Cravatt et al., 2001; Sim, Hatim, Reynolds, & Mohamed, 2013; Sipe, Chiang, Gerber, Beutler, & Cravatt, 2002; Zhou, Huang, Lee, & Kreek, 2016), supporting a potential link between functional abnormalities in eCB signaling and drug/alcohol dependence.

When FAAH is inhibited, another isoform of FAAH, also showing AS and called FAAH-2, cleaves AEA in an alternate route (Wei, Mikkelsen, McKinney, Lander, & Cravatt, 2006). In contrast to FAAH-1, FAAH-2 exhibits a much higher hydrolytic activity with monounsaturated acyl chains and a lower selectivity for polyunsaturated acyl chains (like AEA, that is, C20:4). Interestingly, protease protection assays documented a different cell membrane orientation of these two isozymes. In particular, the C-terminal domain has a cytoplasmic orientation in FAAH-1, whereas it is luminally leaned in FAAH-2, thus suggesting a different contribution of the two enzymes to fatty acid amide catabolism in vivo. Alternatively, AEA can be also degraded by NAE-hydrolyzing acid amidase (NAAA), a cysteine hydrolase that belongs to the N-terminal nucleophile (Ntn) family of enzymes (Tsuboi et al., 2005, 2007; Ueda, Tsuboi, & Uyama, 2010). Molecular cloning demonstrated that NAAA is identical to an acid ceramidase-like protein (Hong et al., 1999; Tsuboi et al., 2005), whose human gene is located in the region of 4q21.1, and contains 11 exons and 10 introns (Hong et al., 1999). NAAA is mainly localized to the lysosomal cellular compartment (Tsuboi, Takezaki, & Ueda, 2007), bears a significant degree of sequence homology with the choloylglycine hydrolases, and shows a strong preference for saturated fatty acid ethanolamides such as PEA (Solorzano et al., 2009; Tsuboi et al., 2005). Like other Ntn enzymes, NAAA is activated by autoproteolysis and displays catalytic activity, due to the cysteine 126 (Cys-126) residue, at acidic pH (Wang et al., 2008). Additionally, cyclooxygenase-2 (COX-2), different lipoxygenase (LOX) isozymes, as well as cytochrome P450 might convert AEA into eicosanoid derivatives like prostaglandin ethanolamides (Hermanson, Gamble-George, Marnett, & Patel, 2014; Kozak et al., 2002; Rouzer & Marnett, 2011), epoxyeicosatrienoic acid ethanolamides (Snider, Walker, & Hollenberg, 2010), and hydroxy-anandamides (Rouzer & Marnett, 2011, van der Stelt et al., 2002).

2-ARACHIDONOYLGLYCEROL METABOLISM: SYNTHESIS

The most important biosynthetic precursors of 2-AG are diacylglycerols (DAGs), which are generated either from the hydrolysis of inositol phospholipids (PI) by a specific PLC or from the hydrolysis of PA by a PA-phosphohydrolase (Bisogno, Melck, De Petrocellis, & Di Marzo, 1999). DAGs are then converted into 2-AG by a sn-1-DAG lipase (DAGL).

Biochemical studies led to the purification, cloning, and characterization of the two sn-1 DAG isoforms: DAGLα and DAGLβ (Bisogno et al., 2003). These two isoforms of this enzyme, DAGLα and DAGLβ, are encoded by different genes (*Dagla* and *Daglb*) located on chromosomes 19 and 5, respectively (*Dagla*: Chr19 6.55 cM, 10245265–10304877 bp, − strand; *Daglb*: Chr5: 82.19 cM, 143472947–143504442 bp, + strand) (Zimmer, 2015). DAGLα, containing 1042 amino acids and with a molecular mass of ~120 kDa, is larger than DAGLβ, which has 672 amino acids and a molecular weight of ~70 kDa. DAGLs have a four transmembrane domain followed by the typical catalytic domain of serine hydrolases; they are plasma membrane proteins that are stimulated by glutathione, are sensitive to but independent of intracellular concentration of Ca^{2+}, and are characterized by a different substrate selectivity. Studies performed by using DAGLα- or DAGLβ-deficient mice highlighted that both enzymes are responsible for most of 2-AG content in the brain, spinal cord, liver, and other tissues, although they differently contribute to 2-AG production within each tissue (Gao et al., 2010). However, there is a general consensus that the former is essential in controlling retrograde synaptic plasticity (Tanimura et al., 2010), depolarization-induced suppression of inhibition in prefrontal cortex neurons (Yoshino et al., 2011), and adult neurogenesis (Gao et al., 2010).

Interestingly, it has been found that the spatial and temporal expression of DAGLs is overlapping with that of CB_1 within the central nervous system, with the CB_1 receptor restricted to the presynaptic terminal and DAGLα to the complementary postsynaptic site, pointing to the importance of DAGL-dependent eCB signaling in brain functioning (Reisenberg, Singh, Williams, & Doherty, 2012).

Much alike NAT, PLC (in particular Cβ) plays a key role in the biosynthesis of 2-AG precursor. Indeed, since its activity is Ca^{2+} dependent, it has been supposed that the enzyme responsible for DAG biosynthesis represents the rate-limiting step in the 2-AG in vivo production.

Also, a secondary pathway for 2-AG synthesis that involves cleavage of the PI precursor by a phospholipase A, followed by hydrolysis of the phosphate ester bond by a lyso-PLC, has been proposed (Bisogno et al., 1999; Carrier et al., 2004), yet its physiological relevance in vivo remains unclear.

2-ARACHIDONOYLGLYCEROL METABOLISM: DEGRADATION

The fate of 2-AG is much similar to that of AEA. Indeed, 2-AG might be hydrolyzed to AA and glycerol by a specific monoacylglycerol lipase (MAGL), yet it can also be oxidized by COX-2 (but not by COX-1), LOXs, and cytochrome P450 enzymes, to generate the corresponding hydroxy-leicosatetraenoyl-glycerols (in the case of LOXs) (van der Stelt et al., 2002), epoxy derivatives (in the case of cytochrome P450), or glyceryl esters (in the case of COXs) (Kozak, Rowlinson, & Marnett, 2001, 2000). The *Mgl* gene encoding MAGL is located on mouse chromosome 6 (39,51 cM, position 88724412–88828360 bp, + strand) (Zimmer, 2015). MAGL is a 33-kDa protein that belongs to the serine hydrolase family, is widely expressed throughout the brain, is localized in the cytosolic compartment of the cell, and often colocalizes with CB_1 receptors in the axon terminals (Savinainen, Saario, & Laitinen, 2012). Studies performed on mouse BV-2 microglial cells, as well as in primary microglia cultures, allow to identify a novel MAGL enzyme (Muccioli et al., 2007). These data led researchers to suppose that the hydrolysis of 2-AG is mainly (~85%) controlled by two MAGLs, differently expressed. Interestingly, while 2-AG levels increase in mice lacking MAGL, on the other hand, 2-AG content is not affected in mice lacking FAAH, suggesting that the complementary localization (pre- and postsynaptic, respectively) of these eCB-hydrolyzing enzymes may be crucial to control the different role played by AEA and 2-AG in healthy and diseased brain.

In addition, two integral membrane proteins belonging to the α/β-hydrolase superfamily, and called α/β-hydrolase-domain-containing protein-6 (ABHD6) and ABHD12, are responsible for the remaining 15% of 2-AG hydrolytic activity. In this context, it has been demonstrated that the suppression of α/β hydrolase domain may be involved both in energy metabolism, pointing to this protein as an interesting therapeutic target in preventing obesity and type 2 diabetes (Zhao et al., 2014, 2016). Indeed, ABHD6, by controlling signal competent 1-MAG levels, positively regulates white adipose browning and food intake process (Zhao et al., 2016), whereas it behaves as a negative modulator of glucose-stimulated insulin secretion in the β cells (Zhao et al., 2014).

On the other hand, ABHD12 is responsible for the degradation of several LPS species (Blankman, Simon, & Cravatt, 2007), and the deregulation of LPS pathway might contribute to the pathogenesis of the autosomal recessive neurodegenerative disease called PHARC (polyneuropathy, hearing loss, ataxia, retinitis pigmentosa, and cataracts), as revealed in experimental studies by using ABHD12 knockout (ABHD12$^{-/-}$) mice (Blankman, Long, Trauger, Siuzdak, & Cravatt, 2013; Fiskerstrand et al., 2010).

All metabolic pathways of AEA and 2-AG are schematically depicted in Fig. 1.2.

MANIPULATION OF THE ENDOCANNABINOID METABOLISM BY PHARMACOLOGICAL TOOLS

The involvement of eCBs in several biological processes supports the idea that alterations of circulating eCB levels is involved in the dysregulation of activity of important physiological systems. In this context, the ability of selective natural or synthetic compounds to inhibit all enzymes described in this chapter and, thus to modulate the metabolic networks regulating eCB actions, has been gaining particular attention. Indeed, selective FAAH and MAGL inhibitors have been shown to tease apart some of the undesirable effects of CB$_1$ activation, opening the gate to the development of eCB-based drugs as therapeutics for the prevention/treatment of a variety of diseases in which a dysregulation of the eCB system is apparent (Fowler, 2015).

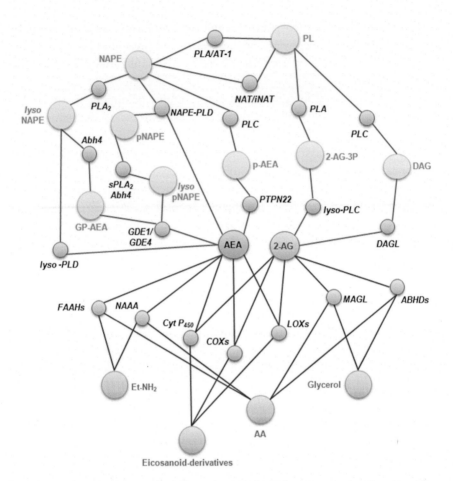

Figure 1.2 Metabolic network of AEA and 2-AG, which shows interactions between biosynthetic (*blue line*) and hydrolytic (*red line*) enzymes. Enzymes and metabolites (detailed in the text) are the *colored circles*, and mutual interactions among them are the lines.

Despite the lack of a selective inhibitor for NAPE-PLD, a remarkable series of potent, specific, and efficacious inhibitors of FAAH have been developed to understand its structure and mechanism of hydrolysis, and to test its possible use in medicine (Bisogno & Maccarrone, 2013). Serine hydrolase inhibitors, including fluorophosphonate and trifluoromethyl ketones, the O-aryl carbamate URB597, the α-ketoheterocycle compound OL-135, the piperidine urea PF-750, the piperazine urea TAK21d (Kono et al., 2014), TC-F2 (Gustin et al., 2011), and SA 57 (Niphakis, Johnson, Ballard, Stiff, & Cravatt, 2012), are some of the most widely used drugs in experimental research aimed at targeting FAAH (Tuo et al., 2016). Big pharmaceutical companies have been evaluating FAAH inhibitors in clinical trials for the treatment of pathologies, such as osteoarthritis pain, depression, Tourette syndrome, and posttraumatic stress disorder.

Several DAGL inhibitors have been identified and used to underline the mechanisms of 2-AG synthesis in cells and tissues. The three lead compounds include tetrahydrolipstatin (THL, Orlistat), RHC80267, and O-3841, which inhibit human DAGLα with apparent IC_{50} values of 60 nM, 60 nM, and 160 nM, respectively (Bisogno et al., 2006; Hoover, Blankman, Niessen, & Cravatt, 2008). THL and RHC-80267, when used at concentrations lower than those required to inhibit other lipases, are able to block DAGL activity without affecting neither MAGL or NAPE-PLD (Bisogno et al., 2006; Lee, Kraemer, & Severson, 1995). However, the use of THL in vivo is limited because of its irreversible binding to DAGL, whereas RHC-80267 interferes with CB_1-mediated effects due to the administration of exogenous 2-AG (Hashimotodani, Ohno-Shosaku, Maejima, Fukami, & Kano, 2008). New analogs of THL and OMDM-188 have been developed, and structure–activity relationship (SAR) studies allow to highlight their ability to inactivate both human and murine DAGLs at nanomolar concentrations (Johnston et al., 2012). The compound O-3841 shows high selectivity for DAGLα over the other proteins of the endocannabinoid system (ECS), but is not suitable for systemic use in vivo due to the lack of stability and poor permeability through the plasma membrane. This problem was partially solved by the design of O-5596, a more stable and slightly more potent and cell-membrane-permeable DAGL inhibitor (Bisogno et al., 2009). A new series of fluorophosphates compounds, O-7458, O-7459, and O-7460, were synthesized, and their pharmacological characterization has been proved both in vitro and in vivo, particularly in the framework of eCB function in the control of food intake and energy homeostasis (Bisogno et al., 2013). Lately, by using SAR and activity-based protein profiling studies, a glycine sulfonamide derivative, LEI-106, has been identified that acts as a potent, dual DAGLα/ABHD6 inhibitor, with promising pharmacological properties for the treatment of diet-induced obesity and metabolic syndrome (Janssen et al., 2014).

General serine hydrolase inhibitors, such as methyl arachidonoylfluorophosphonate, phenylmethanesulfonyl fluoride, as well as sulfhydryl-specific inhibitors, such as 4-chloromercuribenzoic acid and N-ethylmaleimide, are able to attenuate the hydrolyzing action of MAGL. The most preferably used selective MAGL inhibitors are the carbamate compound URB602 and the piperidine carbamate JZL184 (Pan et al., 2009), which are able to increase up to 10-fold the concentration of 2-AG, but not AEA, in the brain. URB602 inhibits MAGL through a noncompetitive, time-independent, and partially reversible mechanism (King et al., 2009), displaying low potency in vivo. A broad series of URB602 derivatives have been designed in an attempt to improve the potency at inhibiting MAGL activity (Szabo, Agostino, Malone, Yuriev, & Capuano, 2011). JZL184 shows high levels of potency and selectivity in inhibiting MAGL over FAAH and has only minimal effect on other monoacylglycerols (Long et al., 2009), and systemic administration of this compound to mice induces a broad array of CB1 receptor–dependent behavioral effects (Long et al., 2009). JJKK048 (4-[bis(1,3-benzodioxol-5-yl) methyl]-1-piperidinyl]-1H-1,2,4-triazol-1-yl-methanone) has been reported as novel molecule that selectively impairs MAGL activity, showing high 630-fold selectivity for MAGL over FAAH and ABHD6, respectively (Aaltonen et al., 2013; Kinsey et al., 2013). Natural terpenoids, such as primesterin (King et al., 2009) and β-amyrin (Chicca, Marazzi, & Gertsch, 2012) have emerged as tools to identify novel scaffolds for MAGL inhibitors able to inactivate this enzyme in a rapid, reversible, and noncompetitive manner (Scalvini, Piomelli, & Mor, 2016).

In conclusion, we should mention that combined FAAH/MAGL inhibitor (i.e., JZL195) exert their action with high efficacy and selectivity in vivo, and by increasing simultaneously brain levels of both AEA and 2-AG, they produce antinociceptive, cataleptic, and hypomotility effects like those produced by direct CB_1 agonists (Fowler, 2015; Long et al., 2009).

The chemical structures and the enzyme kinetic properties of the main inhibitors of AEA and 2-AG metabolism mentioned here are summarized in Table 1.1.

CONCLUSIONS

The search for selective inhibitors of eCB signaling has become a major issue in current drug discovery. As a consequence, many novel and attractive scaffolds with different degrees of potency and selectivity have been discovered. Some of them might be considered promising templates for the development of therapeutic alternatives to CB_1/CB_2 agonists in those pathological conditions that might benefit from an elevation of eCB tone. However, further studies seem necessary to design the first inhibitors (e.g., of FAAH) suitable for clinical exploita-

Table 1.1 Chemical structures and half maximal inhibitory concentrations (IC_{50}) of the main inhibitors of AEA and 2-AG metabolic enzymes

Target enzyme	Compound	Chemical structure	IC_{50} (µM)
DAGL	THL		0.060
	RHC–80267		0.060
	O–3841		0.160
	OMDM–188		0.016
	O–7460		0.69

Continued

Table 1.1 Chemical structures and half maximal inhibitory concentrations (IC_{50}) of the main inhibitors of AEA and 2-AG metabolic enzymes—cont'd

Target enzyme	Compound	Chemical structure	IC_{50} (μM)
	LEI 106		0.018
FAAH	URB 597		0.0041
	OL–135		0.016
	PF–750		0.016

MAGL	TAK21D		0.00072
	TC-F2		0.028
	SA 57		0.001
	URB602		10

Continued

Table 1.1 Chemical structures and half maximal inhibitory concentrations (IC_{50}) of the main inhibitors of AEA and 2-AG metabolic enzymes—cont'd

Target enzyme	Compound	Chemical structure	IC_{50} (µM)
	JZL184		0.002
	JJKK048		0.0004
FAAH/ MAGL	JZL 195		0.002 (FAAH) 0.004 (MAGL)

tion. In particular, research programs aimed at developing new ECS-oriented compounds for human use should take into account the concepts of (1) potency and selectivity versus other proteins belonging to the ECS, as well as versus other potential off-targets; (2) reversibility versus irreversibility; and (3) the emerging differences in the structural and functional properties of ECS elements from different sources (e.g., human vs. rodents).

ACKNOWLEDGMENTS

The authors express their gratitude to all members of Maccarrone's group who have contributed over the years to the investigations into endocannabinoid metabolism. They also thank Dr. Monica Bari and Dr. Filomena Fezza (Tor Vergata University of Rome) for kindly preparing the artwork. This investigation was partially supported by Agenzia Spaziale Italiana (ASI), under 2016–2018 grant to MM.

REFERENCES

Aaltonen, N., Savinainen, J. R., Ribas, C. R., Rönkkö, J., Kuusisto, A., Korhonen, J., … Laitinen, J. T. (2013). Piperazine and piperidine triazole ureas as ultrapotent and highly selective inhibitors of monoacylglycerol lipase. *Chemistry & Biology, 20,* 379–390.

Ben-Shabat, S., Fride, E., Sheskin, T., Tamiri, T., Rhee, M. H., Vogel, Z., … Mechoulam, R. (1998). An entourage effect: Inactive endogenous fatty acid glycerol esters enhance 2-arachidonoyl-glycerol cannabinoid activity. *European Journal of Pharmacology, 353,* 23–31.

Bisogno, T., Burston, J. J., Rai, R., Allarà, M., Saha, B., Mahadevan, A., … Di Marzo, V. (2009). Synthesis and pharmacological activity of a potent inhibitor of the biosynthesis of the endocannabinoid 2-arachidonoylglycerol. *Chemistry and Medicinal Chemistry, 4,* 946–950.

Bisogno, T., Cascio, M. G., Saha, B., Mahadevan, A., Urbani, P., Minassi, A., … Di Marzo, V. (2006). Development of the first potent and specific inhibitors of endocannabinoid biosynthesis. *Biochimica et Biophysica Acta, 1761,* 205–212.

Bisogno, T., Howell, F., Williams, G., Minassi, A., Cascio, M. G., Ligresti, A., … Doherty, P. (2003). Cloning of the first *sn*1-DAG lipases points to the spatial and temporal regulation of endocannabinoid signaling in the brain. *Journal of Cell Biology, 63,* 463–468.

Bisogno, T., & Maccarrone, M. (2013). Latest advances in the discovery of fatty acid amide hydrolase inhibitors. *Expert Opinion on Drug Discovery, 8,* 509–522.

Bisogno, T., Mahadevan, A., Coccurello, R., Chang, J. W., Allarà, M., Chen, Y., … Di Marzo, V. (2013). A novel fluorophosphonate inhibitor of the biosynthesis of the endocannabinoid 2-arachidonoylglycerol with potential anti-obesity effects. *British Journal of Pharmacology, 169,* 784–793.

Bisogno, T., Melck, D., De Petrocellis, L., & Di Marzo, V. (1999). Phosphatidic acid as the biosynthetic precursor of the endocannabinoid 2-arachidonoylglycerol in intact mouse neuroblastoma cells stimulated with ionomycin. *Journal of Neurochemistry, 72,* 2113–2119.

Blankman, J. L., Long, J. Z., Trauger, S. A., Siuzdak, G., & Cravatt, B. F. (2013). ABHD12 controls brain lysophosphatidylserine pathways that are deregulated in a murine model of the neurodegenerative disease PHARC. *Proceedings of the National Academy of Sciences USA, 110*, 1500–1505.

Blankman, J. L., Simon, G. M., & Cravatt, B. F. (2007). A comprehensive profile of brain enzymes that hydrolyze the endocannabinoid 2-arachidonoylglycerol. *Chemistry & Biology, 14*, 1347–1356.

Blednov, Y. A., Cravatt, B. F., Boehm, S. L., Walker, D., & Harris, R. A. (2007). Role of endocannabinoids in alcohol consumption and intoxication: Studies of mice lacking fatty acid amide hydrolase. *Neuropsychopharmacology: Official Publication of the American College of Neuropsychopharmacology, 32*, 1570–1582.

Bracey, M. H., Hanson, M. A., Masuda, K. R., Stevens, R. C., & Cravatt, B. F. (2002). Structural adaptations in a membrane enzyme that terminates endocannabinoid signalling. *Science, 298*, 1793–1796.

Bradshaw, H. B., & Walker, J. M. (2005). The expanding field of cannabimimetic and related lipid mediators. *British Journal of Pharmacology, 144*, 459–465.

Brown, I., Cascio, M. G., Wahle, K. W., Smoum, R., Mechoulam, R., Ross, R. A., … Heys, S. D. (2010). Cannabinoid receptor-dependent and -independent anti-proliferative effects of omega-3 ethanolamides in androgen receptor-positive and -negative prostate cancer cell lines. *Carcinogenesis, 3*, 1584–1591.

Carrier, E. J., Kearn, C. S., Barkmeier, A. J., Breese, N. M., Yang, W., Nithipatikom, K., … Hillard, C. J. (2004). Cultured rat microglial cells synthesize the endocannabinoid 2-arachidonylglycerol, which increases proliferation via a CB2 receptor-dependent mechanism. *Molecular Pharmacology, 65*, 999–1007.

Cascio, M. G. (2013). PUFA-derived endocannabinoids: An overview. *Proceedings of the Nutrition Society, 72*, 451–459.

Cascio, M. G., & Marini, P. (2015). Biosynthesis and fate of endocannabinoids. *Handbook of Experimental Pharmacology, 231*, 39–58.

Chiang, K. P., Gerber, A. L., Sipe, J. C., & Cravatt, B. F. (2004). Reduced cellular expression and activity of the P129T mutant of human fatty acid amide hydrolase: Evidence for a link between defects in the endocannabinoid system and problem drug use. *Human Molecular Genetics, 13*, 2113–2119.

Chicca, A., Marazzi, J., & Gertsch, J. (2012). The antinociceptive triterpene β-amyrin inhibits 2-arachidonoylglycerol (2-AG) hydrolysis without directly targeting cannabinoid receptors. *British Journal of Pharmacology, 167*, 1596–1608.

Costa, B., Comelli, F., Bettoni, I., Colleoni, M., & Giagnoni, G. (2008). The endogenous fatty acid amide, palmitoylethanolamide, has anti-allodynic and anti-hyperalgesic effects in a murine model of neuropathic pain: involvement of CB(1), TRPV1 and PPARgamma receptors and neurotrophic factors. *Pain, 139*, 541–550.

Cravatt, B. F., Demarest, K., Patricelli, M. P., Bracey, M. H., Giang, D. K., Martin, B. R., & Lichtman, A. H. (2001). Supersensitivity to anandamide and enhanced endogenous cannabinoid signaling in mice lacking fatty acid amide hydrolase. *Proceedings of the National Academy of Sciences USA, 98*, 9371–9376.

Cravatt, B. F., Giang, D. K., Mayfield, S. P., Boger, D. L., Lerner, R. A., & Gilula, N. B. (1996). Molecular characterization of an enzyme that degrades neuromodulatory fatty-acid amides. *Nature, 384*, 83–87.

Cravatt, B. F., & Lichtman, A. H. (2003). Fatty acid amide hydrolase: An emerging therapeutic target in the endocannabinoid system. *Current Opinion in Chemical Biology, 7*, 469–475.

Cravatt, B. F., Saghatelian, A., Hawkins, E. G., Clement, A. B., Bracey, M. H., & Lichtman, A. H. (2004). Functional disassociation of the central and peripheral fatty acid amide signaling systems. *Proceedings of the National Academy of Sciences USA, 101*, 10821–10826.

Dainese, E., De Fabritiis, G., Sabatucci, A., Oddi, S., Angelucci, C. B., Di Pancrazio, C., … Maccarrone, M. (2014). Membrane lipids are key modulators of the endocannabinoid-hydrolase FAAH. *Biochemical Journal, 457,* 463–472.

De Petrocellis, L., Davis, J. B., & Di Marzo, V. (2001). Palmitoylethanolamide enhances anandamide stimulation of human vanilloid VR1 receptors. *FEBS Letters, 506,* 253–256.

Di Marzo, V., Fontana, A., Cadas, H., Schinelli, S., Cimino, G., Schwartz, J. C., & Piomelli, D. (1994). Formation and inactivation of endogenous cannabinoid anandamide in central neurons. *Nature, 372,* 686–691.

Fegley, D., Gaetani, S., Duranti, A., Tontini, A., Mor, M., Tarzia, G., & Piomelli, D. (2005). Characterization of the fatty acid amide hydrolase inhibitor cyclohexyl carbamic acid 3′-carbamoyl-biphenyl-3-yl ester (URB597): Effects on anandamide and oleoylethanolamide deactivation. *Journal of Pharmacology and Experimental Therapeutics, 313,* 352–358.

Fezza, F., Bari, M., Florio, R., Talamonti, E., Feole, M., & Maccarrone, M. (2014). Endocannabinoids, related compounds and their metabolic routes. *Molecules: A Journal of Synthetic Chemistry and Natural Product Chemistry, 19,* 17078–17106.

Fezza, F., De Simone, C., Amadio, D., & Maccarrone, M. (2008). Fatty acid amide hydrolase: A gate-keeper of the endocannabinoid system. *Subcellular Biochemistry, 49,* 101–132.

Fiskerstrand, T., H'mida-Ben Brahim, D., Johansson, S., M'zahem, A., Haukanes, B. I., Drouot, N., … Knappskog, P. M. (2010). Mutations in ABHD12 cause the neurodegenerative disease PHARC: An inborn error of endocannabinoid metabolism. *The American Journal of Human Genetics, 87,* 410–417.

Fowler, C. J. (2015). The potential of inhibitors of endocannabinoid metabolism for drug development: A critical review. *Handbook of Experimental Pharmacology, 231,* 95–128.

Gao, Y., Vasilyev, D. V., Goncalves, M. B., Howell, F. V., Hobbs, C., Reisenberg, M., … Doherty, P. (2010). Loss of retrograde endocannabinoid signaling and reduced adult neurogenesis in diacylglycerol lipase knock-out mice. *The Journal of Neuroscience, 30,* 2017–2024.

Giang, D. K., & Cravatt, B. F. (1997). Molecular characterization of human and mouse fatty acid amide hydrolases. *Proceedings of the National Academy of Sciences USA, 94,* 2238–2242.

Gustin, D. J., Ma, Z., Min, X., Li, Y., Hedberg, C., Guimaraes, C., … Kayser, F. (2011). Identification of potent, noncovalent fatty acid amide hydrolase (FAAH) inhibitors. *Bioorganic & Medicinal Chemistry Letters, 21,* 2492–2496.

Hashimotodani, Y., Ohno-Shosaku, T., Maejima, T., Fukami, K., & Kano, M. (2008). Pharmacological evidence for the involvement of diacylglycerol lipase in depolarization-induced endocanabinoid release. *Neuropharmacology, 54,* 58–67.

Hermanson, D. J., Gamble-George, J. C., Marnett, L. J., & Patel, S. (2014). Substrate-selective COX-2 inhibition as a novel strategy for therapeutic endocannabinoid augmentation. *Trends in Pharmacological Sciences, 35,* 358–367.

Ho, W. S., Barrett, D. A., & Randall, M. D. (2008). "Entourage" effects of N-palmitoylethanolamine and N-oleoylethanolamine on vasorelaxation to anandamide occur through TRPV1 receptors. *British Journal of Pharmacology, 155,* 837–846.

Ho, W. S., & Hillard, C. J. (2005). Modulators of endocannabinoid enzymic hydrolysis and membrane transport. *Handbook of Experimental Pharmacology, 168,* 187–207.

Hong, S. B., Li, C. M., Rhee, H. J., Park, J. H., He, X., Levy, B., … Schuchman, E. H. (1999). Molecular cloning and characterization of a human cDNA and gene encoding a novel acid ceramidase-like protein. *Genomics, 62,* 232–241.

Hoover, H. S., Blankman, J. L., Niessen, S., & Cravatt, B. F. (2008). Selectivity of inhibitors of endocannabinoid biosynthesis evaluated by activity-based protein profiling. *Bioorganic & Medicinal Chemistry Letters, 18,* 5838–5841.

Janssen, F. J., Deng, H., Baggelaar, M. P., Allarà, M., van der Wel, T., den Dulk, H., ... van der Stelt, M. (2014). Discovery of glycine sulfonamides as dual inhibitors of sn-1-diacylglycerol lipase α and α/β-hydrolase domain 6. *Journal of Medicinal Chemistry*, *57*, 6610–6622.

Jin, X. H., Okamoto, Y., Morishita, J., Tsuboi, K., Tonai, T., & Ueda, N. (2007). Discovery and characterization of a Ca^{2+}-independent phosphatidylethanolamine *N*-acyltransferase generating the anandamide precursor and its congeners. *Journal of Biological Chemistry*, *282*, 3614–3623.

Jin, X. H., Uyama, T., Wang, J., Okamoto, Y., Tonai, T., & Ueda, N. (2009). cDNA cloning and characterization of human and mouse Ca(2+)-independent phosphatidylethanolamine *N*-acyltransferases. *Biochimica et Biophysica Acta*, *1791*, 32–38.

Johnston, M., Bhatt, S. R., Sikka, S., Mercier, R. W., West, J. M., Makriyannis, A., ... Duclos, R. I., Jr. (2012). Assay and inhibition of diacylglycerol lipase activity. *Bioorganic & Medicinal Chemistry Letters*, *22*, 4585–4592.

King, A. R., Dotsey, E. Y., Lodola, A., Jung, K. M., Ghomian, A., Qiu, Y., ... Piomelli, D. (2009). Discovery of potent and reversible monoacylglycerol lipase inhibitors. *Chemistry & Biology*, *16*, 1045–1052.

Kinsey, S. G., Wise, L. E., Ramesh, D., Abdullah, R., Selley, D. E., Cravatt, B. F., & Lichtman, A. H. (2013). Repeated low-dose administration of the monoacylglycerol lipase inhibitor JZL184 retains cannabinoid receptor type 1-mediated antinociceptive and gastroprotective effects. *Journal of Pharmacology and Experimental Therapeutics*, *345*, 492–501.

Kono, M., Matsumoto, T., Imaeda, T., Kawamura, T., Fujimoto, S., Kosugi, Y., ... Kori, M. (2014). Design, synthesis, and biological evaluation of a series of piperazine ureas as fatty acid amide hydrolase inhibitors. *Bioorganic & Medicinal Chemistry*, *22*, 1468–1478.

Kozak, K. R., Crews, B. C., Morrow, J. D., Wang, L. H., Ma, Y. H., Weinander, R., ... Marnett, L. J. (2002). Metabolism of the endocannabinoids, 2-arachidonoylglycerol and anandamide, into prostaglandin, tromboxane, and prostacyclin glycerol esters and ethanolamides. *Journal of Biological Chemistry*, *277*, 44877–44885.

Kozak, K. R., Crews, B. C., Ray, J. L., Tai, H. H., Morrow, J. D., & Marnett, L. J. (2001). Metabolism of prostaglandin glycerol esters and prostaglandin ethanolamides in vitro and in vivo. *Journal of Biological Chemistry*, *276*, 36993–36998.

Kozak, K. R., Rowlinson, S. W., & Marnett, L. J. (2000). Oxygenation of the endocannabinoid, 2-arachidonylglycerol, to glyceryl prostaglandins by cyclooxygenase-2. *Journal of Biological Chemistry*, *275*, 33744–33749.

Kurahashi, Y., Ueda, N., Suzuki, H., Suzuki, M., & Yamamoto, S. (1997). Reversible hydrolysis and synthesis of anandamide demonstrated by recombinant rat fatty acid amide hydrolase. *Biochemical and Biophysical Research Communications*, *237*, 512–515.

Lee, M. W., Kraemer, F. B., & Severson, D. L. (1995). Characterization of a partially purified diacylglycerol lipase from bovine aorta. *Biochimica et Biophysica Acta*, *1254*, 311–318.

Leung, D., Saghatelian, A., Simon, G. M., & Cravatt, B. F. (2006). Inactivation of *N*-acyl phosphatidylethanolamine phospholipase D reveals multiple mechanisms for the biosynthesis of endocannabinoids. *Biochemistry*, *45*, 4720–4726.

Liu, Q., Tonai, T., & Ueda, N. (2002). Activation of *N*-acylethanolamine-releasing phospholipase D by polyamines. *Chemistry and Physics of Lipids*, *115*, 77–84.

Liu, J., Wang, L., Harvey-White, J., Osei-Hyiaman, D., Razdan, R., Gong, Q., ... Kunos, G. (2006). A biosynthetic pathway for anandamide. *Proceedings of the National Academy of Sciences USA*, *103*, 13345–13350.

Lo Verme, J., Fu, J., Astarita, G., La Rana, G., Russo, R., Calignano, A., & Piomelli, D. (2005). The nuclear receptor peroxisome proliferator-activated receptor-alpha mediates the anti-inflammatory actions of palmitoylethanolamide. *Molecular Pharmacology*, *67*, 15–19.

Long, J. Z., Li, W., Booker, L., Burston, J. J., Kinsey, S. G., Schlosburg, J. E., ... Cravatt, B. F. (2009). Selective blockade of 2-arachidonylglycerol hydrolysis produces cannabinoid behavioral effects. *Nature Chemical Biology*, *5*, 37–44.

Maccarrone, M. (2006). Fatty acid amide hydrolase: A potential target for next generation therapeutics. *Current Pharmaceutical Design*, *12*, 759–772.

Maccarrone, M., Bab, I., Bíró, T., Cabral, G. A., Dey, S. K., Di Marzo, V., ... Zimmer, A. (2015). Endocannabinoid signaling at the periphery: 50 years after THC. *Trends in Pharmacology Sciences*, *36*, 277–296.

Maccarrone, M., Dainese, E., & Oddi, S. (2010). Intracellular trafficking of anandamide: New concepts for signaling. *Trends in Biochemical Sciences*, *35*, 601–608.

Maccarrone, M., Guzmán, M., Mackie, K., Doherty, P., & Harkany, T. (2014). Programming of neural cells by (endo)cannabinoids: From physiological rules to emerging therapies. *Nature Reviews Neuroscience*, *15*, 786–801.

McKinney, M. K., & Cravatt, B. (2005). Structure and function of fatty acid amide hydrolase. *Annual Review of Biochemistry*, *74*, 411–432.

Mei, G., Di Venere, A., Gasperi, V., Nicolai, E., Masuda, K. R., Finazzi-Agrò, A., Cravatt, B. F., & Maccarrone, M. (2007). Closing the gate to the active site: Effect of the inhibitor methoxyarachidonyl fluorophosphonate on the conformation and membrane binding of fatty acid amide hydrolase. *Journal of Biological Chemistry*, *282*, 3829–3836.

Muccioli, G. G., Xu, C., Odah, E., Cudaback, E., Cisneros, J. A., Lambert, D. M., ... Stella, N. (2007). Identification of a novel endocannabinoid-hydrolyzing enzyme expressed by microglial cells. *The Journal of Neuroscience*, *27*, 2883–2889.

Natarajan, V., Schmid, P. C., Reddy, P. V., & Schmid, H. H. (1984). Catabolism of *N*-acylethanolamine phospholipids by dog brain preparations. *Journal of Neurochemistry*, *42*, 1613–1619.

Niphakis, M. J., Johnson, D. S., Ballard, T. E., Stiff, C., & Cravatt, B. F. (2012). O-hydroxyacetamide carbamates as a highly potent and selective class of endocannabinoid hydrolase inhibitors. *ACS Chemical Neuroscience*, *3*, 418–426.

Okamoto, Y., Morishita, J., Tsuboi, K., Tonai, T., & Ueda, N. (2004). Molecular characterization of a phospholipase D generating anandamide and its congeners. *The Journal of Biological Chemistry*, *279*, 5298–5305.

Pan, B., Wang, W., Long, J. Z., Sun, D., Hillard, C. J., Cravatt, B. F., & Liu, Q. S. (2009). Blockade of 2-arachidonylglycerol hydrolysis by selective monoacylglycerol lipase inhibitor 4-nitrophenyl 4-(dibenzo[d][1,3]dioxol-5-yl(hydroxy)methyl)piperidine-1-carboxylate (JZL184) enhances retrograde endocannabinoid signaling. *Journal of Pharmacology and Experimental Therapeutics*, *331*, 591–597.

Petersen, G., & Hansen, H. S. (1999). *N*-acylphosphatidylethanolamine-hydrolysing phospholipase D lacks the ability to transphosphatidylate. *FEBS Letters*, *455*, 41–44.

Petersen, G., Pedersen, A. H., Pickering, D. S., Begtrup, M., & Hansen, H. S. (2009). Effect of synthetic and natural phospholipids on *N*-acylphosphatidylethanolamine-hydrolyzing phospholipase D activity. *Chemistry and Physics of Lipids*, *162*, 53–61.

Reisenberg, M., Singh, P. K., Williams, G., & Doherty, P. (2012). The diacylglycerol lipases: Structure, regulation and roles in and beyond endocannabinoid signaling. *Philosophical Transactions of the Royal Society of London*, *367*, 3264–3275.

Rouzer, C. A., & Marnett, L. J. (2011). Endocannabinoid oxygenation by cyclooxygenases, lipoxygenases, and cytochromes P450: Cross-talk between the eicosanoid and endocannabinoid signaling pathways. *Chemical Reviews*, *111*, 5899–5921.

Savinainen, J. R., Saario, S. M., & Laitinen, J. T. (2012). The serine hydrolases MAGL, ABHD6 and ABHD12 as guardians of 2-arachidonoylglycerol signalling through cannabinoid receptors. *Acta Physiologica*, *204*, 267–276.

Scalvini, L., Piomelli, D., & Mor, M. (2016). Monoglyceride lipase: Structure and inhibitors. *Chemistry and Physics of Lipids*, *197*, 13–24.

Schmid, P. C., Reddy, P. V., Natarajan, V., & Schmid, H. H. (1983). Metabolism of N-acylethanolamine phospholipids by a mammalian phosphodiesterase of the phospholipase D type. *Journal of Biological Chemistry, 258,* 9302–9306.

Sheskin, T., Hanus, L., Slager, J.,Vogelm, Z., & Mechoulamm, R. (1997). Structural requirements for binding of anandamide-type compounds to the brain cannabinoid receptor. *Journal of Medicinal Chemistry, 40,* 659–667.

Sim, M. S., Hatim, A., Reynolds, G. P., & Mohamed, Z. (2013). Association of a functional FAAH polymorphism with methamphetamine-induced symptoms and dependence in a Malaysian population. *Pharmacogenetics, 14,* 505–514.

Simon, G. M., & Cravatt, B. F. (2006). Endocannabinoid biosynthesis proceeding through glycerophospho-N-acyl ethanolamine and a role for alpha/beta-hydrolase 4 in this pathway. *Journal of Biological Chemistry, 281,* 26465–26472.

Simon, G. M., & Cravatt, B. F. (2008). Anandamide biosynthesis catalyzed by the phosphodiesterase GDE1 and detection of glycerophospho-N-acyl ethanolamine precursors in mouse brain. *Journal of Biological Chemistry, 283,* 9341–9349.

Simon, G. M., & Cravatt, B. F. (2010). Characterization of mice lacking candidate N-acyl ethanolamine biosynthetic enzymes provides evidence for multiple pathways that contribute to endocannabinoid production in vivo. *Molecular BioSystems, 6,* 1411–1418.

Sipe, J. C., Chiang, K., Gerber, A. L., Beutler, E., & Cravatt, B. F. (2002). A missense mutation in human fatty acid amide hydrolase associated with problem drug use. *Proceedings of the National Academy of Sciences USA, 99,* 8394–8399.

Snider, N. T., Walker, V. J., & Hollenberg, P. F. (2010). Oxidation of the endogenous cannabinoid arachidonoyl ethanolamide by the cytochrome P450 monooxygenases: Physiological and pharmacological implications. *Pharmacological Reviews, 62,* 136–154.

Solorzano, C., Zhu, C., Battista, N., Astarita, G., Lodola, A., Rivara, S., … Piomelli, D. (2009). Selective N-acylethanolamine-hydrolyzing acid amidase inhibition reveals a key role for endogenous palmitoylethanolamide in inflammation. *Proceedings of the National Academy of Sciences USA, 106,* 20966–20971.

van der Stelt, M., van Kuik, J. A., Bari, M., van Zadelhoff, G., Leeflang, B. R.,Veldink, G. A., … Maccarone, M. (2002). Oxygenated metabolites of anandamide and 2-arachidonoyl glycerol: Conformational analysis and interaction with cannabinoid receptors, membrane transporter and fatty acid amide hydrolase. *Journal of Medicinal Chemistry, 45,* 3709–3720.

Sun, Y. X., Tsuboi, K., Okamoto, Y., Tonai, T., Murakami, M., Kudo, I., & Ueda, N. (2004). Biosynthesis of anandamide and N-palmitoylethanolamine by sequential actions of phospholipase A2 and lysophospholipase D. *Biochemistry Journal, 380,* 749–756.

Szabo, M., Agostino, M., Malone, D. T.,Yuriev, E., & Capuano, B. (2011). The design, synthesis and biological evaluation of novel URB602 analogues as potential monoacylglycerol lipase inhibitors. *Bioorganic & Medicinal Chemistry Letters, 21,* 6782–6787.

Tanimura, A., Yamazaki, M., Hashimotodani, Y., Uchigashima, M., Kawata, S., Abe, M., … Kano, M. (2010). The endocannabinoid 2-arachidonoylglycerol produced by diacylglycerol lipase alpha mediates retrograde suppression of synaptic transmission. *Neuron, 65,* 320–327.

Tsuboi, K., Okamoto, Y., Ikematsu, N., Inoue, M., Shimizu, Y., Uyama, T., … Ueda, N. (2011). Enzymatic formation of N-acylethanolamines from N-acylethanolamine plasmalogen through N-acylphosphatidylethanolamine-hydrolyzing phospholipase D-dependent and -independent pathways. *Biochimica et Biophysica Acta, 1811,* 565–577.

Tsuboi, K., Okamoto, Y., Rahman, I. A., Uyama, T., Inoue, T., Tokumura, A., & Ueda, N. (2015). Glycerophosphodiesterase GDE4 as a novel lysophospholipase D: A possible involvement in bioactive N-acylethanolamine biosynthesis. *Biochimica et Biophysica Acta, 1851,* 537–548.

Tsuboi, K., Sun, Y. X., Okamoto, Y., Araki, N., Tonai, T., & Ueda, N. (2005). Molecular characterization of N-acylethanolamine-hydrolyzing acid amidase, a novel member of the choloylglycine hydrolase family with structural and functional similarity to acid ceramidase. *Journal of Biological Chemistry, 280,* 11082–11092.

Tsuboi, K., Takezaki, N., & Ueda, N. (2007). The N-acylethanolamine-hydrolyzing acid amidase (NAAA). *Chemistry & Biodiversity, 4,* 1914–1925.

Tuo, W., Leleu-Chavain, N., Barczyk, A., Renault, N., Lemaire, L., Chavatte, P., & Millet, R. (2016). Design, synthesis and biological evaluation of potent FAAH inhibitors. *Bioorganic & Medicinal Chemistry Letters, 26,* 2701–2705.

Ueda, N., Kurahashi, Y., Yamamoto, K., Yamamoto, S., & Tokunaga, T. (1996). Enzymes for anandamide biosynthesis and metabolism. *Journal of Lipid Mediators and Cell Signalling, 14,* 57–61.

Ueda, N., Liu, Q., & Yamanaka, K. (2001). Marked activation of the N-acylphosphatidylethanolamine-hydrolyzing phosphodiesterase by divalent cations. *Biochimica et Biophysica Acta, 1532,* 121–127.

Ueda, N., Tsuboi, K., & Uyama, T. (2010). N-acylethanolamine metabolism with special reference to N-acylethanolamine-hydrolyzing acid amidase (NAAA). *Progress in Lipid Research, 49,* 299–315.

Ueda, N., Tsuboi, K., & Uyama, T. (2013). Metabolism of endocannabinoids and related N-acylethanolamines: Canonical and alternative pathways. *FEBS Journal, 280,* 1874–1894.

Ueda, N., Yamanaka, K., & Yamamoto, S. (2001). Purification and characterization of an acid amidase selective for N-palmitoylethanolamine, a putative endogenous anti-inflammatory substance. *Journal of Biological Chemistry, 276,* 35552–35557.

Uyama, T., Ikematsu, N., Inoue, M., Shinohara, N., Jin, X. H., Tsuboi, K., ... Ueda, N. (2012). Generation of N-acylphosphatidylethanolamine by members of the phospholipase A/acyltransferase (PLA/AT) family. *Journal of Biological Chemistry, 287,* 31905–31919.

Uyama, T., Inoue, M., Okamoto, Y., Shinohara, N., Tai, T., Tsuboi, K., ... Ueda, N. (2013). Involvement of phospholipase A/acyltransferase-1 in N-acylphosphatidylethanolamine generation. *Biochimica et Biophysica Acta, 1831,* 1690–1701.

Wang, J., Zhao, L. Y., Uyama, T., Tsuboi, K., Tonai, T., & Ueda, N. (2008). Amino acid residues crucial in pH regulation and proteolytic activation of N-acylethanolamine-hydrolyzing acid amidase. *Biochimica et Biophysica Acta, 1781,* 710–717.

Wei, B. Q., Mikkelsen, T. S., McKinney, M. K., Lander, E. S., & Cravatt, B. F. (2006). A second fatty acid amide hydrolase with variable distribution among placental mammals. *Journal of Biological Chemistry, 281,* 36569–36578.

Yoshino, H., Miyamae, T., Hansen, G., Zambrowicz, B., Flynn, M., Pedicord, D., ... Gonzalez-Burgos, G. J. (2011). Postsynaptic diacylglycerol lipase mediates retrograde endocannabinoid suppression of inhibition in mouse prefrontal cortex. *The Journal of Physiology, 589,* 4857–4884.

Zhao, S., Mugabo, Y., Ballentine, G., Attane, C., Iglesias, J., Poursharifi, P., ... Prentki, M. (2016). α/β-Hydrolase domain 6 deletion induces adipose browning and prevents obesity and type 2 diabetes. *Cell Reports, 14,* 2872–2888.

Zhao, S., Mugabo, Y., Iglesias, J., Xie, L., Delghingaro-Augusto, V., Lussier, R., ... Prentki, M. (2014). α/β-Hydrolase domain-6-accessible monoacylglycerol controls glucose-stimulated insulin secretion. *Cell Metabolism, 19,* 993–1007.

Zhou, Y., Huang, T., Lee, F., & Kreek, M. J. (2016). Involvement of endocannabinoids in alcohol "binge" drinking: Studies of mice with human fatty acid amide hydrolase genetic variation and after CB1 receptor antagonists. *Alcoholism: Clinical and Experimental Research, 40,* 467–473.

Zimmer, A. (2015). Genetic manipulation of the endocannabinoid system. *Handbook of Experimental Pharmacology, 231,* 129–183.

CHAPTER 2

Cannabinoid Receptors and Their Signaling Mechanisms

Balapal S. Basavarajappa[1,2,3]

[1]New York State Psychiatric Institute, New York, NY, United States; [2]Columbia University, New York, NY, United States; [3]Nathan Kline Institute for Psychiatric Research, Orangeburg, NY, United States

ABBREVIATIONS

2-AG 2-Arachidonoylglycerol
ABHD4 Abhydrolase domain 4
AEA Arachidonoylethanolamide (anandamide)
AMPAR α-Amino-3-hydroxy-5-methy l-4-isoxazolepropionic acid receptor
AMT AEA membrane transporter
BDNF Brain-derived neurotrophic factor
CB1 Cannabinoid receptor type 1
CBC Cannabichromene
CBD Cannabidiol
CBDN Cannabinodiol
CBE Cannabielsoin
CBG Cannabigerol
CBL Cannabicyclol
CBN Cannabinol
CBT Cannabitriol
COX Cyclooxygenase
CREB cAMP response-element-binding protein
DAGL Diacylglycerol lipase
EC Endocannabinoid
FAAH Fatty acid amide hydrolase
FAK Focal adhesion kinase
FAS Fetal alcohol syndrome
fEPSCs Field excitatory postsynaptic currents
GABAA Gamma aminobutyric acid A receptor
GDE1 Glycerophosphodiesterase
GPCR G-protein-coupled receptor
LOX Lipoxygenase
MAGL Monoacylglycerol lipase
MAPK Mitogen-activated protein kinase
NADA N-arachidonoyldopamine
NAPE-PLD N-acylphosphatidylethanolamine-specific phospholipase D

The Endocannabinoid System
ISBN 978-0-12-809666-6
http://dx.doi.org/10.1016/B978-0-12-809666-6.00002-2

N-ArPE *N*-arachidonoyl phosphatidylethanolamine
NAT *N*-acyltransferase
NCAM Neural cell adhesion molecule
NMDA *N*-methyl-D-aspartate
PGH2 Prostaglandin H2
PTPN22 Phosphatase
VR1 Vanilloid receptor type 1
Δ⁹-THC Δ⁹-tetrahydrocannabinol

INTRODUCTION

The earliest anthropological evidence of cannabis use comes from the oldest known Neolithic culture in China, where it was cultivated and used by humans in the production of hemp for ropes and textiles and also as an intoxicant for recreational and religious purposes (Kabelik, Krejci, & Santavy, 1960). Cannabis has been used in various cultures for centuries (Basavarajappa, 2007). The primary psychoactive constituent of *Cannabis sativa* (such as marijuana, hashish, and bhang) is Δ⁹-tetrahydrocannabinol (Δ⁹-THC, dronabinol) (Fig. 2.1), which is predominantly responsible for the psychotropic effects of the *Cannabis* plant (Dewey, 1986; Hollister, 1986). Findings have emphasized the potential therapeutic use of many other *Cannabis* components, either alone or in combination. The chemistry of *Cannabis* is very complex, and the biological activity of many of its constituents have yet to be characterized (for review, see Elsohly & Slade, 2005). Analysis of *C. sativa* L. has identified more than 500 natural compounds, many of which

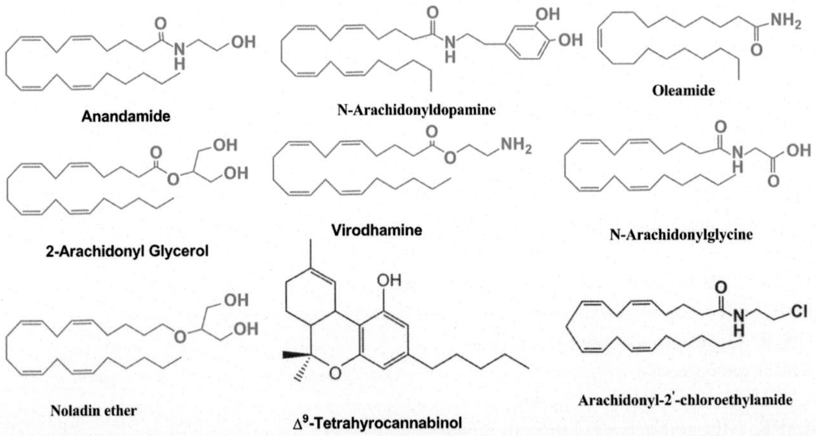

Figure 2.1 *The molecular structure of endogenous and synthetic cannabinoid receptor ligands.*

interact with each other. Thus far, nearly 100 cannabinoids have been characterized in *C. sativa* L. (Elsohly & Slade, 2005; Mehmedic et al., 2010; Ross & ElSohly, 1996; Sharma, Murthy, & Bharath, 2012; Turner, Elsohly, & Boeren, 1980). The term "phytocannabinoids" is used to differentiate the many cannabinoid components of *Cannabis* from endocannabinoids (ECs) and synthetic psychotropic cannabinoids.

Cannabidiol (CBD), cannabigerol, cannabichromene, cannabicyclol, cannabielsoin, cannabinol, cannabinodiol, and cannabitriol, along with 14 other less-studied compounds (for review, see Elsohly & Slade, 2005), have been identified as cannabinoid types. CBD is also enriched in marijuana and was shown to have bioactive and neurological effects. It does not produce a psychotropic effect like Δ^9-THC and has attracted much attention as a potential therapeutic compound for many disorders (Devinsky et al., 2014). In addition, *Cannabis* also contains a broad range of nitrogenous compounds, amino acids, flavonoids, proteins, glycoproteins, sugars, hydrocarbons, ketones and acids, fatty acids, esters, pigments, lactones, steroids, terpenes, vitamins, simple alcohols, and noncannabinoid phenols (Ross & ElSohly, 1996; Turner et al., 1980). Currently, Δ^9-THC and its derivatives are used for the treatment of nausea and vomiting caused by radiotherapy or chemotherapy treatment and wasting syndrome in patients with acquired immune deficiency syndrome. Cannabinoids are also useful for the treatment of pain, spasticity, glaucoma, and other disorders (Watson, Benson, & Joy, 2000). However, the clinical usefulness of Δ^9-THC and its derivatives has been hampered by their profound effects on mental state, such as euphoric or rewarding effects (Maldonado & Rodriguez de Fonseca, 2002), and impairment of attention, working memory (Hampson & Deadwyler, 1999), and executive function (Fried, Watkinson, James, & Gray, 2002). These behavioral effects are consistent with the findings that Δ^9-THC and several synthetic cannabinoids have an abnormal effect on neuronal function in the central nervous system (CNS). The discovery that Δ^9-THC is predominantly responsible for the psychotropic effects of *Cannabis* led to the identification of the first "cannabinoid" receptor (Devane, Dysarz, Johnson, Melvin, & Howlett, 1988) and the EC system (Devane et al., 1992; Mechoulam & Parker, 2013; Sugiura et al., 1996).

In the past two decades, cannabinoid research has received tremendous attention from various researchers. This emerging body of research has elucidated the numerous functions of the EC system in regulating synaptic neurotransmission in different brain regions (Kreitzer & Regehr, 2001;

Ohno-Shosaku, Maejima, & Kano, 2001; Wilson & Nicoll, 2001). An increasing number of studies have delineated the fundamental features of EC signaling in the molecular pathways underlying both short- and long-lasting alterations in synaptic strength (Alger, 2002; Basavarajappa & Arancio, 2008). In fact, the critical involvement of ECs in several molecular mechanisms of synaptic neurotransmission may alter the current cellular models of learning and memory. These models may be essential in understanding and developing a potential treatment for the rewarding and amnestic effects of marijuana and other synthetic cannabinoid drugs. The current chapter summarizes our knowledge of the signaling mechanisms underlying the effects of marijuana and other synthetic cannabinoids in the brain. First, the structure and anatomical distribution of the cannabinoid receptors are described. Finally, the molecular mechanisms of cannabinoid receptor signaling are discussed.

ENDOCANNABINOID SYSTEM

The EC system is now acknowledged as the most widely distributed signaling system not only in the brain but also in the peripheral tissues. Its function is highly specific and localized (Castillo, Younts, Chavez, & Hashimotodani, 2012; Katona & Freund, 2012). The EC system has been described as one of the most widespread and versatile signaling systems ever identified (for review, see Castillo et al., 2012; Maccarrone et al., 2015; Mechoulam & Parker, 2013). The EC system consists of lipid molecules (ECs) that bind to and activate cannabinoid receptors. These lipid molecules are synthesized from phospholipid precursors (Basavarajappa & Hungund, 1999; Basavarajappa, Saito, Cooper, & Hungund, 2000, 2003; Cadas, di Tomaso, & Piomelli, 1997; Di Marzo et al., 1994; Mechoulam, Fride, & Di Marzo, 1998) and are released from the cells following stimulation in a nonvascular manner to act in a paracrine fashion (Basavarajappa & Hungund, 1999, 2000, 2003; Di Marzo et al., 1994; Giuffrida et al., 1999; Mechoulam et al., 1998). Thus the EC system includes cannabinoid receptors type 1 (CB1R) and type 2 (CB2R), their endogenous lipid ligands [2-arachidonoylglycerol (2-AG) and N-arachidonoylethanolamide (AEA) or anandamide] (Fig. 2.1), and EC-synthesizing and EC-degrading enzymes (Katona & Freund, 2012; Mechoulam & Parker, 2013). Additional ECs, such as oleamide, 2-AG, virodhamine, N-arachidonoylglycine, and noladin ether, were also identified in specific brain tissues (Mechoulam & Parker, 2013) (Fig. 2.1). The structures of endogenous and synthetic CB1R-specific agonists are shown in Fig. 2.1.

The EC system functions in a broad range of physiological processes and behaviors, such as neurogenesis, neural development, immune function, metabolism and energy homeostasis, pain, emotional state, arousal, sleep, stress reactivity, synaptic plasticity, learning, and reward processing of many drugs of abuse, including alcohol (Basavarajappa, 2015; Basavarajappa, Nagre, Xie, & Subbanna, 2014; Basavarajappa & Subbanna, 2014; Fernández-Ruiz, Berrendero, Hernandez, & Ramos, 2000; Murillo-Rodriguez et al., 1998, 2003; Murillo-Rodriguez, Blanco-Centurion, Sanchez, Piomelli, & Shiromani, 2003; Murillo-Rodriguez et al., 1998; Pacher & Kunos, 2013; Pertwee, 2001; Ramesh et al., 2011).

Cannabinoid Receptors

Reports on the presence of cannabinoid receptors have been published since the 1980s (Devane et al., 1988; Howlett, Johnson, Melvin, & Milne, 1988). Both CB1R and CB2R have now been cloned. The presence of a third receptor ("CB3" or "anandamide receptor") in the brain and endothelial tissues has been reported in the literature (Breivogel, Griffin, Di Marzo, & Martin, 2001; Di Marzo et al., 2000; Jarai et al., 1999; Wagner, Varga, Jarai, & Kunos, 1999). However, data on the cloning, expression, and characterization of CB3 are not currently available. CB1R and CB2R are heptahelical G-protein-coupled receptors (GPCRs) that couple to $G_{i/o}$ proteins (for more details, see reviews by Basavarajappa, 2015; Mechoulam & Parker, 2013). The CB1Rs are predominantly expressed in the brain and spinal cord and thus are often referred to as the *brain cannabinoid receptor*. CB1R densities are similar to γ-aminobutyric acid (GABA)- and glutamate-gated ion channel levels (Herkenham et al., 1991). The presence of functional CB2Rs in the brain has provoked considerable controversy over the past several years. Currently, the presence of CB2R in the brain is relatively well established, and it has been identified in distinct locations of the brains of many animal species, including humans, in moderate levels (Onaivi et al., 2006; Van Sickle et al., 2005). However, the specific functions of this receptor in the CNS are emerging slowly (Li & Kim, 2015a, 2015b, 2016a; Onaivi et al., 2008, 2006; Zhang et al., 2014, 2016). The structures of CB1R-specific antagonists are shown in Fig. 2.2.

The complementary DNA sequences encoding CB1- or CB2-like receptors have been reported in the rat (Matsuda, Bonner, & Lolait, 1993), humans (Gerard, Mollereau, Vassart, & Parmentier, 1991; Munro, Thomas, & Abu-Shaar, 1993), mouse (Abood, Ditto, Noel, Showalter, & Tao, 1997; Chakrabarti,

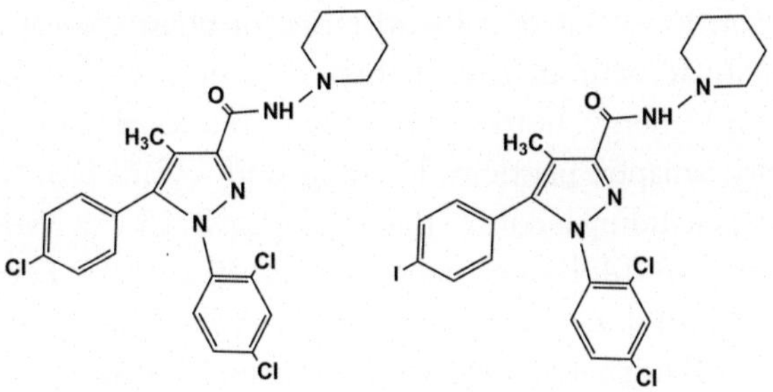

Figure 2.2 The molecular structure of CB1R-specific antagonists.

Onaivi, & Chaudhuri, 1995), cow (Onaivi, Ishiguro, Gu, & Liu, 2012), cat (Gebremedhin, Lange, Campbell, Hillard, & Harder, 1999), pufferfish (Yamaguchi, Macrae, & Brenner, 1996), leech (Stefano, Salzet, & Salzet, 1997), and newt (Soderstrom & Johnson, 2000). Human CB1R and CB2R share 44% overall amino acid identity (for more details, see review by Onaivi et al., 2002). The CB2R shares 81% amino acid identity between rats and mice or humans. Although significant progress has been made in the area of cannabinoid receptor biology, the regulation of cannabinoid receptor genes is poorly understood (D'Addario, Di Francesco, Pucci, Finazzi Agro, & Maccarrone, 2013; Nagre, Subbanna, Shivakumar, Psychoyos, & Basavarajappa, 2015; Subbanna, Nagaraja, Umapathy, Pace, & Basavarajappa, 2015).

CB1 Receptors

The CB1Rs are widely expressed in the developing brain, and their expression patterns parallel neuronal differentiation in the embryo from the most primitive stages. Many studies have examined the CB1R messenger RNA (mRNA) expression pattern and the CB1R distribution in the fetal and neonatal rat brain (Berrendero et al., 1998; Berrendero, Sepe, Ramos, Di Marzo, & Fernández-Ruiz, 1999; Buckley, Hansson, Harta, & Mezey, 1998; Romero et al., 1997). CB1R mRNA expression and receptor binding were found from gestational day (GD) 14 in rats, corresponding to the phenotypic expression pattern of most components of the neurotransmitter system (for review, see Insel, 1995). At this age, CB1Rs were already coupled to GTP-binding proteins, suggesting that they were functional (Berrendero et al., 1998). The developing human and rat brains express higher levels of CB1Rs (Glass, Dragunow, & Faull, 1997; Mailleux & Vanderhaeghen, 1992) compared to the adult brain (Berrendero et al., 1999). However, the distribution of CB1Rs is atypical in the fetal and early neonatal brain, particularly in white matter areas (Romero et al., 1997) and subventricular zones of the forebrain (Berrendero et al., 1998, 1999), compared to the adult brain (Herkenham et al., 1991; Mailleux & Vanderhaeghen, 1992). This distinctive CB1R location was a transient occurrence because the receptors are upregulated during late postnatal development, acquiring the typical distribution pattern observed in the adult brain (Berrendero et al., 1998; Romero et al., 1997). The presence of CB1Rs during early brain development indicates the possible involvement of CB1Rs in cell proliferation, migration, axonal elongation, and later synaptogenesis and myelinogenesis (for review, see Fernández-Ruiz et al., 2000). Therefore, CB1Rs contribute to generating neuronal divergence in particular brain regions during early CNS development. CB1Rs are expressed in the presynaptic area of all brain regions that are central to the processing of learning and memory (hippocampus), fear, anxiety (amygdala), stress (hypothalamic nuclei), depression [long-term potentiation (LTP) prefrontal cortex], and addiction (striatum) activity (Basavarajappa et al., 2014; Basavarajappa & Subbanna, 2014; Cristino et al., 2006; Hungund et al., 2004; Kamprath et al., 2011; Katona et al., 2001, 1999; Marsicano & Lutz, 1999; Puente et al., 2010; Ramikie et al., 2014; Subbanna, Shivakumar, Psychoyos, Xie, & Basavarajappa, 2013).

Gestational exposure to cannabinoids has been shown to impair the development of the neurotransmitter systems and their functions (Fernández-Ruiz, Berrendero, Hernandez, Romero, & Ramos, 1999; Fernández-Ruiz et al., 2000; Fernández-Ruiz, Bonnin, Cebeira, & Ramos, 1994; Fernández-Ruiz, Rodriguez, Navarro, & Ramos, 1992; Fernández-Ruiz, Romero, García-Gil, García-Palomero, & Ramos, 1996). These adverse effects were due to the activation of CB1Rs, which are expressed early in the developing brain (Berrendero et al., 1998, 1999; Fernández-Ruiz et al., 1999, 2000). In the adult brain, the activity of a particular neurotransmitter is the result of a temporally ordered sequence of events that occurs during early brain development. Alterations in this pattern may lead to changes in the functions of this particular neurotransmitter system. For example, advanced or delayed expression of the genes involved in the synthesis of receptors at a specific stage of development can result in altered physiological functions of the receptors. These alterations may also be due to increased or decreased concentrations or modifications in the activity of the CB1R signaling pathways. Exposure to cannabinoids, at doses similar to those found in marijuana consumers, was found to impair neurotransmitter development and may possibly result in neurobehavioral disturbances. Thus adult animals exposed to cannabinoids during the gestation period exhibited long-term alterations in several behaviors. These include male copulatory behavior (Dalterio, 1980), open-field activity (Navarro, Rodriguez de Fonseca, Hernandez, Ramos, & Fernández-Ruiz, 1994), learning ability (Dalterio, 1986), stress response (Mokler, Robinson, Johnson, Hong, & Rosecrans, 1987), pain sensitivity (Vela, Fuentes, Bonnin, Fernández-Ruiz, & Ruiz-Gayo, 1995), social interaction and sexual motivation (Navarro, de Miguel, Rodriguez de Fonseca, Ramos, & Fernández-Ruiz, 1996), drug-seeking behavior (Vela et al., 1998), and neuroendocrine disturbances (Dalterio & Bartke, 1979) (for review, see Dalterio, 1986; Fernández-Ruiz et al., 1994, 1992, 1996). Notably, most of these neurobehavioral effects result from the improper maturation of various neurotransmitter systems caused by the exposure to cannabinoids. These findings suggest that the activation of CB1R during the critical and sensitive period of brain development results in abnormal brain maturation. Interestingly, exposure of postnatal day 7 (P7) mice to a high dose of ethanol, which induced a widespread massive neurodegeneration in neonatal mouse brains, resulted in transcriptional activation of the CB1R gene. We found enhanced levels of CB1R mRNA as well as protein expression in the cortical and hippocampal brain regions (Subbanna et al., 2013). Additionally, we found that postnatal ethanol treatment of mice enhanced CB1R binding and CB1R-ligand-stimulated GTPγS binding (Subbanna et al., 2015). Preadministration of a CB1R-selective antagonist, SR141716A, prior to ethanol treatment prevented neurodegeneration (Subbanna et al., 2015, 2013). The protective effects of CB1R blockade with SR141716A led to normal adult synaptic transmission and learning and memory behavior, including social recognition memory, in mice exposed to ethanol at P7. These observations suggest that the developmental CB1R signaling cascade may be responsible for the synaptic and memory deficits associated with fetal alcohol spectrum disorders (Subbanna et al., 2015, 2013).

CB1Rs may also have a role in glial cells, which play a significant role in brain development. Cannabinoids have been shown to mobilize arachidonic acid (AA) in glial cells, and this activity was inhibited by SR141716A (Shivachar, Martin, & Ellis, 1996). These observations suggest that CB1R may participate in neural–glial signaling in the brain and that anandamide released from the neurons may affect astrocyte function via the activation of CB1R located in these cells. Cannabinoids alter both basal and forskolin-inhibited glucose oxidation and phospholipid synthesis in cortical glial cells (Sanchez, Galve-Roperh, Rueda, & Guzman, 1998) or C6 glioma cells

(Sanchez, Velasco, & Guzman, 1997). Pertussis toxin or SR141716A rescued these activities, thus demonstrating the participation of a G_i/G_o protein-coupled CB1R. During these metabolic alterations, sphingomyelin hydrolysis and mitogen-activated protein kinase (MAPK) stimulation were also observed (Sanchez et al., 1998). Cannabinoids in hippocampal glial cell cultures induced the expression of *kros-24*, which was reversed by SR141716A (Bouaboula, Bourrie et al., 1995), suggesting the involvement of the CB1Rs.

CB1R distribution in the adult brain is highly heterogeneous. High receptor densities were reported in many brain regions, including the basal ganglia, substantia nigra, pars reticulata, and globus pallidus. Additionally, high CB1R binding is present in the hippocampus, particularly in the dentate gyrus, and also in the molecular layer of the cerebellum. In contrast, there are few CB1Rs in the brainstem (Howlett, 2002). The CB1R distribution in humans is similar (Biegon & Kerman, 2001; Glass et al., 1997). The highest densities of CB1R are associated with limbic cortices, with much lower levels found in the primary sensory and motor regions. These findings suggest an important role for CB1R in motivational (limbic) and cognitive (association) information processing. CB1Rs have been shown to be presynaptically localized in GABAergic interneurons and glutamatergic neurons (Basavarajappa, Ninan, & Arancio, 2008; Katona et al., 2001, 1999, 2006). This distribution would be consistent with the proposed role of EC compounds in modulating GABA and glutamate neurotransmission (Basavarajappa et al., 2008; Ohno-Shosaku et al., 2001; Ohno-Shosaku, Sawada, & Kano, 2000; Ohno-Shosaku, Sawada, & Yamamoto, 1998; Ohno-Shosaku et al., 2002; Wilson, Kunos, & Nicoll, 2001; Wilson & Nicoll, 2001, 2002). In the cerebral cortex, hippocampus, and cortical regions of the amygdala, CB1R expression is higher in cholecystokinin-positive GABAergic interneurons, whereas CB1R expression is lacking in calretinin- and parvalbumin-positive interneurons (Katona et al., 1999; Marsicano & Lutz, 1999). Although expression of CB1Rs is lower in the glutamatergic neurons of cortical regions (Katona & Freund, 2012; Katona et al., 2001; Marsicano & Lutz, 1999), it has been found to play a vital role in synaptic plasticity and neuronal excitability (Marinelli, Pacioni, Cannich, Marsicano, & Bacci, 2009; Marsicano et al., 2003; Soltesz et al., 2015). Furthermore, CB1R expression is also found in the cholinergic, noradrenergic, and serotonergic systems (Haring et al., 2015; Haring, Marsicano, Lutz, & Monory, 2007; Kirilly, Hunyady, & Bagdy, 2013), but its physiological function in these neurotransmitter systems has yet to be determined. Additionally, CB1R expression has been found in astrocytes but at low levels (Han et al., 2012).

The structure of the CB1R gene is polymorphic, and these variations have been associated not only with substance abuse but also with other neuropsychiatric disorders. The CB1R gene is intronless and is similar in mice, rats, and humans (Onaivi et al., 2002). Single-exon genes encode the CB1Rs. The CB1R gene is highly conserved among humans, rats, and mice (Griffin, Tao, & Abood, 2000). Like other GPCRs, the CB1R structure is characterized by seven hydrophobic regions of 20–25 amino acids predicted to form transmembrane α helices, which are connected by alternating extracellular and intracellular loops. The 28 N-terminal amino acids in human and rat CB1Rs were similar in the total number of nonpolar, polar, acidic, and basic amino acids. The mouse N-terminal 28 amino acids differed from the rat and human CB1Rs in the number and composition of the total nonpolar and polar amino acids. Furthermore, the molecular weights of human, rat, and mouse CB1Rs are similar. Therefore, the amino acid composition of the mammalian CB1Rs shows high conservation (Onaivi et al., 2002). In CB1Rs, the second extracellular domain contains no cysteines, but two or more cysteines were found in the third extracellular domain. Cysteine residues are suggested to stabilize the tertiary structure of receptors due to their role in forming intramolecular disulfide bridges. In most GPCR receptors, these cysteines occur in the extracellular domains between hydrophobic domains 2 and 3 and hydrophobic domains 4 and 5 (the second and third extracellular domains, on the assumption that the N-terminal domain is also extracellular). One other deviation from most other GPCRs is that CB1Rs lack a highly conserved proline residue in the fifth hydrophobic domain (Matsuda, 1997). The structural features of the CB1R protein that are vital for marijuana or synthetic cannabinoid binding and functional properties have been evaluated in in vivo and in vitro models (Akinshola, Chakrabarti, & Onaivi, 1999; Akinshola, Taylor, Ogunseitan, & Onaivi, 1999). The consensus sites for N-glycosylation are usually present at the N-terminus of the protein. There are three potential N-glycosylation sites that are highly conserved in humans, rats, and mouse. The rodent CB1R protein has an additional potential N-glycosylation site at the C-terminal segment that is absent in the human CB1R protein. One potential N-glycosylation site is present in human and rat CB1Rs, but that site is missing in the mouse CB1R (Onaivi, Chakrabarti, & Chaudhuri, 1996). Whether these potential N-glycosylation sites are naturally glycosylated in CB1R proteins is not well understood. Whether these N-glycosylation sites are essential for CB1R function has yet to be determined. In the N-terminal region, the

CB1R has five potential *N*-glycosylation sites (Onaivi et al., 1996). The biological role of these sites has yet to be determined.

There are four clusters of potential cAMP-dependent protein kinase and Ca^{2+}/calmodulin-dependent protein kinase sites in CB1Rs. These clusters are conserved in the human, rat, and mouse receptor proteins. There is a single potential protein kinase C site that is also conserved in these CB1Rs. The potential N-terminal cAMP cluster in CB1R appears to be conserved. The CB1R does not have any potential protein kinase sites in the C-terminal region. The biological activity of these potential protein phosphorylation sites in CB1R has yet to be determined.

CB1R knockout (KO) mice have been generated. Two groups independently created CB1R gene KO mice. The first CB1R KO mice were generated on a CD1 background (Ledent et al., 1999), and the second KO mouse was established on a C57BL background (Steiner, Bonner, Zimmer, Kitai, & Zimmer, 1999; Zimmer, Zimmer, Hohmann, Herkenham, & Bonner, 1999). The research teams, however, differed in their findings on the baseline motility of the CB1R mutants. The CB1R mutant mice on a CD1 background exhibited higher levels of spontaneous locomotion, even when placed in fear-inducing novel environments (such as elevated plus maze and open field). In contrast, the CB1R mutant mice on a C57BL background displayed reduced activity in the open-field test and increased catalepsy. The basal ganglia, a brain structure with high levels of CB1Rs that is important for sensorimotor and motivational aspects of behavior, displayed significantly increased levels of substance P, dynorphin, enkephalin, and GAD67 expression, which may account for the alterations in spontaneous activity observed in the CB1R mutant mice. Overall, these findings provide many valuable insights into cannabinoid mechanisms, despite some differences in the behavior in these CB1R KO mice. There is a general agreement that the CB1R plays a critical role in mediating most, but not all, CNS effects of cannabinoids. The availability of the CB1R KO mouse provides an excellent opportunity to study the biological roles of these genes, the functions of CB1R, and the mechanism by which cannabinoids elicit their effects.

Furthermore, researchers also have used CB1R KO mice as potential animal models for schizophrenia (Fritzsche, 2000) and other neuropsychiatric conditions because of the ubiquity, diverse functions, and numerous signal transductions involved in the actions of cannabinoids. Mutant mice lacking CB1R specifically in cortical GABAergic neurons (GABA-CB1-KO) or glutamatergic (Glu-CB1-KO) neurons have been developed. These mice served as a tool to understand the specific role of glutamatergic and GABAergic CB1Rs in LTP, spine density and morphology, neuronal network function, and vulnerability to drugs of abuse (Dubreucq et al., 2012; Martin-Garcia et al., 2015; Monory et al., 2006; Ruehle et al., 2013; Sales-Carbonell et al., 2013). Rapid advances in the design of genetically engineered laboratory animals are producing models that are more efficient research tools. As a result, the need for precise genetic characterization of experimental animals is a primary concern.

The genomic location of the human CB1R gene was determined using genetic linkage mapping, and chromosomal in situ hybridization identified the CB1R gene on chromosome 6q14-q15 (Hoehe et al., 1991). The mouse CB1R gene is located on the proximal arm of chromosome 4 (Onaivi et al., 2002; Stubbs, Chittenden, Chakrabarti, & Onaivi, 1996). The location of the rat CB1R gene has not been determined but may be expected to fit the rodent–human homology because the CB1R genes are highly conserved in mammals. The physical and genetic localization of the bovine CB1R genes has been mapped to chromosome 9q22 (Pfister-Genskow, Weesner, Hayes, Eggen, & Bishop, 1997). Emerging evidence has suggested that marijuana and other cannabinoids function in neurobiological events, suggesting the involvement of the CB1R genes in mental and neurological disturbances (Hoenicka et al., 2007; Leroy et al., 2001; Zhang et al., 2004); therefore, the mapping of these genes will undoubtedly enhance our understanding of the linkage and genetic localization of possible cannabinoid abnormalities.

Improved information on CB1R and its allelic variants in humans and rodents can add to our understanding of addiction and other neuropsychiatric disorders. However, little information is available about cannabinoid receptor gene polymorphisms at the molecular level. Different human cannabinoid receptor gene polymorphisms have been reported. A silent mutation resulting from the substitution of G for A at nucleotide position 1359 in codon 453 (Thr) was found to be a common polymorphism in the German GTS patient population (Gadzicki, Muller-Vahl, & Stuhrmann, 1999). A HindIII restriction fragment length polymorphism is located in an approximately 14-kb intron in the 5′ region of the initiation codon of the CB1R gene (Caenazzo et al., 1991). A simple sequence repeat polymorphism consisting of nine alleles containing (AAT) 12–20 repeat sequences in the CB1R gene was linked to mental illness and drug abuse in different populations (Dawson, 1995). Studies assessing the linkage between susceptibility to alcohol or drug dependence and the CB1R gene triplet repeat marker suggested a significant association of the CB1R gene with many diverse types of drug addiction (e.g., cocaine, amphetamines, cannabis) and intravenous drug use. There was no significant association of this marker with alcohol abuse/dependence in non-Hispanic whites (Comings et al., 1997). Furthermore, there was also an essential connection of the triplet repeat marker in the CB1R gene with the P300 event-related potential that has been implicated in substance abuse (Johnson et al., 1997).

CB1R Signaling Mechanisms

CB1R activation facilitates its interaction with GTP-binding proteins, resulting in guanosine diphosphate/guanosine triphosphate exchange and dissociation of the α and $\beta\gamma$ subunit proteins (Fig. 2.3). Furthermore, the dissociated α and $\beta\gamma$ subunit proteins regulate the activity of multiple effector proteins to elicit biological functions. The affinity of CB1Rs to the G_i or G_o protein varies as determined by several receptor-agonist-stimulated and receptor-agonist-stimulated GTPγS binding studies (Breivogel, Sim, & Childers, 1997; Kearn, Greenberg, DiCamelli, Kurzawa, & Hillard, 1999). The CB1R differs from many other GPCR-G proteins because it is constitutively active due to precoupling with G proteins in the absence of exogenously added agonists (Mukhopadhyay, McIntosh, Houston, & Howlett, 2000). Fig. 2.2 depicts a diagram of the cannabinoid-mediated signal transduction pathway. CB1R activation by R-(+)-methanandamide and ECs in N18TG2 cells promoted the inhibition of adenylate cyclase (AC) activity (Childers, Sexton, & Roy, 1994; Howlett & Mukhopadhyay, 2000; Pinto et al., 1994). In some cases, upregulation of AC activity was reported without $G_{i/o}$ coupling (pertussis toxin-sensitive), likely through activation of the G_s proteins (Glass & Felder, 1997). Furthermore, in vitro expression of specific isoforms of AC (I, III, V, VI, or VIII) with coexpression of the CB1R is associated with the suppression of cyclic AMP accumulation. However, the expression of the AC isoforms II, IV, or VII with coexpression of the CB1R is associated with stimulation of cAMP accumulation (Rhee, Bayewitch, Avidor-Reiss, Levy, & Vogel, 1998). Further characterization of the mechanism by which activation of the CB1R can lead to accumulation of G αGTP$\beta\gamma$ heterotrimers, a mechanism that has been proposed for other GPCRs (Digby, Sethi, & Lambert, 2008; Lambert, 2008), would enhance our understanding of CB1R signal transduction. It is also important to characterize the downstream effectors and signaling cascades of these heterotrimer (Gα, G$\beta\gamma$, and G αGTP$\beta\gamma$) proteins.

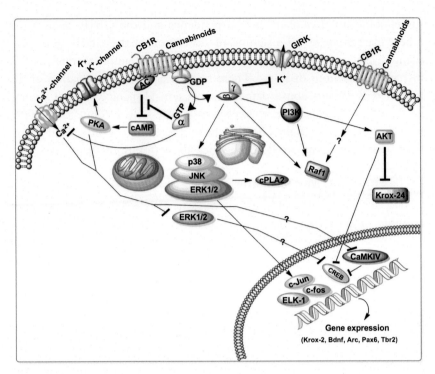

Figure 2.3 *CB1R signaling pathway.* Δ^9-tetrahydrocannabinol (Δ^9-THC) and other cannabinoids elicit their effects by binding to CB1Rs. CB1Rs are seven-transmembrane domain, G protein-coupled receptors located in the cell membrane. The Ca^{2+} channels inhibited by activation of CB1Rs include N-, P/Q-and L-type channels. The actions on Ca^{2+} channels and adenylate cyclase (AC) are thought to be mediated by the G protein α subunits, while GIRK and phosphatidylinositol 3-kinase (PI3K) activation is regulated by the $\beta\gamma$ subunits. The $\beta\gamma$ complex further activates the p38/c-Jun N-terminal kinase (JNK)/ERK1/2 pathways, followed by phosphorylation of several downstream targets, such as cytoplasmic phospholipase A2 (cPLA2), ELK-1, c-fos, c-jun and cAMP response element-binding protein (CREB), leading to the expression of target genes, such as krox-24 and brain-derived neurotrophic factor (BDNF). PI3K mediated the AKT inhibition of CREB activation. Inhibition of AC and the subsequent decrease in cAMP reduces the activation of cAMP-dependent protein kinase (PKA), which results in decreased K^+ channel phosphorylation. Inhibition of ERK1/2 activation followed by inhibition of CaMKIV and CREB phosphorylation was also found in certain conditions, leading to inhibition of Arc expression. Stimulatory effects are shown by ($\rightarrow$) sign and inhibitory effects by ($\perp$) sign.

Cannabinoids have been shown to inhibit N-type voltage-gated channels in several neuronal cells using intracellular Ca^{2+} analysis and whole-cell voltage clamp techniques (Caulfield & Brown, 1992; Mackie & Hille, 1992; Nogueron, Porgilsson, Schneider, Stucky, & Hillard, 2001; Pan, Ikeda, & Lewis, 1996). Cannabinoids have also been shown to inhibit L-type Ca^{2+}

channels in the arterial smooth muscle cells of the cat brain, which express CB1R mRNA and proteins (Gebremedhin et al., 1999) and were blocked by pertussis toxin and SR141716A (Gebremedhin et al., 1999). Stimulation of CB1Rs leads to activation of A-type and inwardly rectifying potassium channels, possibly through AC (Mu, Zhuang, Kirby, Hampson, & Deadwyler, 1999)/$G_{i/o}$ proteins, which leads to inhibition of cAMP-dependent protein kinase A (PKA). Cannabinoids have been shown to regulate potassium current (outward/inward) through PKA-mediated phosphorylation of the potassium channels (Childers & Deadwyler, 1996). Cannabinoids inhibit N-type and P/Q-type calcium channels and D-type potassium channels (Howlett et al., 2002; Howlett & Mukhopadhyay, 2000). Furthermore, cannabinoids can close sodium channels, but whether this effect is receptor-mediated has yet to be determined. In rat hippocampal CA1 pyramidal neurons, CB1Rs are negatively coupled to M-type potassium channels (Schweitzer, 2000). CB1Rs may also mobilize AA and block $5HT_3$ receptor ion channels (Pertwee, 1997). In specific conditions, CB1Rs activate AC (Calandra et al., 1999) and reduce the outward potassium K current via the G_s proteins, possibly through AA-mediated stimulation of protein kinase C (Hampson & Deadwyler, 2000). CB1Rs have also been reported to activate phospholipase C (PLC) through G proteins in COS-7 cells cotransfected with CB1Rs and Gα subunits (Ho, Uezono, Takada, Takase, & Izumi, 1999). Activation of CB1Rs increases N-methyl-D-aspartate (NMDA)-mediated calcium release from inositol 1,4,5-triphosphate-gated intracellular stores in cultured cerebellar granule neurons (Netzeband, Conroy, Parsons, & Gruol, 1999). Activation of CB1Rs by cannabinoid agonists evokes a rapid, transient increase in intracellular free Ca^{2+} in N18TG2 and NG108-15 cells (Sugiura et al., 1997, 1996, 1999).

Cannabinoids have also been shown to regulate neuritogenesis, synaptogenesis and axonal growth. The molecular mechanisms involved in these processes have been rapidly revealed. The regulation of cellular growth is typically associated with activation of tyrosine kinase receptors. However, studies suggest that GPCRs can stimulate the MAPK pathway and thereby induce cellular growth. After the first observation of activation of the MAPK cascade by an EC, AEA (Wartmann, Campbell, Subramanian, Burstein, & Davis, 1995), several studies using both ECs and cannabinoids investigated this pathway in both in vivo and in vitro models. Activation of extracellular signal-regulated kinase 1/2 (ERK1/2) (p42/p44) was observed in nonneuronal U373 MG astrocytoma cells and host cells expressing recombinant CB1Rs and was mediated by CB1R and the $G_{i/o}$ protein (Bouaboula, Poinot-Chazel et al., 1995). Similarly, activation of the $G_{i/o}$ protein via CB1R by Δ^9-THC and HU-210 activated p42/p44 MAPK in C6 glioma and primary astrocyte cultures (Guzman & Sanchez, 1999; Sanchez et al., 1998). In WI-38 fibroblasts, AEA promoted tyrosine phosphorylation and activity of ERK1/2 via CB1R and $G_{i/o}$ (Wartmann et al., 1995). In some cells, CB1R-mediated activation of MAPK involved the phosphatidylinositol 3-kinase (PI3K) pathway (Bouaboula, Poinot-Chazel et al., 1995; Wartmann et al., 1995). EC (AEA) and synthetic cannabinoids (CP 55,940 and WIN 55,212-2) increased the phosphorylation of focal adhesion kinase (FAK)+ 6,7, a neural isoform of FAK, in hippocampal slices and cultured neurons (Derkinderen et al., 1996). Δ^9-THC and ECs stimulated tyrosine phosphorylation of the Tyr-397 residue in the hippocampus, which is crucial for FAK activation (Derkinderen et al., 2001). Activation of CB1Rs by cannabinoid treatment enhanced the phosphorylation of p130 Cas, a protein associated with FAK in the hippocampus. ECs enhanced the association of Fyn, but not Src, with FAK+6,7. These effects were mediated via inhibition of a cAMP pathway. CB1R-stimulated FAK autophosphorylation was shown to function upstream of Src family kinases (Derkinderen et al., 2001). This new mechanism for cannabinoid regulation of the MAP kinase pathway may play a role in the EC-induced regulation of cell migration, neurite remodeling, and synaptic plasticity. Δ^9-THC promoted phosphorylation of Raf-1 and subsequent translocation to the membrane in cortical astrocytes (Sanchez et al., 1998). The CB1R-mediated release of the $\beta\gamma$ subunits leads to activation of PI3K, resulting in tyrosine phosphorylation and activation of Raf-1 and the resulting MAPK phosphorylation. Experiments using CHO cells expressing recombinant CB1Rs (Rueda, Galve-Roperh, Haro, & Guzman, 2000) and human vascular endothelial cells expressing endogenous CB1Rs (Liu et al., 2000) support the activation of p38 MAPK by CB1R stimulation. Furthermore, in CHO cells expressing recombinant CB1Rs, activation of CB1R by Δ^9-THC induced activation of the c-Jun N-terminal kinase (JNK1 and JNK2) (Rueda et al., 2000). The pathway for JNK activation involves CB1R-coupled $G_{i/o}$ protein, PI3K, and Ras (Rueda et al., 2000).

CB1R-mediated activation of MAPK stimulates the Na^+/H^+ exchanger in CHO cells stably expressing CB1R (Bouaboula, Bianchini, McKenzie, Pouyssegur, & Casellas, 1999). Furthermore, EC-stimulated activation of MAPK activity was shown to promote phosphorylation of cytoplasmic phospholipase A2, the release of AA, and the resulting synthesis of prostaglandin E2 in WI-38 cells (Wartmann et al., 1995). MAPK activation by cannabinoids was shown to induce immediate early gene expression (*krox*-24) in U373 MG human astrocytoma cells (Bouaboula, Bourrie et al., 1995). Δ^9-THC-induced the expression of *krox*-24, brain-derived neurotrophic factor (BDNF) and *c-Fos* in the mouse hippocampus (Derkinderen et al., 2003). The CB1R- and MEK-ERK-mediated activation of *krox*-24 is negatively regulated by the PI3K-AKT pathway in Neuro2a cells (Graham et al., 2006). Additionally, EC/CB1R/PKA-mediated MAPK activation results in the inhibition of prolactin and nerve growth factor receptor Trk synthesis (Ben-Shabat et al., 1998). CB1R agonists induce the expression of c-Fos and c-Jun in the brain (Arnold, Topple, Mallet, Hunt, & McGregor, 2001; Mailleux, Verslype, Preud'homme, & Vanderhaeghen, 1994; McGregor, Arnold, Weber, Topple, & Hunt, 1998); whether CB1R-activated MAPK mediates this activity is not known. Δ^9-THC-induced phosphorylation of the transcription factor Elk-1 is mediated by MAPK/ERK (Valjent, Caboche, & Vanhoutte, 2001). Intracerebroventricular injection of ECs induced an increase in the c-Fos protein in rat brains with a similar distribution to that of CB1Rs (Patel, Moldow, Patel, Wu, & Chang, 1998). Δ^9-THC and HU-210 increased glucose metabolism and glycogen synthesis in C6 glioma and astrocyte cultures (Guzman & Sanchez, 1999). In CHO cells expressing recombinant CB1Rs or U373 MG astrocytoma cells, stimulation of the CB1R activated protein kinase B/AKT (isoforms IB) via $G_{i/o}$ and PI3K signaling (Gomez del Pulgar, Velasco, & Guzman, 2000).

The role of the CB1R signaling pathway during brain development has not been well studied. The available evidence supports the participation of ERK1/2 via a mechanism involving the upstream inhibition of Rap1 and B-Raf (for review, see Harkany et al., 2007). Activation of CB1Rs also prevented the recruitment of new synapses by inhibiting the formation of cAMP (Kim & Thayer, 2001). Although the intracellular signaling events involving MAPK coupled to the activation of CB1Rs have been determined in the embryonic developmental stage (Berghuis et al., 2007), they are not well defined during postnatal development. Many studies using different cell lines suggested that MAPK was both up- and downregulated during Δ^9-THC-mediated apoptosis (De Petrocellis et al., 1998; Galve-Roperh et al., 2000). Furthermore, cannabis exposure during brain development also induced a variety of deficits that are similar to several specific human developmental disorders (Stefanis et al., 2004), which were possibly mediated via the activation of CB1Rs.

Studies in P7 mice established a specific role of CB1R-mediated ERK1/2, cAMP response element-binding protein (CREB) phosphorylation, as well as AKT and activity-regulated cytoskeleton-associated protein (Arc) expression in alcohol-induced neurodegeneration. P7 ethanol treatment significantly reduced activation of ERK1/2, AKT, and CREB, followed by suppression of Arc protein expression in the hippocampus and neocortex tissues (Subbanna et al., 2013). Furthermore, activation of ERK1/2, CREB, and Arc protein expression was prevented by SR pretreatment, but AKT activation was not affected. Likewise, CB1R KO mice, which did not show alcohol-induced neurodegeneration, were protected against P7 alcohol-induced inhibition of ERK1/2, CREB activation, and Arc protein expression, but they failed to rescue the inhibition of AKT phosphorylation. Therefore, alcohol-activated CB1R-induced neurodegeneration was regulated by the CB1R/pERK1/2/pCREB/Arc pathway but not by PI3K /AKT signaling in the developing brain (Subbanna et al., 2015, 2013) (Fig. 2.3). CB1R-mediated Arc regulation via the MAP kinase pathway is an important physiological mechanism by which cannabinoids and ECs can modulate synaptic plasticity.

In adult animals, acute activation of CB1R by synthetic cannabinoids failed to alter ERK1/2 activation but impaired activation of CaMKIV and CREB in a CB1R-dependent manner, resulting in LTP and learning and memory deficits (Basavarajappa & Subbanna, 2014). In another study, acute inhibition of FAAH not only enhanced endogenous AEA levels in the hippocampus but also increased ERK1/2 activation and inhibition of CaMKIV and CREB (Basavarajappa et al., 2014). These AEA/CB1R signaling events lead to LTP and learning and memory deficits in a CB1R-dependent manner (Basavarajappa et al., 2014). In several studies, CB1R KO mice exhibited enhanced CREB activation (Basavarajappa et al., 2014; Basavarajappa & Subbanna, 2014; Subbanna et al., 2015), Arc expression (Subbanna et al., 2015), enhanced LTP, and several enhanced behavioral phenotypes, including those involved in learning and memory (Auclair, Otani, Soubrie, & Crepel, 2000; Basavarajappa et al., 2014; Basavarajappa & Subbanna, 2014; Bohme, Laville, Ledent, Parmentier, & Imperato, 2000; Hoffman, Oz, Yang, Lichtman, & Lupica, 2007; Monory et al., 2006; Reibaud et al., 1999; Subbanna et al., 2015, 2013; Terranova et al., 1996; Varvel & Lichtman, 2002).

In other studies, CB1R activation has been associated with the generation of ceramide (Guzman, Galve-Roperh, & Sanchez, 2001). This widespread lipid second messenger is known to play a major role in the control of cell fate in the CNS. Studies showed that cannabinoid-dependent ceramide generation occurs by a G-protein-independent process and involves two different metabolic pathways: sphingomyelin hydrolysis and de novo ceramide synthesis. Ceramide, in turn, mediates cannabinoid-induced apoptosis, as shown by in vitro and in vivo studies. CB1R activation by cannabinoids induces apoptosis via accumulation of ceramide, phosphorylation of p38, depolarization of the mitochondrial membrane, and caspase activation in both mantle-cell lymphoma (MCL) and primary MCL cell lines but not in normal B cells (Gustafsson, Christensson, Sander, & Flygare, 2006). The CB1R of astrocytes is shown to be coupled to sphingomyelin hydrolysis through the adapter protein (FAN) factor associated with neutral sphingomyelinase activation (Sanchez et al., 2001).

CB2 Receptors

As discussed in the previous section, the earlier studies that evaluated the tissue and cell distribution of CB2Rs suggested that this cannabinoid receptor type was solely present in tissues and cells of the immune system. Recent elegant studies have proposed that CB2Rs may be present in the brain even in healthy individuals (Onaivi et al., 2008, 2006) and have provided further evidence for the presence of functional CB2Rs that can influence animal behavior in the brain (Onaivi et al., 2008). Several studies have identified CB2Rs in glial cells, including microglia and astrocytes (Maresz, Carrier, Ponomarev, Hillard, & Dittel, 2005; Nunez et al., 2004; Stella, 2004). CB2Rs were also found in neural (Palazuelos et al., 2006) and oligodendroglial (Molina-Holgado et al., 2002) progenitors and individual neuronal subpopulations in different brain structures of various species, including human samples, using either in vivo or in vitro approaches (for review, see Fernández-Ruiz, Pazos, Garcia-Arencibia, Sagredo, & Ramos, 2008).

The CB2R gene structure has been poorly defined and characterized and is less studied compared to the CB1R gene (Zhang et al., 2004). The mouse CB2R genes are located on the proximal arm of chromosome 4 (Onaivi et al., 2002; Stubbs et al., 1996). The rat CB2R genes both map to chromosome 5. The CB2R gene is intronless, at least in its coding region (Onaivi et al., 2002). Unlike the CB1R gene, which is highly conserved among humans, rats, and mice the CB2R gene is much more divergent (Griffin et al., 2000). Human CB2R is located on chromosome 1p36 and encodes a protein of 360 amino acids with a 44% homology with CB1R, although the homology is greater in the transmembrane domain (approximately 68%). The human CB2R gene consists of a single translated exon

(Sipe, Chiang, Gerber, Beutler, & Cravatt, 2002) and single untranslated exon. The mouse CB2R gene structure is similar to the human gene, but it encodes two transcripts using different first exons (Onaivi et al., 2008). Most regions of the CB2R gene are highly conserved, but at position 63, the human has glutamine and mice and rats have arginine (Sipe et al., 2002). In CB2Rs, no cysteines are found in the second extracellular domain, but the third extracellular domain contains two or more cysteines. Unlike most other GPCRs, the CB2Rs lack a highly conserved proline residue in hydrophobic domain 5 (Matsuda, 1997).

Although the distribution of CB2R in the developing brain is not well characterized, it has been reported that in rats, during development, the CB2R mRNA is exclusively expressed in the liver of the embryo as early as GD13 and lasts throughout gestation (GD21–GD22) (Buckley et al., 1998). CB2R immunoreactivity was detected in neuronal and glial processes but at a much lower level than CB1Rs (Beltramo et al., 2006). Several studies have reported the presence of CB2Rs in the adult brain stem, cortex, cerebellum, dorsal root ganglion, and spinal cord (Ashton, Friberg, Darlington, & Smith, 2006; Beltramo et al., 2006; Gong et al., 2006; Van Sickle et al., 2005). The subcellular localization of CB2Rs using immunoelectron microscopy showed immunostaining in the dendrites from hippocampal areas. This pattern of staining was observed in most hippocampal areas and appears to be predominantly postsynaptic localization of CB2Rs. These studies suggested the presence of CB2Rs at both the pre- and postsynaptic terminals (Onaivi et al., 2008). The functional implication of pre- and postsynaptic localization of CB2Rs requires further investigation using electrophysiological and image analysis methods. Further research is required to characterize the specific CB2R-mediated behavioral effects and their physiological roles in the brain.

Two lines (Buckley et al., 2000 and The Jackson Laboratory) of CB2R KO mice were generated (Buckley et al., 1997). These mice were used to study the function of CB2Rs in various inflammatory disorders, tissue remodeling and fibrosis, metabolic processes, nociception, neurodegenerative disorders, and bone remodeling (for references, see review by Zimmer, 2015). The availability of CB2R KO mice provides an excellent opportunity to study the neurobehavioral roles of these genes.

CB2R KO mice have been used to study the possible involvement of CB2Rs in neuronal signaling, synaptic plasticity, and learning and memory. CB2R KO mice exhibit impaired proliferation of neural progenitor cells in the hippocampal brain region (Palazuelos et al., 2006).

Deletion of CB2R caused schizophrenia-like behaviors (Ortega-Alvaro, Aracil-Fernandez, Garcia-Gutierrez, Navarrete, & Manzanares, 2011), and the mice showed higher levels of aggression in the social interaction test (Rodriguez-Arias et al., 2015) and displayed impaired aversive memory consolidation (Garcia-Gutierrez et al., 2013). Additionally, CB2R KO mice showed impaired hippocampus-dependent, long-term contextual fear memory, whereas hippocampus-independent, cued fear memory was intact (Li & Kim, 2016a). Motor activity and anxiety of CB2R KO mice were intact when they were tested in an open field arena and an elevated zero maze (Li & Kim, 2016a). Consistent with the impaired hippocampal function of the CB1R KO mice, excitatory synaptic transmission, LTP, and dendritic spine density in the hippocampus were also reduced (Li & Kim, 2015a). Additionally, expression of CB2R in cultured autaptic hippocampal neurons obtained from CB1R null mice inhibited synaptic transmission (Atwood, Straiker, & Mackie, 2012). However, CB1R KO mice also displayed improved spatial working memory when tested in a Y-maze, and hippocampal slices prepared from C57BL/6 mice repeatedly treated with JWH-133 (CB2R agonist) (7–9 days) were shown to exhibit enhanced short-term excitatory neurotransmission, an effect that was absent in the CB2R KO mice (Kim & Li, 2014). Although the exact mechanisms underlying these observations have yet to be determined, these results suggest that the CB2R functions are the opposite of those of CB1R on synaptic function in the brain.

Most plant-derived and synthetic cannabinoid agonists activate CB2Rs, although the affinity and potency with which these agonists bind and activate CB2Rs vary compared to CB1Rs (Fernández-Ruiz et al., 2007; Pertwee, 2005). The AEA ligand was shown to be a weaker agonist for CB1Rs and does not significantly bind to CB2Rs (Mechoulam & Hanus, 2000; Sugiura, Kishimoto, Oka, & Gokoh, 2006), except in certain pathological conditions (Eljaschewitsch et al., 2006), although several studies have suggested that 2-AG is an endogenous agonist for CB2Rs (Sugiura et al., 2006). Several selective synthetic CB2R agonists, such as JWH-133 and its analogs HU-308 and AM1241, have been developed and can be used as novel tools to selectively activate CB2Rs without the concomitant activation of CB1Rs. The structures of CB2R-specific synthetic agonists and antagonists are shown in Fig. 2.4. These tools represent a significant step forward because the use of cannabinoid receptor ligands in the clinic is severely limited due to their

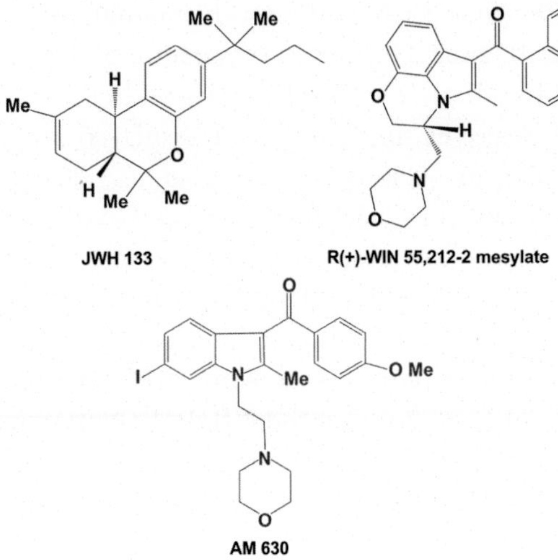

Figure 2.4 *The molecular structure of CB2R-specific synthetic agonists and antagonist.*

psychoactive effects. Therefore, an attractive alternative is to target CB2Rs with selective CB2R-specific agonists that entirely lack psychotropic activity, although they may produce other side effects, such as immune suppression (Pertwee, 2005). The usefulness of several synthetic CB2R agonists for the management of several neurological disorders, including several neuroinflammatory/neurodegenerative diseases, is currently being evaluated in multiple preclinical studies. It has been suggested that CB2R agonists might provide protection against the progression of neuronal damage (see reviews Davis, 2014; Fernández-Ruiz et al., 2008, 2007; Yrjola et al., 2015). Selective antagonists for the CB2R are also currently available. The selective agonists and antagonists along with KO mice constitute essential tools that can be used to elucidate the role of CB2Rs in particular cellular functions.

Mechanisms of CB2Rs Signaling

As discussed in the previous section, CB2Rs belong to the seven transmembrane domain, G protein-coupled class of receptors. Because CB2Rs are also coupled to $G_{i/o}$ proteins, their activation results in the inhibition of AC and the cAMP/PKA-dependent pathway (Howlett, 2002) as observed for CB1Rs. CB2R activation results in the activation of MAPK cascades, specifically the ERK (Bouaboula et al., 1996; Carrier et al.,

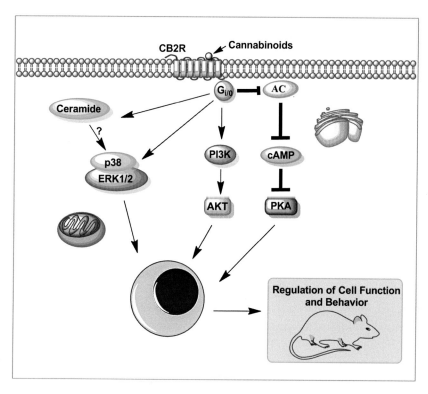

Figure 2.5 *CB2R signaling cascade.* Δ^9-tetrahydrocannabinol and cannabinoids also bind to CB2Rs. CB2Rs are also seven-transmembrane-domain, G protein-coupled receptors located in the cell membrane. Activation of CB2R is coupled to several different cellular pathways, such as adenylate cyclase (AC), cAMP, protein kinase A (PKA), ERK1/2, p38 mitogen-activated protein kinase, and AKT and a pathway for de novo synthesis of ceramide. These signaling cascades may play a role in the regulation of cell function and behavior. Stimulatory effects are shown by ($\rightarrow$) sign and inhibitory effects by ($\perp$) sign.

2004; Palazuelos et al., 2006) and the p38 MAPK pathways (Gertsch, Schoop, Kuenzle, & Suter, 2004; Herrera, Carracedo, Diez-Zaera, Guzman, & Velasco, 2005) (Fig. 2.5). Additionally, the stimulation of CB2Rs has been linked to the activation of several intracellular molecules, including the PI3K/AKT pathway (Molina–Holgado et al., 2002; Palazuelos et al., 2006; Samson et al., 2003). These pathways have been associated with prosurvival effects and the de novo synthesis of the sphingolipid messenger ceramide (Carracedo, Gironella et al., 2006; Carracedo, Lorente et al., 2006; Herrera et al., 2006; Sanchez et al., 2001), which has been linked to the proapoptotic mechanism of cannabinoids. In early brain injury, which involves vasogenic edema and apoptotic cell death that are known to cause subarachnoid hemorrhage (SAH) pathophysiology, a CB2R-specific agonist attenuated apoptosis and SAH by increasing Bcl-2 levels and CREB activation (Fujii et al., 2014). CREB is a transcription factor that promotes neuronal survival by upregulating Bcl-2 expression and inhibiting apoptosis (Meller et al., 2005; Tokudome et al., 2004; Wilson, Mochon, & Boxer, 1996). In another study, activation of CB2R by a specific agonist, trans-caryophyllene, rescued ischemic injury through upregulation of AMP-activated protein kinase and CREB activation followed by enhanced expression of the CREB target gene product, BDNF (Choi et al., 2013).

In summary, the research on the signaling mechanisms of cannabinoid receptors and their relation to a variety of brain disorders has received intense attention after the identification of the cannabinoid receptors and their endogenous ligands. In addition to well-known euphoria and elation, marijuana binding to the CB1Rs can also cause anxiety, short-term memory problems, attention deficits, and many other cognitive, affective and psychomotor effects. The extent to which it disturbs brain development and functions is unknown in young marijuana users, who have been shown to have increased risk of cannabinoid dependence. Although the acute effects of cannabinoids are well studied, studies on the chronic abuse of cannabis drugs are still insufficient and require further research. The acute effects of cannabinoids are probably related to the activation of pre-synaptic CB1/CB2R-mediated signaling cascades and the inhibition of the release of neurotransmitters in the brain. In chronic cannabis users, there may be adaptive changes in the cannabinoid-receptor-mediated signaling cascades and the related neurotransmitter systems, which may result in adverse or therapeutic effects of cannabinoids. Understanding the structure–function relationship and signaling mechanisms of cannabinoid receptors during ontogeny may identify potential therapeutic targets. These results may also help to develop compounds to treat several illnesses, including disorders induced by clinical insufficiency of the EC system during development.

ACKNOWLEDGMENTS

This work was supported by National Institute of Alcohol and Alcoholism grant (R01 AA019443).

REFERENCES

Abood, M. E., Ditto, K. E., Noel, M. A., Showalter, V. M., & Tao, Q. (1997). Isolation and expression of a mouse CB1 cannabinoid receptor gene. Comparison of binding properties with those of native CB1 receptors in mouse brain and N18TG2 neuroblastoma cells. *Biochemical Pharmacology, 53*, 207–214.

Akinshola, B. E., Chakrabarti, A., & Onaivi, E. S. (1999). In-vitro and in-vivo action of cannabinoids. *Neurochemical Research, 24*, 1233–1240.

Akinshola, B. E., Taylor, R. E., Ogunseitan, A. B., & Onaivi, E. S. (1999). Anandamide inhibition of recombinant AMPA receptor subunits in *Xenopus* oocytes is increased by forskolin and 8-bromo-cyclic AMP. *Naunyn-Schmiedeberg's Archives of Pharmacology, 360*, 242–248.

Alger, B. E. (2002). Retrograde signaling in the regulation of synaptic transmission: Focus on endocannabinoids. *Progress in Neurobiology, 68*, 247–286.

Arnold, J. C., Topple, A. N., Mallet, P. E., Hunt, G. E., & McGregor, I. S. (2001). The distribution of cannabinoid-induced Fos expression in rat brain: Differences between the Lewis and Wistar strain. *Brain Research, 921*, 240–255.

Ashton, J. C., Friberg, D., Darlington, C. L., & Smith, P. F. (2006). Expression of the cannabinoid CB2 receptor in the rat cerebellum: An immunohistochemical study. *Neuroscience Letters, 396*, 113–116.

Atwood, B. K., Straiker, A., & Mackie, K. (2012). CB(2) cannabinoid receptors inhibit synaptic transmission when expressed in cultured autaptic neurons. *Neuropharmacology, 63*, 514–523.

Auclair, N., Otani, S., Soubrie, P., & Crepel, F. (2000). Cannabinoids modulate synaptic strength and plasticity at glutamatergic synapses of rat prefrontal cortex pyramidal neurons. *Journal of Neurophysiology, 83*, 3287–3293.

Basavarajappa, B. S. (2007). Neuropharmacology of the endocannabinoid signaling system-molecular mechanisms, biological actions and synaptic plasticity. *Current Neuropharmacology, 5*, 81–97.

Basavarajappa, B. S. (2015). Fetal alcohol spectrum disorder: Potential role of endocannabinoids signaling. *Brain Science, 5*, 456–493.

Basavarajappa, B. S., & Arancio, O. (2008). Synaptic plasticity: Emerging role for endocannabinoid system. In T. F. Kaiser, & F. J. Peters (Eds.), *Synaptic plasticity: New research* (pp. 77–112). NY, USA: Nova Science Publishers, Inc.

Basavarajappa, B. S., & Hungund, B. L. (1999). Chronic ethanol increases the cannabinoid receptor agonist, anandamide and its precursor N-arachidonyl phosphatidyl ethanolamine in SK-N-SH cells. *Journal of Neurochemistry, 72*, 522–528.

Basavarajappa, B. S., Nagre, N. N., Xie, S., & Subbanna, S. (2014). Elevation of endogenous anandamide impairs LTP, learning, and memory through CB1 receptor signaling in mice. *Hippocampus, 24*, 808–818.

Basavarajappa, B. S., Ninan, I., & Arancio, O. (2008). Acute ethanol suppresses glutamatergic neurotransmission through endocannabinoids in hippocampal neurons. *Journal of Neurochemistry, 107*, 1001–1013.

Basavarajappa, B. S., Saito, M., Cooper, T. B., & Hungund, B. L. (2000). Stimulation of cannabinoid receptor agonist 2-arachidonylglycerol by chronic ethanol and its modulation by specific neuromodulators in cerebellar granule neurons. *Biochemica Biophysica Acta, 1535*, 78–86.

Basavarajappa, B. S., Saito, M., Cooper, T. B., & Hungund, B. L. (2003). Chronic ethanol inhibits the anandamide transport and increases extracellular anandamide levels in cerebellar granule neurons. *European Journal of Pharmacology, 466*, 73–83.

Basavarajappa, B. S., & Subbanna, S. (2014). CB1 receptor-mediated signaling underlies the hippocampal synaptic, learning and memory deficits following treatment with JWH-081, a new component of spice/K2 preparations. *Hippocampus, 24*, 178–188.

Beltramo, M., Bernardini, N., Bertorelli, R., Campanella, M., Nicolussi, E., Fredduzzi, S., et al. (2006). CB2 receptor-mediated antihyperalgesia: Possible direct involvement of neural mechanisms. *The European Journal of Neuroscience, 23*, 1530–1538.

Ben-Shabat, S., Fride, E., Sheskin, T., Tamiri, T., Rhee, M. H., Vogel, Z., et al. (1998). An entourage effect: Inactive endogenous fatty acid glycerol esters enhance 2-arachidonoyl-glycerol cannabinoid activity. *European Journal of Pharmacology, 353*, 23–31.

Berghuis, P., Rajnicek, A. M., Morozov, Y. M., Ross, R. A., Mulder, J., Urban, G. M., et al. (2007). Hardwiring the brain: Endocannabinoids shape neuronal connectivity. *Science, 316*, 1212–1216.

Berrendero, F., Garcia-Gil, L., Hernandez, M. L., Romero, J., Cebeira, M., de Miguel, R., et al. (1998). Localization of mRNA expression and activation of signal transduction mechanisms for cannabinoid receptor in rat brain during fetal development. *Development, 125*, 3179–3188.

Berrendero, F., Sepe, N., Ramos, J. A., Di Marzo, V., & Fernández-Ruiz, J. J. (1999). Analysis of cannabinoid receptor binding and mRNA expression and endogenous cannabinoid contents in the developing rat brain during late gestation and early postnatal period. *Synapse, 33*, 181–191.

Biegon, A., & Kerman, I. A. (2001). Autoradiographic study of pre- and postnatal distribution of cannabinoid receptors in human brain. *Neuroimage, 14*, 1463–1468.

Bohme, G. A., Laville, M., Ledent, C., Parmentier, M., & Imperato, A. (2000). Enhanced long-term potentiation in mice lacking cannabinoid CB1 receptors. *Neuroscience, 95*, 5–7.

Bouaboula, M., Bianchini, L., McKenzie, F. R., Pouyssegur, J., & Casellas, P. (1999). Cannabinoid receptor CB1 activates the Na^+/H^+ exchanger NHE-1 isoform via Gi-mediated mitogen activated protein kinase signaling transduction pathways. *FEBS Letters, 449*, 61–65.

Bouaboula, M., Bourrie, B., Rinaldi-Carmona, M., Shire, D., Le Fur, G., & Casellas, P. (1995). Stimulation of cannabinoid receptor CB1 induces krox-24 expression in human astrocytoma cells. *The Journal of Biological Chemistry, 270*, 13973–13980.

Bouaboula, M., Poinot-Chazel, C., Bourrie, B., Canat, X., Calandra, B., Rinaldi-Carmona, M., et al. (1995). Activation of mitogen-activated protein kinases by stimulation of the central cannabinoid receptor CB1. *The Biochemical Journal, 312*, 637–641.

Bouaboula, M., Poinot-Chazel, C., Marchand, J., Canat, X., Bourrie, B., Rinaldi-Carmona, M., et al. (1996). Signaling pathway associated with stimulation of CB2 peripheral cannabinoid receptor. Involvement of both mitogen-activated protein kinase and induction of Krox-24 expression. *European Journal of Biochemistry, 237*, 704–711.

Breivogel, C. S., Griffin, G., Di Marzo, V., & Martin, B. R. (2001). Evidence for a new G protein-coupled cannabinoid receptor in mouse brain. *Molecular Pharmacology, 60*, 155–163.

Breivogel, C. S., Sim, L. J., & Childers, S. R. (1997). Regional differences in cannabinoid receptor/G-protein coupling in rat brain. *The Journal of Pharmacology and Experimental Therapeutics, 282*, 1632–1642.

Buckley, N. E., Hansson, S., Harta, G., & Mezey, E. (1998). Expression of the CB1 and CB2 receptor messenger RNAs during embryonic development in the rat. *Neuroscience, 82*, 1131–1149.

Buckley, N. E., McCoy, K. L., Mezey, E., Bonner, T., Zimmer, A., Felder, C. C., et al. (2000). Immunomodulation by cannabinoids is absent in mice deficient for the cannabinoid CB(2) receptor. *European Journal of Pharmacology, 396*, 141–149.

Cadas, H., di Tomaso, E., & Piomelli, D. (1997). Occurrence and biosynthesis of endogenous cannabinoid precursor, N-arachidonoyl phosphatidylethanolamine, in rat brain. *The Journal of Neuroscience, 17*, 1226–1242.

Caenazzo, L., Hoehe, M. R., Hsieh, W.-T., Berrettini, W. H., Bonner, T. I., & Gershon, E. S. (1991). HindIII identifies a two allele DNA polymorphism of the human cannabinoid receptor gene (CNR). *Nucleic Acids Research, 19*, 4798.

Calandra, B., Portier, M., Kerneis, A., Delpech, M., Carillon, C., Le Fur, G., et al. (1999). Dual intracellular signaling pathways mediated by the human cannabinoid CB1 receptor. *European Journal of Pharmacology, 374*, 445–455.

Carracedo, A., Gironella, M., Lorente, M., Garcia, S., Guzman, M., Velasco, G., et al. (2006). Cannabinoids induce apoptosis of pancreatic tumor cells via endoplasmic reticulum stress-related genes. *Cancer Research, 66*, 6748–6755.

Carracedo, A., Lorente, M., Egia, A., Blazquez, C., Garcia, S., Giroux, V., et al. (2006). The stress-regulated protein p8 mediates cannabinoid-induced apoptosis of tumor cells. *Cancer Cell, 9*, 301–312.

Carrier, E. J., Kearn, C. S., Barkmeier, A. J., Breese, N. M., Yang, W., Nithipatikom, K., et al. (2004). Cultured rat microglial cells synthesize the endocannabinoid 2-arachidonylglycerol, which increases proliferation via a CB2 receptor-dependent mechanism. *Molecular Pharmacology, 65*, 999–1007.

Castillo, P. E., Younts, T. J., Chavez, A. E., & Hashimotodani, Y. (2012). Endocannabinoid signaling and synaptic function. *Neuron, 76*, 70–81.

Caulfield, M. P., & Brown, D. A. (1992). Cannabinoid receptor agonists inhibit Ca current in NG108-15 neuroblastoma cells via a pertussis toxin-sensitive mechanism. *British Journal of Pharmacology, 106*, 231–232.

Chakrabarti, A., Onaivi, E. S., & Chaudhuri, G. (1995). Cloning and sequencing of a cDNA encoding the mouse brain-type cannabinoid receptor protein. *DNA Sequence, 6*, 385–388.

Childers, S. R., & Deadwyler, S. A. (1996). Role of cyclic AMP in the actions of cannabinoid receptors. *Biochemical Pharmacology, 52*, 819–827.

Childers, S. R., Sexton, T., & Roy, M. B. (1994). Effects of anandamide on cannabinoid receptors in rat brain membranes. *Biochemical Pharmacology, 47*, 711–715.

Choi, I. Y., Ju, C., Anthony Jalin, A. M., Lee da, I., Prather, P. L., & Kim, W. K. (2013). Activation of cannabinoid CB2 receptor-mediated AMPK/CREB pathway reduces cerebral ischemic injury. *The American Journal of Pathology, 182*, 928–939.

Comings, D. E., Muhleman, D., Gade, R., Johnson, J. P., Verde, R., & Saucier, G. (1997). Cannabinoid receptor gene (CNR1): Association with i.v. drug use. *Molecular Psychiatry, 2*, 161–168.

Cristino, L., de Petrocellis, L., Pryce, G., Baker, D., Guglielmotti, V., & Di Marzo, V. (2006). Immunohistochemical localization of cannabinoid type 1 and vanilloid transient receptor potential vanilloid type 1 receptors in the mouse brain. *Neuroscience, 139*, 1405–1415.

D'Addario, C., Di Francesco, A., Pucci, M., Finazzi Agro, A., & Maccarrone, M. (2013). Epigenetic mechanisms and endocannabinoid signalling. *FEBS Journal, 280*, 1905–1917.

Dalterio, S. L. (1980). Perinatal or adult exposure to cannabinoids alters male reproductive functions in mice. *Pharmacology, Biochemistry, and Behavior, 12*, 143–153.

Dalterio, S. L. (1986). Cannabinoid exposure: Effects on development. *Neurobehavioral Toxicology and Teratology, 8*, 345–352.

Dalterio, S., & Bartke, A. (1979). Perinatal exposure to cannabinoids alters male reproductive function in mice. *Science, 205*, 1420–1422.

Davis, M. P. (2014). Cannabinoids in pain management: CB1, CB2 and non-classic receptor ligands. *Expert Opinion on Investigational Drugs, 23*, 1123–1140.

Dawson, E. (1995). Identification of a polymorphic triplet marker for the brain cannabinoid receptor gene: Use in linkage and association studies of schizophrenia. *Psychiatric Genetics, 5*, S50.

De Petrocellis, L., Melck, D., Palmisano, A., Bisogno, T., Laezza, C., Bifulco, M., et al. (1998). The endogenous cannabinoid anandamide inhibits human breast cancer cell proliferation. *Proceedings of the National Academy of Sciences of the United States of America, 95*, 8375–8380.

Derkinderen, P., Toutant, M., Burgaya, F., Le Bert, M., Siciliano, J. C., de Franciscis, V., et al. (1996). Regulation of a neuronal form of focal adhesion kinase by anandamide. *Science*, *273*, 1719–1722.

Derkinderen, P., Toutant, M., Kadare, G., Ledent, C., Parmentier, M., & Girault, J. A. (2001). Dual role of Fyn in the regulation of FAK+6,7 by cannabinoids in hippocampus. *The Journal of Biological Chemistry*, *276*, 38289–38296.

Derkinderen, P., Valjent, E., Toutant, M., Corvol, J. C., Enslen, H., Ledent, C., et al. (2003). Regulation of extracellular signal-regulated kinase by cannabinoids in hippocampus. *The Journal of Neuroscience*, *23*, 2371–2382.

Devane, W. A., Dysarz, F. A. I., Johnson, M. R., Melvin, L. S., & Howlett, A. C. (1988). Determination and characterization of a cannabinoid receptor in rat brain. *Molecular Pharmacology*, *34*, 605–613.

Devane, W. A., Hanus, L., Breuer, A., Pertwee, R. G., Stevenson, L. A., Griffin, G., et al. (1992). Isolation and structure of a brain constituent that binds to the cannabinoid receptor. *Science*, *258*, 1946–1949.

Development of a CB2 knockout mouse. In N. E. Buckley, E. Mezey, T. Bonner, A. Zimmer, C. C. Felder, M. Glass, et al. (Eds.), (1997). *Proceedings of the symposium on the cannabinoids*. Berlington, Vermont: International Cannabinoid Research Society.

Devinsky, O., Cilio, M. R., Cross, H., Fernández-Ruiz, J., French, J., Hill, C., et al. (2014). Cannabidiol: Pharmacology and potential therapeutic role in epilepsy and other neuropsychiatric disorders. *Epilepsia*, *55*, 791–802.

Dewey, W. L. (1986). Cannabinoid pharmacology. *Pharmacological Reviews*, *38*, 151–178.

Di Marzo, V., Breivogel, C. S., Tao, Q., Bridgen, D. T., Razdan, R. K., Zimmer, A. M., et al. (2000). Levels, metabolism, and pharmacological activity of anandamide in CB(1) cannabinoid receptor knockout mice: Evidence for non-CB(1), non-CB(2) receptor-mediated actions of anandamide in mouse brain. *Journal of Neurochemistry*, *75*, 2434–2444.

Di Marzo, V., Fontana, A., Cadas, H., Schinelli, S., Cimino, G., Schwartz, J. C., et al. (1994). Formation and inactivation of endogenous cannabinoid anandamide in central neurons. *Nature*, *372*, 686–691.

Digby, G. J., Sethi, P. R., & Lambert, N. A. (2008). Differential dissociation of G protein heterotrimers. *Journal of Physiology*, *586*, 3325–3335.

Dubreucq, S., Matias, I., Cardinal, P., Haring, M., Lutz, B., Marsicano, G., et al. (2012). Genetic dissection of the role of cannabinoid type-1 receptors in the emotional consequences of repeated social stress in mice. *Neuropsychopharmacology*, *37*, 1885–1900.

Eljaschewitsch, E., Witting, A., Mawrin, C., Lee, T., Schmidt, P. M., Wolf, S., et al. (2006). The endocannabinoid anandamide protects neurons during CNS inflammation by induction of MKP-1 in microglial cells. *Neuron*, *49*, 67–79.

Elsohly, M. A., & Slade, D. (2005). Chemical constituents of marijuana: The complex mixture of natural cannabinoids. *Life Sciences*, *78*, 539–548.

Fernández-Ruiz, J. J., Berrendero, F., Hernandez, M. L., Romero, J., & Ramos, J. A. (1999). Role of endocannabinoids in brain development. *Life Sciences*, *65*, 725–736.

Fernández-Ruiz, J., Berrendero, F., Hernandez, M. L., & Ramos, J. A. (2000). The endogenous cannabinoid system and brain development. *Trends in Neurosciences*, *23*, 14–20.

Fernández-Ruiz, J. J., Bonnin, A., Cebeira, M., & Ramos, J. A. (1994). Ontogenic and adults changes in the activity of hypothalamic and extrahypothalamic dopaminergic neurons after perinatal cannabinoid exposure. In T. Palomo, & T. Archer (Eds.), *Strategies for studying brain disorders* (Vol. 1) (pp. 357–390). England: Farrand Press.

Fernández-Ruiz, J., Pazos, M. R., Garcia-Arencibia, M., Sagredo, O., & Ramos, J. A. (2008). Role of CB2 receptors in neuroprotective effects of cannabinoids. *Molecular and Cellular Endocrinology*, *286*, S91–S96.

Fernández-Ruiz, J. J., Rodriguez, F., Navarro, M., & Ramos, J. A. (1992). Maternal cannabinoid exposure and brain development: Changes in the ontogeny of dopaminergic neurons. In L. L. Bartke AaM (Ed.), *Neurobiology and neurophysiology of cannabinoids: Biochemistry and physiology of substance abuse* (Vol. 4), (pp. 119–164). Boca Raton, FL: CRC Press.

Fernández-Ruiz, J. J., Romero, J., García-Gil, L., García-Palomero, E., & Ramos, J. A. (1996). Dopaminergic neurons as neurochemical substrates of neurobehavioral effects of marihuana: Developmental and adult studies. In R. J. Beninger, T. Archer, & T. Palomo (Eds.), *Dopamine disease states* (pp. 359–387). Madrid, Spain: CYM Press.

Fernández-Ruiz, J., Romero, J., Velasco, G., Tolon, R. M., Ramos, J. A., & Guzman, M. (2007). Cannabinoid CB2 receptor: A new target for controlling neural cell survival? *Trends in Pharmacological Sciences, 28*, 39–45.

Fried, P., Watkinson, B., James, D., & Gray, R. (2002). Current and former marijuana use: Preliminary findings of a longitudinal study of effects on IQ in young adults. *CMAJ, 166*, 887–891

Fritzsche, M. (2000). Are cannabinoid receptor knockout mice animal models of schizophrenia. *Medical Hypotheses, 56*, 638–643.

Fujii, M., Sherchan, P., Soejima, Y., Hasegawa, Y., Flores, J., Doycheva, D., et al. (2014). Cannabinoid receptor type 2 agonist attenuates apoptosis by activation of phosphorylated CREB-Bcl-2 pathway after subarachnoid hemorrhage in rats. *Experimental Neurology, 261*, 396–403.

Gadzicki, D., Muller-Vahl, K., & Stuhrmann, M. (1999). A frequent polymorphism in the coding exon of the human cannabinoid receptor (CNR1) gene. *Molecular and Cellular Probes, 13*, 321–323.

Galve-Roperh, I., Sanchez, C., Cortes, M. L., Gomez del Pulgar, T., Izquierdo, M., & Guzman, M. (2000). Anti-tumoral action of cannabinoids: Involvement of sustained ceramide accumulation and extracellular signal-regulated kinase activation. *Nature Medicine, 6*, 313–319.

Garcia-Gutierrez, M. S., Ortega-Alvaro, A., Busquets-Garcia, A., Perez-Ortiz, J. M., Caltana, L., Ricatti, M. J., et al. (2013). Synaptic plasticity alterations associated with memory impairment induced by deletion of CB2 cannabinoid receptors. *Neuropharmacology, 73*, 388–396.

Gebremedhin, D., Lange, A. R., Campbell, W. B., Hillard, C. J., & Harder, D. R. (1999). Cannabinoid CB1 receptor of cat cerebral arterial muscle functions to inhibit L-type Ca^{2+} channel current. *American Journal of Physiology, 276*, H2085–H2093.

Gerard, C. M., Mollereau, C., Vassart, G., & Parmentier, M. (1991). Molecular cloning of a human cannabinoid receptor which is also expressed in testis. *The Biochemical Journal, 279*, 129–134.

Gertsch, J., Schoop, R., Kuenzle, U., & Suter, A. (2004). Echinacea alkylamides modulate TNF-alpha gene expression via cannabinoid receptor CB2 and multiple signal transduction pathways. *FEBS Letters, 577*, 563–569.

Giuffrida, A., Parsons, L. H., Kerr, T. M., Rodriguez de Fonseca, F., Navarro, M., & Piomelli, D. (1999). Dopamine activation of endogenous cannabinoid signaling in dorsal striatum. *Nature Neuroscience, 2*, 358–363.

Glass, M., Dragunow, M., & Faull, R. L. (1997). Cannabinoid receptors in the human brain: A detailed anatomical and quantitative autoradiographic study in the fetal, neonatal and adult human brain. *Neuroscience, 77*, 299–318.

Glass, M., & Felder, C. C. (1997). Concurrent stimulation of cannabinoid CB1 and dopamine D2 receptors augments cAMP accumulation in striatal neurons: Evidence for a Gs linkage to the CB1 receptor. *The Journal of Neuroscience, 17*, 5327–5333.

Gomez del Pulgar, T., Velasco, G., & Guzman, M. (2000). The CB1 cannabinoid receptor is coupled to the activation of protein kinase B/Akt. *The Biochemical Journal, 347*, 369–373.

Gong, J. P., Onaivi, E. S., Ishiguro, H., Liu, Q. R., Tagliaferro, P. A., Brusco, A., et al. (2006). Cannabinoid CB2 receptors: Immunohistochemical localization in rat brain. *Brain Research, 1071*, 10–23.

Graham, E. S., Ball, N., Scotter, E. L., Narayan, P., Dragunow, M., & Glass, M. (2006). Induction of Krox-24 by endogenous cannabinoid type 1 receptors in Neuro2A cells is mediated by the MEK-ERK MAPK pathway and is suppressed by the phosphatidylinositol 3-kinase pathway. *The Journal of Biological Chemistry, 281*, 29085–29095.

Griffin, G., Tao, Q., & Abood, M. E. (2000). Cloning and pharmacological characterization of the rat CB(2) cannabinoid receptor. *The Journal of Pharmacology and Experimental Therapeutics, 292*, 886–894.

Gustafsson, K., Christensson, B., Sander, B., & Flygare, J. (2006). Cannabinoid receptor-mediated apoptosis induced by R(+)-methanandamide and Win55,212-2 is associated with ceramide accumulation and p38 activation in mantle cell lymphoma. *Molecular Pharmacology, 70*, 1612–1620.

Guzman, M., & Sanchez, C. (1999). Effects of cannabinoids on energy metabolism. *Life Sciences, 65*, 657–664.

Guzman, M., Galve-Roperh, I., & Sanchez, C. (2001). Ceramide: A new second messenger of cannabinoid action. *Trends in Pharmacological Sciences, 22*, 19–22.

Hampson, R. E., & Deadwyler, S. A. (1999). Cannabinoids, hippocampal function and memory. *Life Sciences, 65*, 715–723.

Hampson, R. E., & Deadwyler, S. A. (2000). Cannabinoids reveal the necessity of hippocampal neural encoding for short-term memory in rats. *The Journal of Neuroscience, 20*, 8932–8942.

Han, J., Kesner, P., Metna-Laurent, M., Duan, T., Xu, L., Georges, F., et al. (2012). Acute cannabinoids impair working memory through astroglial CB1 receptor modulation of hippocampal LTD. *Cell, 148*, 1039–1050.

Haring, M., Enk, V., Aparisi Rey, A., Loch, S., Ruiz de Azua, I., Weber, T., et al. (2015). Cannabinoid type-1 receptor signaling in central serotonergic neurons regulates anxiety-like behavior and sociability. *Frontiers in Behavioral Neuroscience, 9*, 235.

Haring, M., Marsicano, G., Lutz, B., & Monory, K. (2007). Identification of the cannabinoid receptor type 1 in serotonergic cells of raphe nuclei in mice. *Neuroscience, 146*, 1212–1219.

Harkany, T., Guzman, M., Galve-Roperh, I., Berghuis, P., Devi, L. A., & Mackie, K. (2007). The emerging functions of endocannabinoid signaling during CNS development. *Trends in Pharmacological Sciences, 28*, 83–92.

Herkenham, M., Lynn, A. B., Johnson, M. R., Melvin, L. S., de Cost, B. R., & Rice, K. C. (1991). Characterization and localization of cannabinoid receptors in rat brain: A quantitative in vitro autoradiographic study. *The Journal of Neuroscience, 16*, 8057–8066.

Herrera, B., Carracedo, A., Diez-Zaera, M., Gomez del Pulgar, T., Guzman, M., & Velasco, G. (2006). The CB2 cannabinoid receptor signals apoptosis via ceramide-dependent activation of the mitochondrial intrinsic pathway. *Experimental Cell Research, 312*, 2121–2131.

Herrera, B., Carracedo, A., Diez-Zaera, M., Guzman, M., & Velasco, G. (2005). p38 MAPK is involved in CB2 receptor-induced apoptosis of human leukaemia cells. *FEBS Letters, 579*, 5084–5088.

Ho, B. Y., Uezono, Y., Takada, S., Takase, I., & Izumi, F. (1999). Coupling of the expressed cannabinoid CB1 and CB2 receptors to phospholipase C and G protein-coupled inwardly rectifying K$^+$ channels. *Receptors Channels, 6*, 363–374.

Hoehe, M. R., Caenazzo, L., Martinez, M. M., Hsieh, W. T., Modi, W. S., Gershon, E. S., et al. (1991). Genetic and physical mapping of the human cannabinoid receptor gene to chromosome 6q14-q15. *New Biology, 3*, 880–885.

Hoenicka, J., Ponce, G., Jimenez-Arriero, M. A., Ampuero, I., Rodriguez-Jimenez, R., Rubio, G., et al. (2007). Association in alcoholic patients between psychopathic traits and the additive effect of allelic forms of the CNR1 and FAAH endocannabinoid genes, and the 3′ region of the DRD2 gene. *Neurotox Research, 11,* 51–60.

Hoffman, A. F., Oz, M., Yang, R., Lichtman, A. H., & Lupica, C. R. (2007). Opposing actions of chronic {Delta}9-tetrahydrocannabinol and cannabinoid antagonists on hippocampal long-term potentiation. *Learning & Memory, 14,* 63–74.

Hollister, L. E. (1986). Health aspects of cannabis. *Pharmacological Reviews, 38,* 1–20.

Howlett, A. C. (2002). The cannabinoid receptors. *Prostaglandins & Other Lipid Mediators, 68-69,* 619–631.

Howlett, A. C., Barth, F., Bonner, T. I., Cabral, G., Casellas, P., Devane, W. A., et al. (2002). International union of pharmacology. XXVII. Classification of cannabinoid receptors. *Pharmacological Reviews, 54,* 161–202.

Howlett, A. C., Johnson, M. R., Melvin, L. S., & Milne, G. M. (1988). Nonclassical cannabinoid analgetics inhibit adenylate cyclase: Development of a cannabinoid receptor model. *Molecular Pharmacology, 33,* 297–302.

Howlett, A. C., & Mukhopadhyay, S. (2000). Cellular signal transduction by anandamide and 2-arachidonoylglycerol. *Chemistry and Physics of Lipids, 108,* 53–70.

Hungund, B. L., Vinod, K. Y., Kassir, S. A., Basavarajappa, B. S., Yalamanchili, R., Cooper, T. B., et al. (2004). Upregulation of CB1 receptors and agonist-stimulated [35S]GTPgammaS binding in the prefrontal cortex of depressed suicide victims. *Molecular Psychiatry, 9,* 184–190.

Insel, T. R. (1995). The development of brain and behavior. In F. E. Bloom, & D. J. Kupfer (Eds.), *Psychopharmacology: The four generation of progress* (pp. 683–694). New York: Raven Press.

Jarai, Z., Wagner, J. A., Varga, K., Lake, K. D., Compton, D. R., Martin, B. R., et al. (1999). Cannabinoid-induced mesenteric vasodilation through an endothelial site distinct from CB1 or CB2 receptors. *Proceedings of the National Academy of Sciences of the United States of America, 96,* 14136–14141.

Johnson, J. P., Muhleman, D., MacMurray, J., Gade, R., Verde, R., Ask, M., et al. (1997). Association between the cannabinoid receptor gene (CNR1) and the P300 event-related potential. *Molecular Psychiatry, 2,* 169–171.

Kabelik, J., Krejci, Z., & Santavy, F. (1960). Cannabis as a medicant. *Bulletin on Narcotics, 12,* 5–23.

Kamprath, K., Romo-Parra, H., Haring, M., Gaburro, S., Doengi, M., Lutz, B., et al. (2011). Short-term adaptation of conditioned fear responses through endocannabinoid signaling in the central amygdala. *Neuropsychopharmacology, 36,* 652–663.

Katona, I., & Freund, T. F. (2012). Multiple functions of endocannabinoid signaling in the brain. *Annual Review of Neuroscience, 35,* 529–558.

Katona, I., Rancz, E. A., Acsady, L., Ledent, C., Mackie, K., Hajos, N., et al. (2001). Distribution of CB1 cannabinoid receptors in the amygdala and their role in the control of GABAergic transmission. *The Journal of Neuroscience, 21,* 9506–9518.

Katona, I., Sperlagh, B., Sik, A., Kafalvi, A., Vizi, E. S., Mackie, K., et al. (1999). Presynaptically located CB1 cannabinoid receptors regulate GABA release from axon terminals of specific hippocampal interneurons. *The Journal of Neuroscience, 19,* 4544–4558.

Katona, I., Urban, G. M., Wallace, M., Ledent, C., Jung, K. M., Piomelli, D., et al. (2006). Molecular composition of the endocannabinoid system at glutamatergic synapses. *The Journal of Neuroscience, 26,* 5628–5637.

Kearn, C. S., Greenberg, M. J., DiCamelli, R., Kurzawa, K., & Hillard, C. J. (1999). Relationships between ligand affinities for the cerebellar cannabinoid receptor CB1 and the induction of GDP/GTP exchange. *Journal of Neurochemistry, 72,* 2379–2387.

Kim, D., & Thayer, S. A. (2001). Cannabinoids inhibit the formation of new synapses between hippocampal neurons in culture. *The Journal of Neuroscience, 21*, RC146.

Kim, J., & Li, Y. (2014). Chronic activation of CB2 cannabinoid receptors in the hippocampus increases excitatory synaptic transmission. *Journal of Physiology, 493*, 871–886.

Kirilly, E., Hunyady, L., & Bagdy, G. (2013). Opposing local effects of endocannabinoids on the activity of noradrenergic neurons and release of noradrenaline: Relevance for their role in depression and in the actions of CB(1) receptor antagonists. *Journal of Neural Transmission (Vienna), 120*, 177–186.

Kreitzer, A. C., & Regehr, W. G. (2001). Retrograde inhibition of presynaptic calcium influx by endogenous cannabinoids at excitatory synapses onto Purkinje cells. *Neuron, 29*, 717–727.

Lambert, N. A. (2008). Dissociation of heterotrimeric g proteins in cells. *Science Signaling, 1*, re5.

Ledent, C., Valverde, O., Cossu, G., Petitet, F., Aubert, J. F., Beslot, F., et al. (1999). Unresponsiveness to cannabinoids and reduced addictive effects of opiates in CB1 receptor knockout mice. *Science, 283*, 401–404.

Leroy, S., Griffon, N., Bourdel, M. C., Olie, J. P., Poirier, M. F., & Krebs, M. O. (2001). Schizophrenia and the cannabinoid receptor type 1 (CB1): Association study using a single-base polymorphism in coding exon 1. *American Journal of Medical Genetics, 105*, 749–752.

Li, Y., & Kim, J. (2015a). Deletion of CB2 cannabinoid receptors reduces synaptic transmission and long-term potentiation in the mouse hippocampus. *Hippocampus, 26*(3), 275–281.

Li, Y., & Kim, J. (2015b). Neuronal expression of CB2 cannabinoid receptor mRNAs in the mouse hippocampus. *Neuroscience, 311*, 253–267.

Li, Y., & Kim, J. (2016a). CB2 cannabinoid receptor knockout in mice impairs contextual long-term memory and enhances spatial working memory. *Neural Plasticity, 2016*, 9817089.

Liu, J., Gao, B., Mirshahi, F., Sanyal, A. J., Khanolkar, A. D., Makriyannis, A., et al. (March 15, 2000). Functional CB1 cannabinoid receptors in human vascular endothelial cells. *The Biochemical Journal, 346*(Pt 3), 835–840.

Maccarrone, M., Bab, I., Biro, T., Cabral, G. A., Dey, S. K., Di Marzo, V., et al. (2015). Endocannabinoid signaling at the periphery: 50 years after THC. *Trends in Pharmacological Sciences, 36*, 277–296.

Mackie, K., & Hille, B. (1992). Cannabinoids inhibit N-type calcium channels in neuroblastoma-glioma cells. *Proceedings of the National Academy of Sciences of the United States of America, 89*, 3825–3829.

Mailleux, P., & Vanderhaeghen, J. J. (1992). Distribution of neuronal cannabinoid receptor in the adult rat brain: A comparative receptor binding radioautography and in situ hybridization histochemistry. *Neuroscience, 48*, 655–668.

Mailleux, P., Verslype, M., Preud'homme, X., & Vanderhaeghen, J. J. (1994). Activation of multiple transcription factor genes by tetrahydrocannabinol in rat forebrain. *Neuroreport, 5*, 1265–1268.

Maldonado, R., & Rodriguez de Fonseca, F. (2002). Cannabinoid addiction: Behavioral models and neural correlates. *The Journal of Neuroscience, 22*, 3326–3331.

Maresz, K., Carrier, E. J., Ponomarev, E. D., Hillard, C. J., & Dittel, B. N. (2005). Modulation of the cannabinoid CB2 receptor in microglial cells in response to inflammatory stimuli. *Journal of Neurochemistry, 95*, 437–445.

Marinelli, S., Pacioni, S., Cannich, A., Marsicano, G., & Bacci, A. (2009). Self-modulation of neocortical pyramidal neurons by endocannabinoids. *Nature Neuroscience, 12*, 1488–1490.

Marsicano, G., Goodenough, S., Monory, K., Hermann, H., Eder, M., Cannich, A., et al. (2003). CB1 cannabinoid receptors and on-demand defense against excitotoxicity. *Science, 302*, 84–88.

Marsicano, G., & Lutz, B. (1999). Expression of the cannabinoid receptor CB1 in distinct neuronal subpopulations in the adult mouse forebrain. *The European Journal of Neuroscience, 11*, 4213–4225.

Martin-Garcia, E., Bourgoin, L., Cathala, A., Kasanetz, F., Mondesir, M., Gutierrez-Rodriguez, A., et al. (2015). Differential control of cocaine self-administration by GABAergic and glutamatergic CB1 cannabinoid receptors. *Neuropsychopharmacology, 41*(9), 2192–2205.

Matsuda, L. A. (1997). Molecular aspects of cannabinoid receptors. *Critical Review Neurobiology, 11*, 143–166.

Matsuda, L. A., Bonner, T. I., & Lolait, S. J. (1993). Localization of cannabinoid receptor mRNA in rat brain. *Journal of Comparative Neurology, 327*, 535–550.

McGregor, I. S., Arnold, J. C., Weber, M. F., Topple, A. N., & Hunt, G. E. (1998). A comparison of delta 9-THC and anandamide induced c-fos expression in the rat forebrain. *Brain Research, 802*, 19–26.

Mechoulam, R., Fride, E., & Di Marzo, V. (1998). Endocannabinoids. *European Journal of Pharmacology, 359*, 1–18.

Mechoulam, R., & Hanus, L. (2000). A historical overview of chemical research on cannabinoids. *Chemistry and Physics of Lipids, 108*, 1–13.

Mechoulam, R., & Parker, L. A. (2013). The endocannabinoid system and the brain. *Annual Review of Psychology, 64*, 21–47.

Mehmedic, Z., Chandra, S., Slade, D., Denham, H., Foster, S., Patel, A. S., et al. (2010). Potency trends of Delta9-THC and other cannabinoids in confiscated cannabis preparations from 1993 to 2008. *Journal of Forensic Sciences, 55*, 1209–1217.

Meller, R., Minami, M., Cameron, J. A., Impey, S., Chen, D., Lan, J. Q., et al. (2005). CREB-mediated Bcl-2 protein expression after ischemic preconditioning. *Journal of Cerebral Blood Flow & Metabolism, 25*, 234–246.

Mokler, D. J., Robinson, S. E., Johnson, J. H., Hong, J. S., & Rosecrans, J. A. (1987). Neonatal administration of delta-9-tetrahydrocannabinol (THC) alters the neurochemical response to stress in the adult Fischer-344 rat. *Neurotoxicology and Teratology, 9*, 321–327.

Molina-Holgado, E., Vela, J. M., Arevalo-Martin, A., Almazan, G., Molina-Holgado, F., Borrell, J., et al. (2002). Cannabinoids promote oligodendrocyte progenitor survival: Involvement of cannabinoid receptors and phosphatidylinositol-3 kinase/Akt signaling. *The Journal of Neuroscience, 22*, 9742–9753.

Monory, K., Massa, F., Egertova, M., Eder, M., Blaudzun, H., Westenbroek, R., et al. (2006). The endocannabinoid system controls key epileptogenic circuits in the hippocampus. *Neuron, 51*, 455–466.

Mu, J., Zhuang, S. Y., Kirby, M. T., Hampson, R. E., & Deadwyler, S. A. (1999). Cannabinoid receptors differentially modulate potassium A and D currents in hippocampal neurons in culture. *The Journal of Pharmacology and Experimental Therapeutics, 291*, 893–902.

Mukhopadhyay, S., McIntosh, H. H., Houston, D. B., & Howlett, A. C. (2000). The CB(1) cannabinoid receptor juxtamembrane C-terminal peptide confers activation to specific G proteins in brain. *Molecular Pharmacology, 57*, 162–170.

Munro, S., Thomas, K. L., & Abu-Shaar, M. (1993). Molecular characterization of a peripheral receptor for cannabinoids. *Nature, 365*, 61–65.

Murillo-Rodriguez, E., Blanco-Centurion, C., Sanchez, C., Piomelli, D., & Shiromani, P. J. (2003). Anandamide enhances extracellular levels of adenosine and induces sleep: An in vivo microdialysis study. *Sleep, 26*, 943–947.

Murillo-Rodriguez, E., Sanchez-Alavez, M., Navarro, L., Martinez-Gonzalez, D., Drucker-Colin, R., & Prospero-Garcia, O. (1998). Anandamide modulates sleep and memory in rats. *Brain Research, 812,* 270–274.

Nagre, N. N., Subbanna, S., Shivakumar, M., Psychoyos, D., & Basavarajappa, B. S. (2015). CB1-Receptor knockout neonatal mice are protected against ethanol-induced impairments of DNMT1, DNMT3A and DNA methylation. *Journal of Neurochemistry, 132,* 429–442.

Navarro, M., de Miguel, R., Rodriguez de Fonseca, F., Ramos, J. A., & Fernández-Ruiz, J. J. (1996). Perinatal cannabinoid exposure modifies the sociosexual approach behavior and the mesolimbic dopaminergic activity of adult male rats. *Behavioural Brain Research, 75,* 91–98.

Navarro, M., Rodriguez de Fonseca, F., Hernandez, M. L., Ramos, J. A., & Fernández-Ruiz, J. J. (1994). Motor behavior and nigrostriatal dopaminergic activity in adult rats perinatally exposed to cannabinoids. *Pharmacology, Biochemistry, and Behavior, 47,* 47–58.

Netzeband, J. G., Conroy, S. M., Parsons, K. L., & Gruol, D. L. (1999). Cannabinoids enhance NMDA-elicited Ca^{2+} signals in cerebellar granule neurons in culture. *The Journal of Neuroscience, 19,* 8765–8777.

Nogueron, M. I., Porgilsson, B., Schneider, W. E., Stucky, C. L., & Hillard, C. J. (2001). Cannabinoid receptor agonists inhibit depolarization-induced calcium influx in cerebellar granule neurons. *Journal of Neurochemistry, 79,* 371–381.

Nunez, E., Benito, C., Pazos, M. R., Barbachano, A., Fajardo, O., Gonzalez, S., et al. (2004). Cannabinoid CB2 receptors are expressed by perivascular microglial cells in the human brain: An immunohistochemical study. *Synapse, 53,* 208–213.

Ohno-Shosaku, T., Maejima, T., & Kano, M. (2001). Endogenous cannabinoids mediate retrograde signals from depolarized postsynaptic neurons to presynaptic terminal. *Neuron, 29,* 729–738.

Ohno-Shosaku, T., Sawada, S., & Kano, M. (2000). Heterosynaptic expression of depolarization-induced suppression of inhibition (DSI) in rat hippocampal cultures. *Neuroscience Research, 36,* 67–71.

Ohno-Shosaku, T., Sawada, S., & Yamamoto, C. (1998). Properties of depolarization-induced suppression of inhibitory transmission in cultured rat hippocampal neurons. *Pflugers Archiv, 435,* 273–279.

Ohno-Shosaku, T., Tsubokawa, H., Mizushima, I., Yoneda, N., Zimmer, A., & Kano, M. (2002). Presynaptic cannabinoid sensitivity is a major determinant of depolarization-induced retrograde suppression at hippocampal synapses. *The Journal of Neuroscience, 22,* 3864–3872.

Onaivi, E. S., Chakrabarti, A., & Chaudhuri, G. (1996). Cannabinoid receptor genes. *Progress in Neurobiology, 48,* 275–305.

Onaivi, E. S., Ishiguro, H., Gong, J. P., Patel, S., Meozzi, P. A., Myers, L., et al. (2008). Brain neuronal CB2 cannabinoid receptors in drug abuse and depression: From mice to human subjects. *PLoS One, 3,* e1640.

Onaivi, E. S., Ishiguro, H., Gong, J. P., Patel, S., Perchuk, A., Meozzi, P. A., et al. (2006). Discovery of the presence and functional expression of cannabinoid CB2 receptors in brain. *Annals of the New York Academy of Sciences, 1074,* 514–536.

Onaivi, E. S., Ishiguro, H., Gu, S., & Liu, Q. R. (2012). CNS effects of CB2 cannabinoid receptors: beyond neuroimmuno-cannabinoid activity. *Journal of Psychopharmacology, 26,* 92–103.

Onaivi, E. S., Leonard, C. M., Ishiguro, H., Zhang, P. W., Lin, Z., Akinshola, B. E., et al. (2002). Endocannabinoids and cannabinoid receptor genetics. *Progress in Neurobiology, 66,* 307–344.

Ortega-Alvaro, A., Aracil-Fernandez, A., Garcia-Gutierrez, M. S., Navarrete, F., & Manzanares, J. (2011). Deletion of CB2 cannabinoid receptor induces schizophrenia-related behaviors in mice. *Neuropsychopharmacology, 36,* 1489–1504.

Pacher, P., & Kunos, G. (2013). Modulating the endocannabinoid system in human health and disease—successes and failures. *FEBS Journal, 280*, 1918–1943.

Palazuelos, J., Aguado, T., Egia, A., Mechoulam, R., Guzman, M., & Galve-Roperh, I. (2006). Non-psychoactive CB2 cannabinoid agonists stimulate neural progenitor proliferation. *FASEB Journal, 20*, 2405–2407.

Pan, X., Ikeda, S. R., & Lewis, D. L. (1996). Rat brain cannabinoid receptor modulates N-type Ca^{2+} channels in a neuronal expression system. *Molecular Pharmacology, 49*, 707–714.

Patel, N. A., Moldow, R. L., Patel, J. A., Wu, G., & Chang, S. L. (1998). Arachidonylethanolamide (AEA) activation of FOS proto-oncogene protein immunoreactivity in the rat brain. *Brain Research, 797*, 225–233.

Pertwee, R. G. (1997). Pharmacology of cannabinoid CB1 and CB2 receptors. *Pharmacology & Therapeutics, 74*, 129–180.

Pertwee, R. G. (2001). Cannabinoid receptors and pain. *Progress in Neurobiology, 63*, 569–611.

Pertwee, R. G. (2005). Pharmacological actions of cannabinoids. *Handbook of Experimental Pharmacology*, 1–51.

Pfister-Genskow, M., Weesner, G. D., Hayes, H., Eggen, A., & Bishop, M. D. (1997). Physical and genetic localization of the bovine cannabinoid receptor (CNR1) gene to bovine chromosome 9. *Mammalian Genome, 8*, 301–302.

Pinto, J. C., Potie, F., Rice, K. C., Boring, D., Johnson, M. R., Evans, D. M., et al. (1994). Cannabinoid receptor binding and agonist activity of amides and esters of arachidonic acid. *Molecular Pharmacology, 46*, 516–522.

Puente, N., Elezgarai, I., Lafourcade, M., Reguero, L., Marsicano, G., Georges, F., et al. (2010). Localization and function of the cannabinoid CB1 receptor in the anterolateral bed nucleus of the stria terminalis. *PLoS One, 5*, e8869.

Ramesh, D., Ross, G. R., Schlosburg, J. E., Owens, R. A., Abdullah, R. A., Kinsey, S. G., et al. (2011). Blockade of endocannabinoid hydrolytic enzymes attenuates precipitated opioid withdrawal symptoms in mice. *The Journal of Pharmacology and Experimental Therapeutics, 339*, 173–185.

Ramikie, T. S., Nyilas, R., Bluett, R. J., Gamble-George, J. C., Hartley, N. D., Mackie, K., et al. (2014). Multiple mechanistically distinct modes of endocannabinoid mobilization at central amygdala glutamatergic synapses. *Neuron, 81*, 1111–1125.

Reibaud, M., Obinu, M. C., Ledent, C., Parmentier, M., Bohme, G. A., & Imperato, A. (1999). Enhancement of memory in cannabinoid CB1 receptor knock-out mice. *European Journal of Pharmacology, 379*, R1–R2.

Rhee, M. H., Bayewitch, M., Avidor-Reiss, T., Levy, R., & Vogel, Z. (1998). Cannabinoid receptor activation differentially regulates the various adenylyl cyclase isozymes. *Journal of Neurochemistry, 71*, 1525–1534.

Rodriguez-Arias, M., Navarrete, F., Blanco-Gandia, M. C., Arenas, M. C., Aguilar, M. A., Bartoll-Andres, A., et al. (2015). Role of CB2 receptors in social and aggressive behavior in male mice. *Psychopharmacology (Berlin), 232*, 3019–3031.

Romero, J., Garcia-Palomero, E., Berrendero, F., Garcia-Gil, L., Hernandez, M. L., Ramos, J. A., et al. (1997). Atypical location of cannabinoid receptors in white matter areas during rat brain development. *Synapse, 26*, 317–323.

Ross, S. A., & ElSohly, M. A. (1996). The volatile oil composition of fresh and air-dried buds of *Cannabis sativa*. *Journal of Natural Products, 59*, 49–51.

Rueda, D., Galve-Roperh, I., Haro, A., & Guzman, M. (2000). The CB(1) cannabinoid receptor is coupled to the activation of c-Jun N-terminal kinase. *Molecular Pharmacology, 58*, 814–820.

Ruehle, S., Remmers, F., Romo-Parra, H., Massa, F., Wickert, M., Wortge, S., et al. (2013). Cannabinoid CB1 receptor in dorsal telencephalic glutamatergic neurons: Distinctive sufficiency for hippocampus-dependent and amygdala-dependent synaptic and behavioral functions. *The Journal of Neuroscience, 33*, 10264–10277.

Sales-Carbonell, C., Rueda-Orozco, P. E., Soria-Gomez, E., Buzsaki, G., Marsicano, G., & Robbe, D. (2013). Striatal GABAergic and cortical glutamatergic neurons mediate contrasting effects of cannabinoids on cortical network synchrony. *Proceedings of the National Academy of Sciences of the United States of America, 110*, 719–724.

Samson, M. T., Small-Howard, A., Shimoda, L. M., Koblan-Huberson, M., Stokes, A. J., & Turner, H. (2003). Differential roles of CB1 and CB2 cannabinoid receptors in mast cells. *The Journal of Immunology, 170*, 4953–4962.

Sanchez, C., Galve-Roperh, I., Rueda, D., & Guzman, M. (1998). Involvement of sphingomyelin hydrolysis and the mitogen-activated protein kinase cascade in the Delta9-tetrahydrocannabinol-induced stimulation of glucose metabolism in primary astrocytes. *Molecular Pharmacology, 54*, 834–843.

Sanchez, C., Rueda, D., Segui, B., Galve-Roperh, I., Levade, T., & Guzman, M. (2001). The CB(1) cannabinoid receptor of astrocytes is coupled to sphingomyelin hydrolysis through the adaptor protein fan. *Molecular Pharmacology, 59*, 955–959.

Sanchez, C., Velasco, G., & Guzman, M. (1997). Delta9-tetrahydrocannabinol stimulates glucose utilization in C6 glioma cells. *Brain Research, 767*, 64–71.

Schweitzer, P. (2000). Cannabinoids decrease the K(+) M-current in hippocampal CA1 neurons. *The Journal of Neuroscience, 20*, 51–58.

Sharma, P., Murthy, P., & Bharath, M. M. (2012). Chemistry, metabolism, and toxicology of cannabis: Clinical implications. *Iranian Journal of Psychiatry, 7*, 149–156.

Shivachar, A. C., Martin, B. R., & Ellis, E. F. (1996). Anandamide- and delta9-tetrahydrocannabinol-evoked arachidonic acid mobilization and blockade by SR141716A [N-(Piperidin-1-yl)-5-(4-chlorophenyl)-1-(2,4-dichlorophenyl)-4-methyl-1H-pyrazole-3-carboximide hydrochloride]. *Biochemical Pharmacology, 51*, 669–676.

Sipe, J. C., Chiang, K., Gerber, A. L., Beutler, E., & Cravatt, B. F. (2002). A missense mutation in human fatty acid amide hydrolase associated with problem drug use. *Proceedings of the National Academy of Sciences of the United States of America, 99*, 8394–8399.

Soderstrom, K., & Johnson, F. (2000). CB1 cannabinoid receptor expression in brain regions associated with zebra finch song control. *Brain Research, 857*, 151–157.

Soltesz, I., Alger, B. E., Kano, M., Lee, S. H., Lovinger, D. M., Ohno-Shosaku, T., et al. (2015). Weeding out bad waves: Towards selective cannabinoid circuit control in epilepsy. *Nature Reviews Neuroscience, 16*, 264–277.

Stefanis, N. C., Delespaul, P., Henquet, C., Bakoula, C., Stefanis, C. N., & Van Os, J. (2004). Early adolescent cannabis exposure and positive and negative dimensions of psychosis. *Addiction, 99*, 1333–1341.

Stefano, G. B., Salzet, B., & Salzet, M. (1997). Identification and characterization of the leech CNS cannabinoid receptor: Coupling to nitric oxide release. *Brain Research, 753*, 219–224.

Steiner, H., Bonner, T. I., Zimmer, A. M., Kitai, S. T., & Zimmer, A. (1999). Altered gene expression in striatal projection neurons in CB1 cannabinoid receptor knockout mice. *Proceedings of the National Academy of Sciences of the United States of America, 96*, 5786–5790.

Stella, N. (2004). Cannabinoid signaling in glial cells. *Glia, 48*, 267–277.

Stubbs, L., Chittenden, L., Chakrabarti, A., & Onaivi, E. (1996). The gene encoding the central cannabinoid receptor is located in proximal mouse chromosome 4. *Mammalian Genome, 7*, 165–166.

Subbanna, S., Nagaraja, N. N., Umapathy, N. S., Pace, B. S., & Basavarajappa, B. S. (2015). Ethanol exposure induces neonatal neurodegeneration by enhancing CB1R Exon1 Histone H4K8 acetylation and up-regulating CB1R function causing neurobehavioral abnormalities in adult mice. *International Journal of Neuropsychopharmacology, 18*(5), (in press).

Subbanna, S., Shivakumar, M., Psychoyos, D., Xie, S., & Basavarajappa, B. S. (2013). Anandamide-CB1 receptor signaling contributes to postnatal ethanol-induced neonatal neurodegeneration, adult synaptic and memory deficits. *Journal of Neuroscience, 33*, 6350–6366.

Sugiura, T., Kishimoto, S., Oka, S., & Gokoh, M. (2006). Biochemistry, pharmacology and physiology of 2-arachidonoylglycerol, an endogenous cannabinoid receptor ligand. *Progress in Lipid Research, 45*, 405–446.

Sugiura, T., Kodaka, T., Kondo, S., Nakane, S., Kondo, H., Waku, K., et al. (1997). Is the cannabinoid CB1 receptor a 2-arachidonoylglycerol receptor? Structural requirements for triggering a Ca^{2+} transient in NG108-15 cells. *Journal of Biochemistry (Tokyo), 122*, 890–895.

Sugiura, T., Kodaka, T., Kondo, S., Tonegawa, T., Nakane, S., Kishimoto, S., et al. (1996). 2-Arachidonoylglycerol, a putative endogenous cannabinoid receptor ligand, induces rapid, transient elevation of intracellular free Ca^{2+} in neuroblastoma x glioma hybrid NG108-15 cells. *Biochemical and Biophysical Research Communications, 229*, 58–64.

Sugiura, T., Kodaka, T., Nakane, S., Miyashita, T., Kondo, S., Suhara, Y., et al. (1999). Evidence that the cannabinoid CB1 receptor is a 2-arachidonylglycerol receptor. *The Journal of Biological Chemistry, 274*, 2794–2801.

Terranova, J. P., Storme, J. J., Lafon, N., Perio, A., Rinaldi-Carmona, M., Le Fur, G., et al. (1996). Improvement of memory in rodents by the selective CB1 cannabinoid receptor antagonist, SR 141716. *Psychopharmacology (Berlin), 126*, 165–172.

Tokudome, T., Horio, T., Fukunaga, M., Okumura, H., Hino, J., Mori, K., et al. (2004). Ventricular nonmyocytes inhibit doxorubicin-induced myocyte apoptosis: Involvement of endogenous endothelin-1 as a paracrine factor. *Endocrinology, 145*, 2458–2466.

Turner, C. E., Elsohly, M. A., & Boeren, E. G. (1980). Constituents of *Cannabis sativa* L. XVII. A review of the natural constituents. *Journal of Natural Products, 43*, 169–234.

Valjent, E., Caboche, J., & Vanhoutte, P. (2001). Mitogen-activated protein kinase/extracellular signal-regulated kinase induced gene regulation in brain: A molecular substrate for learning and memory? *Molecular Neurobiology, 23*, 83–99.

Van Sickle, M. D., Duncan, M., Kingsley, P. J., Mouihate, A., Urbani, P., Mackie, K., et al. (2005). Identification and functional characterization of brainstem cannabinoid CB2 receptors. *Science, 310*, 329–332.

Varvel, S. A., & Lichtman, A. H. (2002). Evaluation of CB1 receptor knockout mice in the Morris water maze. *The Journal of Pharmacology and Experimental Therapeutics, 301*, 915–924.

Vela, G., Fuentes, J. A., Bonnin, A., Fernández-Ruiz, J., & Ruiz-Gayo, M. (1995). Perinatal exposure to delta 9-tetrahydrocannabinol (delta 9-THC) leads to changes in opioid-related behavioral patterns in rats. *Brain Research, 680*, 142–147.

Vela, G., Martin, S., Garcia-Gil, L., Crespo, J. A., Ruiz-Gayo, M., Fernández-Ruiz, J. J., et al. (1998). Maternal exposure to delta9-tetrahydrocannabinol facilitates morphine self-administration behavior and changes regional binding to central mu opioid receptors in adult offspring female rats. *Brain Research, 807*, 101–109.

Wagner, J. A., Varga, K., Jarai, Z., & Kunos, G. (1999). Mesenteric vasodilation mediated by endothelial anandamide receptors. *Hypertension, 33*, 429–434.

Wartmann, M., Campbell, D., Subramanian, A., Burstein, S. H., & Davis, R. J. (1995). The MAP kinase signal transduction pathway is activated by the endogenous cannabinoid anandamide. *FEBS Letters, 2–3*, 133–136.

Watson, S. J., Benson, J. A., Jr., & Joy, J. E. (2000). Marijuana and medicine: Assessing the science base: A summary of the 1999 Institute of Medicine report. *Archives of General Psychiatry, 57*, 547–552.

Wilson, B. E., Mochon, E., & Boxer, L. M. (1996). Induction of bcl-2 expression by phosphorylated CREB proteins during B-cell activation and rescue from apoptosis. *Molecular and Cellular Biology, 16*, 5546–5556.

Wilson, R. I., Kunos, G., & Nicoll, R. A. (2001). Presynaptic specificity of endocannabinoid signaling in the hippocampus. *Neuron, 31*, 453–462.

Wilson, R. I., & Nicoll, R. A. (2001). Endogenous cannabinoids mediate retrograde signalling at hippocampal synapses. *Nature, 410*, 588 592.

Wilson, R. I., & Nicoll, R. A. (2002). Endocannabinoid signaling in the brain. *Science, 296,* 678–682.

Yamaguchi, F., Macrae, A. D., & Brenner, S. (1996). Molecular cloning of two cannabinoid type 1-like receptor genes from the puffer fish *Fugu rubripes*. *Genomics, 35,* 603–605.

Yrjola, S., Sarparanta, M., Airaksinen, A. J., Hytti, M., Kauppinen, A., Pasonen-Seppanen, S., et al. (2015). Synthesis, in vitro and in vivo evaluation of 1,3,5-triazines as cannabinoid CB2 receptor agonists. *European Journal of Pharmaceutical Sciences, 67,* 85–96.

Zhang, H.Y., Gao, M., Liu, Q. R., Bi, G. H., Li, X., Yang, H. J., et al. (2014). Cannabinoid CB2 receptors modulate midbrain dopamine neuronal activity and dopamine-related behavior in mice. *Proceedings of the National Academy of Sciences of the United States of America, 111,* E5007–E5015.

Zhang, H.Y., Gao, M., Shen, H., Bi, G. H., Yang, H. J., Liu, Q. R., et al. (2016). Expression of functional cannabinoid CB receptor in VTA dopamine neurons in rats. *Addiction Biology*.

Zhang, P. W., Ishiguro, H., Ohtsuki, T., Hess, J., Carillo, F., Walther, D., et al. (2004). Human cannabinoid receptor 1:5′ exons, candidate regulatory regions, polymorphisms, haplotypes and association with polysubstance abuse. *Molecular Psychiatry, 9,* 916–931.

Zimmer, A. (2015). Genetic manipulation of the endocannabinoid system. *Handbook of Experimental Pharmacology, 231,* 129–183.

Zimmer, A., Zimmer, A. M., Hohmann, A. G., Herkenham, M., & Bonner, T. I. (1999). Increased mortality, hypoactivity, and hypoalgesia in cannabinoid CB1 receptor knockout mice. *Proceedings of the National Academy of Sciences of the United States of America, 96,* 5780–5785.

CHAPTER 3

The Endocannabinoid System and Parkinson Disease

Andrea Giuffrida, Alex Martinez
University of Texas Health Science Center, San Antonio, TX, United States

INTRODUCTION: THE ENDOCANNABINOID SYSTEM

The endocannabinoid (EC) system is composed of several lipid-signaling molecules (ECs), the enzymes responsible for their synthesis and degradation, and the metabotropic, ionotropic, and nuclear receptors activated by these ligands (Bouaboula et al., 2005; De Petrocellis & Di Marzo, 2010; Gasperi et al., 2007; Zygmunt et al., 1999).

Two metabotropic G-protein-coupled cannabinoid (CB) receptors have been cloned to date, CB1 (Matsuda, Lolait, Brownstein, Young, & Bonner, 1990) and CB2 (Munro, Thomas, & Abu-Shaar, 1993). The former are the most abundant G-protein coupled receptors within the central nervous system (CNS), and given their localization at GABAergic and glutamatergic synapses, their activation can modulate inhibitory and excitatory neurotransmission (Gerdeman & Lovinger, 2001; Herkenham, Lynn, de Costa, & Richfield, 1991; Katona et al., 1999). High concentrations of CB1 receptors are found in the caudate putamen, substantia nigra, globus pallidus, hippocampus, and the molecular layer of the cerebellum (Herkenham et al., 1990), which is consistent with the profound effects exerted by the major ingredient of marijuana, Δ^9-tetrahydrocannabinol (THC) on cognition and motor activity.

CB2 receptors are primarily found on peripheral immunocompetent cells and lymphoid organs (Berdyshev, 2000; Sugiura & Waku, 2000); however, studies have shown that they are also present in the CNS, although in smaller concentrations than CB1 (Gong et al., 2006; Onaivi et al., 2006). CB2 receptors modulate microglia activation and migration (Walter et al., 2003) and consequently play a role in neuroinflammation, including the inflammatory response observed in animal models of neurodegenerative disorders (Carta, Pisanu, & Carboni, 2011; Price et al., 2009).

The Endocannabinoid System
ISBN 978-0-12-809666-6
http://dx.doi.org/10.1016/B978-0-12-809666-6.00003-4

Anandamide (AEA) and 2-arachidonoylglycerol (2-AG) are the two most studied ECs. Unlike classical neurotransmitters, ECs are not stored in synaptic vesicles, but are synthesized and released on demand from postsynaptic neurons to act on presynaptic CB receptors in a retrograde fashion (Kreitzer & Regehr, 2001).

AEA, which is a partial agonist at CB1 and CB2 receptors, is synthesized from the phospholipid precursor N-arachidonoyl phosphatidylethanolamine (NAPE) by a phospholipase D (PLD) (Cadas, Gaillet, Beltramo, Venance, & Piomelli, 1996; Okamoto, Morishita, Tsuboi, Tonai, & Ueda, 2004). However, NAPE-PLD knockout mice show unaltered AEA brain levels (Leung, Saghatelian, Simon, & Cravatt, 2006) suggesting that alternative biosynthetic pathways may lead to its production. Indeed, non-NAPE-PLD biosynthesis of AEA has been shown to be initiated via phospholipase C (PLC)/phosphatase and αβ-hydrolase 4 (Abhd4) (Liu et al., 2008; Simon & Cravatt, 2006). AEA metabolism occurs via enzymatic hydrolysis by fatty acid amide hydrolase (FAAH) and, to a lesser extent, by cyclooxygenase-2 (Giuffrida & Seillier, 2012), which converts AEA into a prostaglandin-like molecule.

In addition to CB receptors, AEA can activate members of the transient receptor potential channel family (TRP channels), specifically TRPV1 (Zygmunt et al., 1999), as well as different peroxisome proliferator-activated receptor (PPAR) subtypes (Bouaboula et al., 2005).

TRPV1 are nonselective cation channels (Szallasi & Blumberg, 1999), primarily known for their role in processing inflammatory and thermal pain stimuli (Starowicz et al., 2007). They are expressed in sensory nerves, as well as in basal ganglia regions, such as the striatum, globus pallidus, and substantia nigra (Cristino et al., 2006; Mezey et al., 2000; Moran, Xu, & Clapham, 2004). Within the substantia nigra, activation of TRPV1 channels has been shown to increase the frequency of excitatory postsynaptic currents of dopamine neurons (Marinelli et al., 2003). Although AEA has lower affinity for TRPV1 than CB receptors, it can act as a full/partial agonist at TRPV1 under certain conditions (Di Marzo, Bisogno, & De Petrocellis, 2001; Ross, 2003), such as upon pharmacological blockade of FAAH, which elevates AEA intracellular concentrations. There is also evidence of a cross talk between CB1 and TRPV1 receptors as CB1 activation can modulate TRPV1 phosphorylation and consequently affect its functional state (Patwardhan et al., 2006; Starowicz et al., 2007).

The second major EC, 2-AG is a full agonist at CB receptors and is mobilized from biological membranes via the hydrolysis of inositol phospholipids by PLC, leading to the production of diacylglycerol (DAG) and its conversion into 2-AG by DAG lipase (Stella, Schweitzer, & Piomelli, 1997).

Metabolism of 2-AG primarily occurs via the hydrolyzing enzyme mono-acylglycerol lipase (MAGL) (Beltramo & Piomelli, 2000; Dinh et al., 2002); however, knockdown of the serine hydrolase alpha-beta-hydrolase domain 6 (ABHD6) has been shown to decrease 2-AG hydrolysis and affect 2-AG-induced stimulation of cell migration in vitro (Marrs et al., 2010).

MAGL genetic ablation can alter synaptic plasticity in mouse hippo-campus and cerebellum via 2-AG mobilization and persistent activation of CB1 receptors, leading to their desensitization (Pan et al., 2011; Zhong et al., 2011). Similarly, inhibition of ABHD6 can increase CB1-dependent long-term depression (LTD) in mouse cortical excitatory synapses, suggest-ing that not only MAGL but also ABHD6 may control the amount of 2-AG reaching presynaptic CB1 receptors (Marrs et al., 2010).

EC can also serve as allosteric modulators of other metabotropic recep-tors, such as GPR55 (Begg et al., 2005; Henstridge et al., 2011; Sharir et al., 2012)—an orphan receptor activated by the CB1 antagonists rimonabant and AM251 (Johns et al., 2007)—and GPR18 (McHugh et al., 2010). However, neither AEA nor 2-AG have shown consistent pharmacological effects upon stimulation of these receptors, and their functional roles have not been yet elucidated (Yin et al., 2009).

ROLE OF THE ENDOCANNABINOID SYSTEM IN PARKINSON DISEASE

According to the classical model of basal ganglia circuitry (Fig. 3.1), striatal medium spiny neurons (MSNs) receive excitatory glutamatergic projec-tions from the cerebral cortex. MSNs are in turn modulated by nigrostriatal dopaminergic afferents that exert excitatory or inhibitory effects on "direct" and "indirect" striatofugal pathways via stimulation of dopamine D1 and D2 receptors, respectively (Lanciego, Luquin, & Obeso, 2012).

Although CB1 are not present on dopaminergic neurons (Matyas et al., 2006), they indirectly affect dopamine output in the dorsal striatum by modulat-ing neurotransmitter release from projecting inhibitory and excitatory terminals via stimulation of CB1 receptors (Gerdeman & Lovinger, 2001; Julian et al., 2003; Martin et al., 2008; Morera-Herreras, Ruiz-Ortega, Gómez-Urquijo, & Ugedo, 2008; Sidlo, Reggio, & Rice, 2008; Szabo, Wallmichrath, Mathonia, & Pfreundtner, 2000). The overall effect of CB1 stimulation on dopamine release in the caudate putamen remains controversial, as some studies have shown a decrease (Sidlo et al., 2008), an increase (Malone & Taylor, 1999), or no effect at all (Kofalvi et al., 2005; Szabo, Muller, & Koch, 1999). AEA and EC-enhancing drugs, such as FAAH inhibitors, can also modulate nigrostriatal dopamine

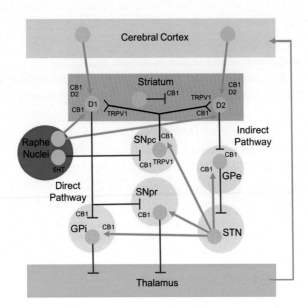

Figure 3.1 Basal ganglia organization.

transmission by acting at TRPV1 (de Lago, de Miguel, Lastres-Becker, Ramos, & Fernández-Ruiz, 2004; Marinelli et al., 2003, 2007; Tzavara et al., 2006) or PPAR receptors (Melis et al., 2008).

Striatal ECs are regulated in opposite ways on stimulation of dopamine D1- (which decrease AEA) and D2-like receptors (which increase AEA) (Centonze et al., 2004; Ferrer, Asbrock, Kathuria, Piomelli, & Giuffrida, 2003; Giuffrida et al., 1999; Morgese, Cassano, Cuomo, & Giuffrida, 2007). These effects may depend on the ability of D1 and D2 agonists to enhance or diminish spontaneous excitatory postsynaptic currents in striatal MSN, respectively, and suggest a dopamine-mediated control of EC mobilization, which in turn can affect glutamate release differently by acting at presynaptic CB1 receptors (Andre et al., 2010).

Changes in striatal synaptic plasticity, specifically LTD, are regulated by a decrease in glutamate release probability via D2-receptor activation (Tang, Low, Grandy, & Lovinger, 2001). Several groups have shown that EC-mediated striatal LTD is compromised after striatal dopamine denervation (Kreitzer & Malenka, 2005) or blockade of D2 receptors (Kreitzer & Malenka, 2005; Tang et al., 2001; Wang et al., 2006), and that this form of plasticity is rescued by D2 receptor agonism or inhibition of EC metabolism (Kreitzer & Malenka, 2007; Tang et al., 2001). These findings are in line with the observation that EC-mediated LTD involves cortical glutamatergic projections onto the D2 receptor expressing MSNs of the indirect pathway (Kreitzer & Malenka, 2007; Shen, Flajolet, Greengard, & Surmeier, 2008). Either AEA or 2-AG has been implicated in striatal-EC-mediated plasticity, depending on the stimulation frequency applied to the glutamatergic afferents (Ade & Lovinger, 2007; Adermark, Talani, & Lovinger, 2009; Lerner & Kreitzer, 2012).

Interestingly, EC-mediated LTD at corticostriatal synapses is completely lost in dyskinetic—but not in nondyskinetic—parkinsonian rats treated with L-DOPA (Picconi et al., 2011).

Finally, EC, in particular AEA, also mediates synaptic depression of GABAergic afferents projecting onto striatal MSNs (Adermark et al., 2009; Centonze et al., 2004; Narushima, Uchigashima, Hashimoto, Watanabe, & Kano, 2006; Szabo, Dörner, Pfreundtner, Nörenberg, & Starke, 1998) leading to their disinhibition. Together, these observations suggest that dopamine denervation affects EC transmission leading to maladaptive plasticity in the striatum.

Behavioral studies indicate that the AEA elevation observed after administration of dopaminergic agonists may serve as an inhibitory feedback signal to offset dopamine-induced hyperactivity (Beltramo et al., 2000; Giuffrida et al., 1999; Masserano, Karoum, & Wyatt, 1999). Thus abnormalities in dopamine and EC transmission may disrupt this feedback mechanism and contribute to motor disturbances, particularly upon long-term activation of dopamine receptors, as in the case of chronic L-DOPA administration.

Recent studies have added a further level of complexity to this scenario, showing a competitive interaction between dopamine D2 and adenosine A2A receptors in the induction of EC-mediated plasticity, such that activation of dopamine D2 receptors would promote LTD, whereas A2A receptor activation would lead to LTP (Shen et al., 2008; Tozzi et al., 2011). Also, coadministration of A2A and CB1 agonists have been shown to partially inhibit the CB1-dependent decrease of glutamate transmission (Martire et al., 2011). As A2A receptors are expressed on glutamatergic terminals projecting to MNS spines (Quiroz et al., 2009), these synapses may represent the anatomical substrates for these complex interactions.

In the 6-OHDA rodent model of Parkinson disease (PD), dopamine denervation has been shown to increase CB1 messenger RNA (mRNA) in striatal neurons (Zeng et al., 1999) and decrease CB1 receptor expression in the substantia nigra pars reticulata (Walsh et al., 2010), although no changes in CB1 receptor binding were observed (Romero et al., 2000). Postmortem studies have confirmed an increase of striatal CB1 mRNA levels in patients with PD chronically treated with L-DOPA (Lastres-Becker et al., 2001); however, other reports have shown that CB1 mRNA is actually decreased in the striatum and globus pallidus of these patients (Hurley, Mash, & Jenner, 2003). An in vivo study reported regional difference in the availability of CB1 receptors in patients with PD, with a decrease within the substantia nigra and increases in the nigrostriatal, mesolimbic, and mesocortical regions (Van Laere et al., 2012). Conflicting data are not only limited to studies on CB receptor expression but also include EC measurements across the basal ganglia. In the globus pallidus of reserpine-treated rats, increased 2-AG levels were associated with decreased locomotor activity, which was fully reversed by coadministration of a dopamine D2 agonist and a CB1 receptor antagonist (Di Marzo, Hill, Bisogno, Crossman, & Brotchie, 2000). Some studies carried out in 6-OHDA lesioned rats have shown an increase of AEA and a decrease of the AEA-metabolic enzyme FAAH, which was reversed by L-DOPA administration (De Petrocellis & Di Marzo, 2010; Maccarone et al., 2003). However, in the same animal model, other groups have shown either no change or a decrease of AEA within the basal ganglia, and no effect of acute L-DOPA administration on EC levels (Ferrer et al., 2003; Morgese et al., 2007). Also, there is no electrophysiological evidence for increased EC tone in 6-OHDA rats before or after L-DOPA administration (Kreitzer & Malenka, 2007).

To date, there is no general consensus whether CB-based therapies can ameliorate PD symptoms, and studies on the effects of either CB agonists or antagonists in animal models have produced conflicting results (Cao et al., 2007; Meschler, Howlett, & Madras, 2001; Mesnage et al., 2004; Papa, 2008; van der Stelt et al., 2005). These discrepancies might be attributable to specific differences across PD models, to the overall physiological state of the animals at the time of the experiments, or other variables (such as stress levels, light conditions, circadian rhythms), which may all affect EC mobilization.

Nevertheless, increasing experimental evidence suggests that CB drugs may delay the underlying neurodegenerative process of PD by modulating neuro-inflammation and the brain immune responses via CB-receptor-dependent and CB-receptor-independent mechanisms (Molina-Holgado et al., 2003; Ramírez, Blázquez, Gómez del Pulgar, Guzmán, & de Ceballos, 2005; Sancho, Calzado, Di Marzo, Appendino, & Muñoz, 2003; Walter & Stella, 2004).

CANNABINOIDS AND NEUROPROTECTION IN PARKINSON DISEASE MODELS

The antiinflammatory properties of CBs are well documented (Klein, 2005; Nagarkatti et al., 2009; Rettori, De Laurentiis, Zorrilla Zubilete, Rettori, & Elverdin, 2012), and several studies have shown significant effects on neuroinflammation and modulation of astroglial and microglia cells (Banati, Daniel, &

Blunt, 1998; Block, Zecca, & Hong, 2007; Hirsch et al., 2003;Vila et al., 2001), suggesting a possible beneficial action against neurodegeneration.

In 1-methyl-4-phenyl-1,2,3,6-tetrahydropyridine (MPTP)-treated mice, administration of WIN or of the CB2 agonist JWH015 protected against MPTP-induced loss of tyrosine-hydroxylase-positive neurons in the substantia nigra pars compacta, reduced MPTP-induced microglial activation via a CB2-dependent mechanism, and reversed MPTP-associated motor deficits (Price et al., 2009). On the other hand, genetic ablation of CB2 receptors exacerbated the systemic toxicity of MPTP (Price et al., 2009). In line with these observations, other studies have shown that activation of CB2, rather than CB1, receptors may be a more effective pharmacological strategy to slow down nigrostriatal degeneration (Marsicano et al., 2003; Nagayama, Pandey, Rieder, Hegde, & Nagarkatti, 1999; Panikashvili et al., 2001). Other groups, however, have shown that treatment with WIN and HU210 protected nigrostriatal dopamine neurons against MPTP-induced neurotoxicity and reduced microglia activation via a CB1-dependent mechanism (Chung et al., 2011). Also, MAGL inhibition has been shown to protect against MPTP-induced neuronal loss in a CB1/CB2-independent manner, possibly by preventing the conversion of 2-AG into proinflammatory prostaglandins (Nomura et al., 2011). Finally, CBs can also exert neuroprotective actions due to their antioxidant properties (Garcia et al., 2011; Mounsey et al., 2015), or by promoting the differentiation of progenitor cells in neurogenic areas (Galve-Roperh, Aguado, Palazuelos, & Guzmán, 2007; Lastres-Becker, Molina-Holgado, Ramos, Mechoulam, & Fernández-Ruiz, 2005; Marsicano et al., 2002).

THE ENDOCANNABINOID SYSTEM AND L-DOPA-INDUCED DYSKINESIA

Chronic L-DOPA administration, the gold standard treatment for PD motor symptoms, can produce disabling chorealike involuntary movements—known as L-DOPA-induced dyskinesia (LID)—in as many as 90% of patients who have received the drug for at least 5 years (Ahlskog & Muenter, 2001). Experimental evidence suggests that the EC system is a potential target for LID pharmacotherapy. In 6-OHDA-lesioned rats, subchronic administration of the CB agonist WIN55212-2 (WIN) at doses that do not affect general locomotor activity or dopamine-elicited motor responses reduced L-DOPA-induced abnormal involuntary movements (AIM)—a behavioral correlate of clinical LID—via a CB1-dependent mechanism (Morgese et al., 2007, 2009). In the same model, the

WIN-induced decrease of AIM also reduced L-DOPA-induced protein kinase A (PKA) hyperactivity, which was positively correlated with dyskinesia severity (Martinez, Macheda, Morgese, Trabace, & Giuffrida, 2012).

Elevation of AEA via pharmacological blockade of FAAH by URB597 has been shown to reduce AIM only when coadministered with the TRPV1 antagonist capsazepine, suggesting that stimulation of TRPV1 receptors has a prodyskinetic effect (Martinez et al., 2015; Morgese et al., 2007). However, recent reports indicate that administration of oleoylethanolamide—a structural analog of AEA that it elevated upon FAAH blockade and has no activity at CB1 receptors—may reduce AIM via a TRPV1-dependent mechanism (Gonzalez-Aparicio & Moratalla, 2013). Similarly, administration of URB597 alone, or stimulation of TRPV1 receptors by capsaicin, has been shown to attenuate L-DOPA-induced hyperactivity in reserpine-treated rats (Lee, Di Marzo, & Brotchie, 2006). The explanation of these discrepancies requires further investigations and may be due to the use of different animal models of PD and behavioral outcomes (vertical motor activity vs. AIM).

As previously mentioned, studies from our laboratory have shown that 6-OHDA-treated rats have lower striatal AEA than sham controls, and treatment with L-DOPA does not affect AEA levels (Ferrer et al., 2003; Morgese et al., 2007). As AEA counteracts dopamine-induced motor activation (Giuffrida et al., 1999), and given the deficit in AEA mobilization in 6-OHDA rats following L-DOPA administration (Morgese et al., 2007), it is plausible to hypothesize that stimulation of CB receptors may indeed ameliorate LID by correcting this deficit. In addition, increased glutamatergic neurotransmission has been implicated in the development and maintenance of LID, and L-DOPA administration increases glutamate transmission and the expression of glutamate receptors in dopamine-denervated animals (Calon et al., 2002; Jonkers, Sarre, Ebinger, & Michotte, 2002). Thus the antidyskinetic effect of CB agonists may be attributed to their ability to reduce striatal glutamatergic activity (Gubellini et al., 2002).

However, the reduction of dyskinesia via increased AEA tone may not entirely depend on CB1 as it is fully reversed by a nonselective antagonist for PPARs. In addition, it is important to note that CB1 receptors are expressed on serotonergic raphe striatal fibers (Hermann, Marsicano, & Lutz, 2002) (Fig. 3.1), which are able to (1) convert L-DOPA into dopamine and release it as a "false neurotransmitter," thus contributing to LID development (Carta, Carlsson, Kirik, & Björklund, 2007); (2) influence nigrostriatal dopamine release (Abdallah et al., 2009); and consequently (3) affect the dopamine- and CB1-mediated control of glutamate release (Mathur, Capik, Alvarez, & Lovinger, 2011).

Thus we could speculate that CB agents may exert their antidyskinetic effects by dampening the ectopic dopamine release from serotonergic terminals, and/or by controlling dopamine transmission indirectly via inhibition of serotonin release (Haring, Grieb, Monory, Lutz, & Moreira, 2013; Nakazi, Bauer, Nickel, Kathmann, & Schlicker, 2000).

Nevertheless, the use of CBs as antidyskinetic agents remains controversial. Cannabis consumption failed to decrease LID in a randomized, double-blind, placebo-controlled crossover trial (Carroll et al., 2004). We should, however, keep in mind that cannabis has a more complex pharmacological profile than synthetic CB agonists, which may account for the lack of effect in this study.

An important point to consider is the abuse liability associated with the use of CB drugs (Justinova, Goldberg, Heishman, & Tanda, 2005; Tanda & Goldberg, 2003; Tanda, Munzar, & Goldberg, 2000). However, THC and AEA are self-administered in squirrel monkeys (Justinova et al., 2005), whereas URB597 is not, and this drug does not share discriminative stimulus effects with THC in either nonhuman primates or rats (Giuffrida & McMahon, 2010; Justinova, Tanda, Redhi, & Goldberg, 2003; Solinas et al., 2007). Another potential limitation of CBs for the treatment of LID is the development of tolerance over time, which has been demonstrated with respect to CB-induced antinociception, hypolocomotion, catalepsy, and neuroendocrine effects to name a few (D'Souza et al., 2008; Karler, Calder, & Turkanis, 1984; Martin, 1985; Miczek & Dixit, 1980; Pertwee, 1974; Ramaekers, Kauert, Theunissen, Toennes, & Moeller, 2009). Also, chronic THC decreases WIN-induced stimulation of CB1 receptors and CB1 binding in rat cerebellum, hippocampus, caudate putamen, and globus pallidus (Breivogel et al., 1999).

Overall, more research is needed to determine the role of the EC system in PD and assess the potential use of CB-based drugs for the treatment of LID.

CONCLUSIONS

The central role played by the EC system in regulating dopamine transmission, synaptic plasticity, motor activity, and inflammation makes CBs ideal tools to study the pathophysiology of neurodegenerative disorders and explore their potential for PD pharmacotherapy and neuroprotection.

With PD being the second most frequent age-related neurodegenerative disease and a growing aging population, the need to find effective treatments for this pathology is of great importance. Although it is still unclear whether stimulation/inhibition of CB receptors can ameliorate PD motor symptoms, direct

CB agonists have shown antidyskinetic properties in preclinical studies. Thus, the possibility of using synthetic CBs as an adjunctive therapy to control LID while preserving L-DOPA beneficial effects is very attractive. Nevertheless, the complexity of CB pharmacology may pose several challenges for drug development. The biggest hurdle is the development of undesired side effects and tolerance. Given their low abuse liability, FAAH or MAGL inhibitors may represent a valuable alternative to synthetic CBs or cannabis extracts. Indeed, as their pharmacological action is limited to sites engaged in EC mobilization, drug-induced side effects should be smaller than those observed after the brainwide activation/inhibition of CB receptors achieved with synthetic CBs.

Finally, further studies are required to explore the therapeutic potentials of other EC targets, such as PPAR receptors, which may offer an opportunity to develop novel approaches to control the neuroinflammatory response underlying PD progression.

REFERENCES

Abdallah, L., Bonasera, S. J., Hopf, F. W., O'Dell, L., Giorgetti, M., Jongsma, M., ... Tecott, L. H. (2009). Impact of serotonin 2C receptor null mutation on physiology and behavior associated with nigrostriatal dopamine pathway function. *The Journal of Neuroscience: The Official Journal of the Society for Neuroscience, 29*(25), 8156–8165.

Ade, K. K., & Lovinger, D. M. (2007). Anandamide regulates postnatal development of long-term synaptic plasticity in the rat dorsolateral striatum. *The Journal of Neuroscience: The Official Journal of the Society for Neuroscience, 27*(9), 2403–2409.

Adermark, L., Talani, G., & Lovinger, D. M. (2009). Endocannabinoid-dependent plasticity at GABAergic and glutamatergic synapses in the striatum is regulated by synaptic activity. *The European Journal of Neuroscience, 29*(1), 32–41.

Ahlskog, J. E., & Muenter, M. D. (2001). Frequency of levodopa-related dyskinesias and motor fluctuations as estimated from the cumulative literature. *Movement Disorders: Official Journal of the Movement Disorder Society, 16*(3), 448–458.

Andre, V. M., Cepeda, C., Cummings, D. M., Jocoy, E. L., Fisher, Y. E., William Yang, X., & Levine, M. S. (2010). Dopamine modulation of excitatory currents in the striatum is dictated by the expression of D1 or D2 receptors and modified by endocannabinoids. *The European Journal of Neuroscience, 31*(1), 14–28.

Banati, R. B., Daniel, S. E., & Blunt, S. B. (1998). Glial pathology but absence of apoptotic nigral neurons in long-standing Parkinson's disease. *Movement Disorders: Official Journal of the Movement Disorder Society, 13*(2), 221–227.

Begg, M., Pacher, P., Bátkai, S., Osei-Hyiaman, D., Offertáler, L., Mo, F. M., ... Kunos, G. (2005). Evidence for novel cannabinoid receptors. *Pharmacology & Therapeutics, 106*(2), 133–145.

Beltramo, M., de Fonseca, F. R., Navarro, M., Calignano, A., Gorriti, M. A., Grammatikopoulos, G., ... Piomelli, D. (2000). Reversal of dopamine D(2) receptor responses by an anandamide transport inhibitor. *The Journal of Neuroscience: The Official Journal of the Society for Neuroscience, 20*(9), 3401–3407.

Beltramo, M., & Piomelli, D. (2000). Carrier-mediated transport and enzymatic hydrolysis of the endogenous cannabinoid 2-arachidonylglycerol. *Neuroreport, 11*(6), 1231–1235.

Berdyshev, E. V. (2000). Cannabinoid receptors and the regulation of immune response. *Chemistry and Physics of Lipids, 108*(1–2), 169–190.

Block, M. L., Zecca, L., & Hong, J. S. (2007). Microglia-mediated neurotoxicity: Uncovering the molecular mechanisms. *Nature Reviews Neuroscience, 8*(1), 57–69.

Bouaboula, M., Hilairet, S., Marchand, J., Fajas, L., Le Fur, G., & Casellas, P. (2005). Anandamide induced PPARgamma transcriptional activation and 3T3-L1 preadipocyte differentiation. *European Journal of Pharmacology, 517*(3), 174–181.

Breivogel, C. S., Childers, S. R., Deadwyler, S. A., Hampson, R. E., Vogt, L. J., & Sim-Selley, L. J. (1999). Chronic delta9-tetrahydrocannabinol treatment produces a time-dependent loss of cannabinoid receptors and cannabinoid receptor-activated G proteins in rat brain. *Journal of Neurochemistry, 73*(6), 2447–2459.

Cadas, H., Gaillet, S., Beltramo, M., Venance, L., & Piomelli, D. (1996). Biosynthesis of an endogenous cannabinoid precursor in neurons and its control by calcium and cAMP. *The Journal of Neuroscience: The Official Journal of the Society for Neuroscience, 16*(12), 3934–3942.

Calon, F., Morissette, M., Ghribi, O., Goulet, M., Grondin, R., Blanchet, P. J., ... Di Paolo, T. (2002). Alteration of glutamate receptors in the striatum of dyskinetic 1-methyl-4-phenyl-1,2,3,6-tetrahydropyridine-treated monkeys following dopamine agonist treatment. *Progress in Neuro-psychopharmacology & Biological Psychiatry, 26*(1), 127–138.

Cao, X., Liang, L., Hadcock, J. R., Iredale, P. A., Griffith, D. A., Menniti, F. S., ... Papa, S. M. (2007). Blockade of cannabinoid type 1 receptors augments the antiparkinsonian action of levodopa without affecting dyskinesias in 1-methyl-4-phenyl-1,2,3,6-tetrahydropyridine-treated rhesus monkeys. *The Journal of Pharmacology and Experimental Therapeutics, 323*(1), 318–326.

Carroll, C. B., Bain, P. G., Teare, L., Liu, X., Joint, C., Wroath, C., ... Zajicek, J. P. (2004). Cannabis for dyskinesia in Parkinson disease: A randomized double-blind crossover study. *Neurology, 63*(7), 1245–1250.

Carta, M., Carlsson, T., Kirik, D., & Björklund, A. (2007). Dopamine released from 5-HT terminals is the cause of L-DOPA-induced dyskinesia in parkinsonian rats. *Brain: A Journal of Neurology, 130*(Pt 7), 1819–1833.

Carta, A. R., Pisanu, A., & Carboni, E. (2011). Do PPAR-gamma agonists have a future in Parkinson's disease therapy? *Parkinsons Disease, 2011*, 689181.

Centonze, D., Battista, N., Rossi, S., Mercuri, N. B., Finazzi-Agrò, A., Bernardi, G., ... Maccarrone, M. (2004). A critical interaction between dopamine D2 receptors and endocannabinoids mediates the effects of cocaine on striatal gabaergic transmission. *Neuropsychopharmacology: Official Publication of the American College of Neuropsychopharmacology, 29*(8), 1488–1497.

Chung, Y. C., et al. (2011). Cannabinoid receptor type 1 protects nigrostriatal dopaminergic neurons against MPTP neurotoxicity by inhibiting microglial activation. *The Journal of Immunology: Official Journal of the American Association of Immunologists, 187*(12), 6508–6517.

Cristino, L., de Petrocellis, L., Pryce, G., Baker, D., Guglielmotti, V., & Di Marzo, V. (2006). Immunohistochemical localization of cannabinoid type 1 and vanilloid transient receptor potential vanilloid type 1 receptors in the mouse brain. *Neuroscience, 139*(4), 1405–1415.

De Petrocellis, L., & Di Marzo, V. (2010). Non-CB1, non-CB2 receptors for endocannabinoids, plant cannabinoids, and synthetic cannabimimetics: Focus on G-protein-coupled receptors and transient receptor potential channels. *Journal of Neuroimmune Pharmacology: The Official Journal of the Society on Neuroimmune Pharmacology, 5*(1), 103–121.

Di Marzo, V., Bisogno, T., & De Petrocellis, L. (2001). Anandamide: Some like it hot. *Trends in Pharmacological Sciences, 22*(7), 346–349.

Di Marzo, V., Hill, M. P., Bisogno, T., Crossman, A. R., & Brotchie, J. M. (2000). Enhanced levels of endogenous cannabinoids in the globus pallidus are associated with a reduction in movement in an animal model of Parkinson's disease. *FASEB Journal: Official Publication of the Federation of American Societies for Experimental Biology, 14*(10), 1432–1438.

Dinh, T. P., Carpenter, D., Leslie, F. M., Freund, T. F., Katona, I., Sensi, S. L., … Piomelli, D. (2002). Brain monoglyceride lipase participating in endocannabinoid inactivation. *Proceedings of the National Academy of Sciences of the United States of America, 99*(16), 10819–10824.

D'Souza, D. C., Ranganathan, M., Braley, G., Gueorguieva, R., Zimolo, Z., Cooper, T., … Krystal, J. (2008). Blunted psychotomimetic and amnestic effects of delta-9-tetrahydrocannabinol in frequent users of cannabis. *Neuropsychopharmacology: Official Publication of the American College of Neuropsychopharmacology, 33*(10), 2505–2516.

Ferrer, B., Asbrock, N., Kathuria, S., Piomelli, D., & Giuffrida, A. (2003). Effects of levodopa on endocannabinoid levels in rat basal ganglia: Implications for the treatment of levodopa-induced dyskinesias. *The European Journal of Neuroscience, 18*(6), 1607–1614.

Galve-Roperh, I., Aguado, T., Palazuelos, J., & Guzmán, M. (2007). The endocannabinoid system and neurogenesis in health and disease. *Neuroscientist, 13*(2), 109–114.

Garcia, C., Palomo-Garo, C., García-Arencibia, M., Ramos, J., Pertwee, R., & Fernández-Ruiz, J. (2011). Symptom-relieving and neuroprotective effects of the phytocannabinoid Delta(9)-THCV in animal models of Parkinson's disease. *British Journal of Pharmacology, 163*(7), 1495–1506.

Gasperi, V., Fezza, F., Pasquariello, N., Bari, M., Oddi, S., Agrò, A. F., & Maccarrone, M. (2007). Endocannabinoids in adipocytes during differentiation and their role in glucose uptake. *Cellular and Molecular Life Sciences: CMLS, 64*(2), 219–229.

Gerdeman, G., & Lovinger, D. M. (2001). CB1 cannabinoid receptor inhibits synaptic release of glutamate in rat dorsolateral striatum. *Journal of Neurophysiology, 85*(1), 468–471.

Giuffrida, A., & McMahon, L. R. (2010). In vivo pharmacology of endocannabinoids and their metabolic inhibitors: Therapeutic implications in Parkinson's disease and abuse liability. *Prostaglandins & Other Lipid Mediators, 91*(3–4), 90–103.

Giuffrida, A., Parsons, L. H., Kerr, T. M., Rodríguez de Fonseca, F., Navarro, M., & Piomelli, D. (1999). Dopamine activation of endogenous cannabinoid signaling in dorsal striatum. *Nature Neuroscience, 2*(4), 358–363.

Giuffrida, A., & Seillier, A. (2012). New insights on endocannabinoid transmission in psychomotor disorders. *Progress in Neuro-psychopharmacology & Biological Psychiatry, 38*(1), 51–58.

Gong, J. P., Onaivi, E. S., Ishiguro, H., Liu, Q. R., Tagliaferro, P. A., Brusco, A., & Uhl, G. R. (2006). Cannabinoid CB2 receptors: Immunohistochemical localization in rat brain. *Brain Research, 1071*(1), 10–23.

Gonzalez-Aparicio, R., & Moratalla, R. (2013). Oleoylethanolamide reduces L-DOPA-induced dyskinesia via TRPV1 receptor in a mouse model of Parkinson s disease. *Neurobiology of Disease, 62*, 416–425.

Gubellini, P., Picconi, B., Bari, M., Battista, N., Calabresi, P., Centonze, D., … Maccarrone, M. (2002). Experimental parkinsonism alters endocannabinoid degradation: Implications for striatal glutamatergic transmission. *The Journal of Neuroscience: The Official Journal of the Society for Neuroscience, 22*(16), 6900–6907.

Haring, M., Grieb, M., Monory, K., Lutz, B., & Moreira, F. A. (2013). Cannabinoid CB(1) receptor in the modulation of stress coping behavior in mice: The role of serotonin and different forebrain neuronal subpopulations. *Neuropharmacology, 65*, 83–89.

Henstridge, C. M., Balenga, N. A., Kargl, J., Andradas, C., Brown, A. J., Irving, A., … Waldhoer, M. (2011). Minireview: Recent developments in the physiology and pathology of the lysophosphatidylinositol-sensitive receptor GPR55. *Molecular Endocrinology, 25*(11), 1835–1848.

Herkenham, M., Lynn, A. B., de Costa, B. R., & Richfield, E. K. (1991). Neuronal localization of cannabinoid receptors in the basal ganglia of the rat. *Brain Research, 547*(2), 267–274.

Herkenham, M., Lynn, A. B., Little, M. D., Johnson, M. R., Melvin, L. S., de Costa, B. R., & Rice, K. C. (1990). Cannabinoid receptor localization in brain. *Proceedings of the National Academy of Sciences of the United States of America, 87*(5), 1932–1936.

Hermann, H., Marsicano, G., & Lutz, B. (2002). Coexpression of the cannabinoid receptor type 1 with dopamine and serotonin receptors in distinct neuronal subpopulations of the adult mouse forebrain. *Neuroscience, 109*(3), 451–460.

Hirsch, E. C., Breidert, T., Rousselet, E., Hunot, S., Hartmann, A., & Michel, P. P. (2003). The role of glial reaction and inflammation in Parkinson's disease. *Annals of the New York Academy of Sciences, 991*, 214–228.

Hurley, M. J., Mash, D. C., & Jenner, P. (2003). Expression of cannabinoid CB1 receptor mRNA in basal ganglia of normal and parkinsonian human brain. *Journal of Neural Transmission, 110*(11), 1279–1288.

Johns, D. G., Behm, D. J., Walker, D. J., Ao, Z., Shapland, E. M., Daniels, D. A., … Douglas, S. A. (2007). The novel endocannabinoid receptor GPR55 is activated by atypical cannabinoids but does not mediate their vasodilator effects. *British Journal of Pharmacology, 152*(5), 825–831.

Jonkers, N., Sarre, S., Ebinger, G., & Michotte, Y. (2002). MK801 suppresses the L-DOPA-induced increase of glutamate in striatum of hemi-Parkinson rats. *Brain Research, 926*(1–2), 149–155.

Julian, M. D., Martin, A. B., Cuellar, B., Rodriguez De Fonseca, F., Navarro, M., Moratalla, R., & Garcia-Segura, L. M. (2003). Neuroanatomical relationship between type 1 cannabinoid receptors and dopaminergic systems in the rat basal ganglia. *Neuroscience, 119*(1), 309–318.

Justinova, Z., Goldberg, S. R., Heishman, S. J., & Tanda, G. (2005). Self-administration of cannabinoids by experimental animals and human marijuana smokers. *Pharmacology, Biochemistry, and Behavior, 81*(2), 285–299.

Justinova, Z., Tanda, G., Redhi, G. H., & Goldberg, S. R. (2003). Self-administration of delta9-tetrahydrocannabinol (THC) by drug naive squirrel monkeys. *Psychopharmacology, 169*(2), 135–140.

Karler, R., Calder, L. D., & Turkanis, S. A. (1984). Changes in CNS sensitivity to cannabinoids with repeated treatment: Tolerance and auxoesthesia. *NIDA Research Monograph, 54*, 312–322.

Katona, I., Sperlágh, B., Sík, A., Käfalvi, A., Vizi, E. S., Mackie, K., & Freund, T. F. (1999). Presynaptically located CB1 cannabinoid receptors regulate GABA release from axon terminals of specific hippocampal interneurons. *The Journal of Neuroscience: The Official Journal of the Society for Neuroscience, 19*(11), 4544–4558.

Klein, T. W. (2005). Cannabinoid-based drugs as anti-inflammatory therapeutics. *Nature Reviews Immunology, 5*(5), 400–411.

Kofalvi, A., Rodrigues, R. J., Ledent, C., Mackie, K., Vizi, E. S., Cunha, R. A., & Sperlágh, B. (2005). Involvement of cannabinoid receptors in the regulation of neurotransmitter release in the rodent striatum: A combined immunochemical and pharmacological analysis. *The Journal of Neuroscience: The Official Journal of the Society for Neuroscience, 25*(11), 2874–2884.

Kreitzer, A. C., & Malenka, R. C. (2005). Dopamine modulation of state-dependent endocannabinoid release and long-term depression in the striatum. *The Journal of Neuroscience: The Official Journal of the Society for Neuroscience, 25*(45), 10537–10545.

Kreitzer, A. C., & Malenka, R. C. (2007). Endocannabinoid-mediated rescue of striatal LTD and motor deficits in Parkinson's disease model. *Nature, 445*(7128), 643–647.

Kreitzer, A. C., & Regehr, W. G. (2001). Cerebellar depolarization-induced suppression of inhibition is mediated by endogenous cannabinoids. *The Journal of Neuroscience: The Official Journal of the Society for Neuroscience, 21*(20), RC174.

de Lago, E., de Miguel, R., Lastres-Becker, I., Ramos, J. A., & Fernández-Ruiz, J. (2004). Involvement of vanilloid-like receptors in the effects of anandamide on motor behavior and nigrostriatal dopaminergic activity: In vivo and in vitro evidence. *Brain Research, 1007*(1–2), 152–159.

Lanciego, J. L., Luquin, N., & Obeso, J. A. (2012). Functional neuroanatomy of the basal ganglia. *Cold Spring Harbor Perspectives in Medicine [Electronic Resource], 2*(12), a009621.

Lastres-Becker, I., Cebeira, M., de Ceballos, M. L., Zeng, B. Y., Jenner, P., Ramos, J. A., & Fernández-Ruiz, J. J. (2001). Increased cannabinoid CB1 receptor binding and activation of GTP-binding proteins in the basal ganglia of patients with Parkinson's syndrome and of MPTP-treated marmosets. *The European Journal of Neuroscience, 14*(11), 1827–1832.

Lastres-Becker, I., Molina-Holgado, F., Ramos, J. A., Mechoulam, R., & Fernández-Ruiz, J. (2005). Cannabinoids provide neuroprotection against 6-hydroxydopamine toxicity in vivo and in vitro: Relevance to Parkinson's disease. *Neurobiology of Disease, 19*(1–2), 96–107.

Lee, J., Di Marzo, V., & Brotchie, J. M. (2006). A role for vanilloid receptor 1 (TRPV1) and endocannabinoid signalling in the regulation of spontaneous and L-DOPA induced locomotion in normal and reserpine-treated rats. *Neuropharmacology, 51*(3), 557–565.

Lerner, T. N., & Kreitzer, A. C. (2012). RGS4 is required for dopaminergic control of striatal LTD and susceptibility to parkinsonian motor deficits. *Neuron, 73*(2), 347–359.

Leung, D., Saghatelian, A., Simon, G. M., & Cravatt, B. F. (2006). Inactivation of N-acyl phosphatidylethanolamine phospholipase D reveals multiple mechanisms for the biosynthesis of endocannabinoids. *Biochemistry, 45*(15), 4720–4726.

Liu, J., Wang, L., Harvey-White, J., Huang, B. X., Kim, H. Y., Luquet, S., ... Kunos, G. (2008). Multiple pathways involved in the biosynthesis of anandamide. *Neuropharmacology, 54*(1), 1–7.

Maccarrone, M., Gubellini, P., Bari, M., Picconi, B., Battista, N., Centonze, D., ... Calabresi, P. (2003). Levodopa treatment reverses endocannabinoid system abnormalities in experimental parkinsonism. *Journal of Neurochemistry, 85*(4), 1018–1025.

Malone, D. T., & Taylor, D. A. (1999). Modulation by fluoxetine of striatal dopamine release following Delta9-tetrahydrocannabinol: A microdialysis study in conscious rats. *British Journal of Pharmacology, 128*(1), 21–26.

Marinelli, S., Di Marzo, V., Berretta, N., Matias, I., Maccarrone, M., Bernardi, G., & Mercuri, N. B. (2003). Presynaptic facilitation of glutamatergic synapses to dopaminergic neurons of the rat substantia nigra by endogenous stimulation of vanilloid receptors. *The Journal of Neuroscience: The Official Journal of the Society for Neuroscience, 23*(8), 3136–3144.

Marinelli, S., Di Marzo, V., Florenzano, F., Fezza, F., Viscomi, M. T., van der Stelt, M., ... Mercuri, N. B. (2007). N-arachidonoyl-dopamine tunes synaptic transmission onto dopaminergic neurons by activating both cannabinoid and vanilloid receptors. *Neuropsychopharmacology: Official Publication of the American College of Neuropsychopharmacology, 32*(2), 298–308.

Marrs, W. R., Blankman, J. L., Horne, E. A., Thomazeau, A., Lin, Y. H., Coy, J., ... Stella, N. (2010). The serine hydrolase ABHD6 controls the accumulation and efficacy of 2-AG at cannabinoid receptors. *Nature Neuroscience, 13*(8), 951–957.

Marsicano, G., Goodenough, S., Monory, K., Hermann, H., Eder, M., Cannich, A., ... Lutz, B. (2003). CB1 cannabinoid receptors and on-demand defense against excitotoxicity. *Science, 302*(5642), 84–88.

Marsicano, G., Wotjak, C. T., Azad, S. C., Bisogno, T., Rammes, G., Cascio, M. G., ... Lutz, B. (2002). The endogenous cannabinoid system controls extinction of aversive memories. *Nature, 418*(6897), 530–534.

Martin, A. B., Fernandez-Espejo, E., Ferrer, B., Gorriti, M. A., Bilbao, A., Navarro, M., ... Moratalla, R. (2008). Expression and function of CB1 receptor in the rat striatum: Localization and effects on D1 and D2 dopamine receptor-mediated motor behaviors. *Neuropsychopharmacology: Official Publication of the American College of Neuropsychopharmacology, 33*(7), 1667–1679.

Martin, B. R. (1985). Structural requirements for cannabinoid-induced antinociceptive activity in mice. *Life Sciences, 36*(16), 1523–1530.

Martinez, A., Macheda, T., Morgese, M. G., Trabace, L., & Giuffrida, A. (2012). The cannabinoid agonist WIN55212-2 decreases L-DOPA-induced PKA activation and dyskinetic behavior in 6-OHDA-treated rats. *Neuroscience Research, 72*(3), 236–242.

Martinez, A. A., Morgese, M. G., Pisanu, A., Macheda, T., Paquette, M. A., Seillier, A., … Giuffrida, A. (2015). Activation of PPAR gamma receptors reduces levodopa-induced dyskinesias in 6-OHDA-lesioned rats. *Neurobiology of Disease, 74*, 295–304.

Martire, A., Tebano, M. T., Chiodi, V., Ferreira, S. G., Cunha, R. A., Köfalvi, A., & Popoli, P. (2011). Pre-synaptic adenosine A2A receptors control cannabinoid CB1 receptor-mediated inhibition of striatal glutamatergic neurotransmission. *Journal of Neurochemistry, 116*(2), 273–280.

Masserano, J. M., Karoum, F., & Wyatt, R. J. (1999). SR 141716A, a CB1 cannabinoid receptor antagonist, potentiates the locomotor stimulant effects of amphetamine and apomorphine. *Behavioural Pharmacology, 10*(4), 429–432.

Mathur, B. N., Capik, N. A., Alvarez, V. A., & Lovinger, D. M. (2011). Serotonin induces long-term depression at corticostriatal synapses. *The Journal of Neuroscience: The Official Journal of the Society for Neuroscience, 31*(20), 7402–7411.

Matsuda, L. A., Lolait, S. J., Brownstein, M. J., Young, A. C., & Bonner, T. I. (1990). Structure of a cannabinoid receptor and functional expression of the cloned cDNA. *Nature, 346*(6284), 561–564.

Matyas, F., Yanovsky, Y., Mackie, K., Kelsch, W., Misgeld, U., & Freund, T. F. (2006). Subcellular localization of type 1 cannabinoid receptors in the rat basal ganglia. *Neuroscience, 137*(1), 337–361.

McHugh, D., Hu, S. S., Rimmerman, N., Juknat, A., Vogel, Z., Walker, J. M., & Bradshaw, H. B. (2010). N-arachidonoyl glycine, an abundant endogenous lipid, potently drives directed cellular migration through GPR18, the putative abnormal cannabidiol receptor. *BMC Neuroscience, 11*, 44.

Melis, M., Pillolla, G., Luchicchi, A., Muntoni, A. L., Yasar, S., Goldberg, S. R., & Pistis, M. (2008). Endogenous fatty acid ethanolamides suppress nicotine-induced activation of mesolimbic dopamine neurons through nuclear receptors. *The Journal of Neuroscience: The Official Journal of the Society for Neuroscience, 28*(51), 13985–13994.

Meschler, J. P., Howlett, A. C., & Madras, B. K. (2001). Cannabinoid receptor agonist and antagonist effects on motor function in normal and 1-methyl-4-phenyl-1,2,5,6-tetrahydropyridine (MPTP)-treated non-human primates. *Psychopharmacology, 156*(1), 79–85.

Mesnage, V., Houeto, J. L., Bonnet, A. M., Clavier, I., Arnulf, I., Cattelin, F., … Agid, Y. (2004). Neurokinin B, neurotensin, and cannabinoid receptor antagonists and Parkinson disease. *Clinical Neuropharmacology, 27*(3), 108–110.

Mezey, E., Tóth, Z. E., Cortright, D. N., Arzubi, M. K., Krause, J. E., Elde, R., … Szallasi, A. (2000). Distribution of mRNA for vanilloid receptor subtype 1 (VR1), and VR1-like immunoreactivity, in the central nervous system of the rat and human. *Proceedings of the National Academy of Sciences of the United States of America, 97*(7), 3655–3660.

Miczek, K. A., & Dixit, B. N. (1980). Behavioral and biochemical effects of chronic delta 9-tetrahydrocannabinol in rats. *Psychopharmacology, 67*(2), 195–202.

Molina-Holgado, F., Pinteaux, E., Moore, J. D., Molina-Holgado, E., Guaza, C., Gibson, R. M., & Rothwell, N. J. (2003). Endogenous interleukin-1 receptor antagonist mediates anti-inflammatory and neuroprotective actions of cannabinoids in neurons and glia. *The Journal of Neuroscience: The Official Journal of the Society for Neuroscience, 23*(16), 6470–6474.

Moran, M. M., Xu, H., & Clapham, D. E. (2004). TRP ion channels in the nervous system. *Current Opinion in Neurobiology, 14*(3), 362–369.

Morera-Herreras, T., Ruiz-Ortega, J. A., Gómez-Urquijo, S., & Ugedo, L. (2008). Involvement of subthalamic nucleus in the stimulatory effect of Delta(9)-tetrahydrocannabinol on dopaminergic neurons. *Neuroscience, 151*(3), 817–823.

Morgese, M. G., Cassano, T., Cuomo, V., & Giuffrida, A. (2007). Anti-dyskinetic effects of cannabinoids in a rat model of Parkinson's disease: Role of CB(1) and TRPV1 receptors. *Experimental Neurology, 208*(1), 110–119.

Morgese, M. G., Cassano, T., Gaetani, S., Macheda, T., Laconca, L., Dipasquale, P., … Giuffrida, A. (2009). Neurochemical changes in the striatum of dyskinetic rats after administration of the cannabinoid agonist WIN55,212-2. *Neurochemistry International, 54*(1), 56–64.

Mounsey, R. B., Mustafa, S., Robinson, L., Ross, R. A., Riedel, G., Pertwee, R. G., & Teismann, P. (2015). Increasing levels of the endocannabinoid 2-AG is neuroprotective in the 1-methyl-4-phenyl-1,2,3,6-tetrahydropyridine mouse model of Parkinson's disease. *Experimental Neurology, 273*, 36–44.

Munro, S., Thomas, K. L., & Abu-Shaar, M. (1993). Molecular characterization of a peripheral receptor for cannabinoids. *Nature, 365*(6441), 61–65.

Nagarkatti, P., et al. (2009). Cannabinoids as novel anti-inflammatory drugs. *Future Medicinal Chemistry, 1*(7), 1333–1349.

Nagayama, T., Pandey, R., Rieder, S. A., Hegde, V. L., & Nagarkatti, M. (1999). Cannabinoids and neuroprotection in global and focal cerebral ischemia and in neuronal cultures. *The Journal of Neuroscience: The Official Journal of the Society for Neuroscience, 19*(8), 2987–2995.

Nakazi, M., Bauer, U., Nickel, T., Kathmann, M., & Schlicker, E. (2000). Inhibition of serotonin release in the mouse brain via presynaptic cannabinoid CB1 receptors. *Naunyn-Schmiedeberg's Archives of Pharmacology, 361*(1), 19–24.

Narushima, M., Uchigashima, M., Hashimoto, K., Watanabe, M., & Kano, M. (2006). Depolarization-induced suppression of inhibition mediated by endocannabinoids at synapses from fast-spiking interneurons to medium spiny neurons in the striatum. *The European Journal of Neuroscience, 24*(8), 2246–2252.

Nomura, D. K., Morrison, B. E., Blankman, J. L., Long, J. Z., Kinsey, S. G., Marcondes, M. C., … Cravatt, B. F. (2011). Endocannabinoid hydrolysis generates brain prostaglandins that promote neuroinflammation. *Science, 334*(6057), 809–813.

Okamoto, Y., Morishita, J., Tsuboi, K., Tonai, T., & Ueda, N. (2004). Molecular characterization of a phospholipase D generating anandamide and its congeners. *The Journal of Biological Chemistry, 279*(7), 5298–5305.

Onaivi, E. S., Ishiguro, H., Gong, J. P., Patel, S., Perchuk, A., Meozzi, P. A., … Uhl, G. R. (2006). Discovery of the presence and functional expression of cannabinoid CB2 receptors in brain. *Annals of the New York Academy of Sciences, 1074*, 514–536.

Pan, J. P., Zhang, H. Q., Wei-Wang, Guo, Y. F., Na-Xiao, Cao, X. H., & Liu, L. J. (2011). Some subtypes of endocannabinoid/endovanilloid receptors mediate docosahexaenoic acid-induced enhanced spatial memory in rats. *Brain Research, 1412*, 18–27.

Panikashvili, D., Simeonidou, C., Ben-Shabat, S., Hanus, L., Breuer, A., Mechoulam, R., & Shohami, E. (2001). An endogenous cannabinoid (2-AG) is neuroprotective after brain injury. *Nature, 413*(6855), 527–531.

Papa, S. M. (2008). The cannabinoid system in Parkinson's disease: Multiple targets to motor effects. *Experimental Neurology, 211*(2), 334–338.

Patwardhan, A. M., Jeske, N. A., Price, T. J., Gamper, N., Akopian, A. N., & Hargreaves, K. M. (2006). The cannabinoid WIN 55,212-2 inhibits transient receptor potential vanilloid 1 (TRPV1) and evokes peripheral antihyperalgesia via calcineurin. *Proceedings of the National Academy of Sciences of the United States of America, 103*(30), 11393–11398.

Pertwee, R. G. (1974). Tolerance to the effect of delta1-tetrahydrocannabinol on corticosterone levels in mouse plasma produced by repeated administration of cannabis extract or delta1-tetrahydrocannabinol. *British Journal of Pharmacology, 51*(3), 391–397.

Picconi, B., Bagetta, V., Ghiglieri, V., Paillè, V., Di Filippo, M., Pendolino, V., … Calabresi, P. (2011). Inhibition of phosphodiesterases rescues striatal long-term depression and reduces levodopa-induced dyskinesia. *Brain: A Journal of Neurology, 134*(Pt 2), 375–387.

Price, D. A., Martinez, A. A., Seillier, A., Koek, W., Acosta, Y., Fernandez, E., … Giuffrida, A. (2009). WIN55,212-2, a cannabinoid receptor agonist, protects against nigrostriatal cell loss in the 1-methyl-4-phenyl-1,2,3,6-tetrahydropyridine mouse model of Parkinson's disease. *The European Journal of Neuroscience, 29*(11), 2177–2186.

Quiroz, C., Luján, R., Uchigashima, M., Simoes, A. P., Lerner, T. N., Borycz, J., … Ferré, S. (2009). Key modulatory role of presynaptic adenosine A2A receptors in cortical neurotransmission to the striatal direct pathway. *The Scientific World Journal [Electronic Resource], 9*, 1321–1344.

Ramaekers, J. G., Kauert, G., Theunissen, E. L., Toennes, S. W., & Moeller, M. R. (2009). Neurocognitive performance during acute THC intoxication in heavy and occasional cannabis users. *Journal of Psychopharmacology, 23*(3), 266–277.

Ramirez, B. G., Blázquez, C., Gómez del Pulgar, T., Guzmán, M., & de Ceballos, M. L. (2005). Prevention of Alzheimer's disease pathology by cannabinoids: Neuroprotection mediated by blockade of microglial activation. *The Journal of Neuroscience: The Official Journal of the Society for Neuroscience, 25*(8), 1904–1913.

Rettori, E., De Laurentiis, A., Zorrilla Zubilete, M., Rettori, V., & Elverdin, J. C. (2012). Anti-inflammatory effect of the endocannabinoid anandamide in experimental periodontitis and stress in the rat. *Neuroimmunomodulation, 19*(5), 293–303.

Romero, J., Berrendero, F., Pérez-Rosado, A., Manzanares, J., Rojo, A., Fernández-Ruiz, J. J., … Ramos, J. A. (2000). Unilateral 6-hydroxydopamine lesions of nigrostriatal dopaminergic neurons increased CB1 receptor mRNA levels in the caudate-putamen. *Life Sciences, 66*(6), 485–494.

Ross, R. A. (2003). Anandamide and vanilloid TRPV1 receptors. *British Journal of Pharmacology, 140*(5), 790–801.

Sancho, R., Calzado, M. A., Di Marzo, V., Appendino, G., & Muñoz, E. (2003). Anandamide inhibits nuclear factor-kappaB activation through a cannabinoid receptor-independent pathway. *Molecular Pharmacology, 63*(2), 429–438.

Sharir, H., Console-Bram, L., Mundy, C., Popoff, S. N., Kapur, A., & Abood, M. E. (2012). The endocannabinoids anandamide and virodhamine modulate the activity of the candidate cannabinoid receptor GPR55. *Journal of Neuroimmune Pharmacology: The Official Journal of the Society on Neuroimmune Pharmacology, 7*(4), 856–865.

Shen, W., Flajolet, M., Greengard, P., & Surmeier, D. J. (2008). Dichotomous dopaminergic control of striatal synaptic plasticity. *Science, 321*(5890), 848–851.

Sidlo, Z., Reggio, P. H., & Rice, M. E. (2008). Inhibition of striatal dopamine release by CB1 receptor activation requires nonsynaptic communication involving GABA, H_2O_2, and KATP channels. *Neurochemistry International, 52*(1–2), 80–88.

Simon, G. M., & Cravatt, B. F. (2006). Endocannabinoid biosynthesis proceeding through glycerophospho-N-acyl ethanolamine and a role for alpha/beta-hydrolase 4 in this pathway. *Journal of Biological Chemistry, 281*(36), 26465–26472.

Solinas, M., Tanda, G., Justinova, Z., Wertheim, C. E., Yasar, S., Piomelli, D., … Goldberg, S. R. (2007). The endogenous cannabinoid anandamide produces delta-9-tetrahydrocannabinol-like discriminative and neurochemical effects that are enhanced by inhibition of fatty acid amide hydrolase but not by inhibition of anandamide transport. *The Journal of Pharmacology and Experimental Therapeutics, 321*(1), 370–380.

Starowicz, K., Maione, S., Cristino, L., Palazzo, E., Marabese, I., Rossi, F., … Di Marzo, V. (2007). Tonic endovanilloid facilitation of glutamate release in brainstem descending antinociceptive pathways. *The Journal of Neuroscience: The Official Journal of the Society for Neuroscience, 27*(50), 13739–13749.

Stella, N., Schweitzer, P., & Piomelli, D. (1997). A second endogenous cannabinoid that modulates long-term potentiation. *Nature, 388*(6644), 773–778.

van der Stelt, M., Fox, S. H., Hill, M., Crossman, A. R., Petrosino, S., Di Marzo, V., & Brotchie, J. M. (2005). A role for endocannabinoids in the generation of parkinsonism and levodopa-induced dyskinesia in MPTP-lesioned non-human primate models of Parkinson's disease. *FASEB Journal: Official Publication of the Federation of American Societies for Experimental Biology, 19*(9), 1140–1142.

Sugiura, T., & Waku, K. (2000). 2-Arachidonoylglycerol and the cannabinoid receptors. *Chemistry and Physics of Lipids, 108*(1–2), 89–106.

Szabo, B., Dörner, L., Pfreundtner, C., Nörenberg, W., & Starke, K. (1998). Inhibition of GABAergic inhibitory postsynaptic currents by cannabinoids in rat corpus striatum. *Neuroscience, 85*(2), 395–403.

Szabo, B., Muller, T., & Koch, H. (1999). Effects of cannabinoids on dopamine release in the corpus striatum and the nucleus accumbens in vitro. *Journal of Neurochemistry, 73*(3), 1084–1089.

Szabo, B., Wallmichrath, I., Mathonia, P., & Pfreundtner, C. (2000). Cannabinoids inhibit excitatory neurotransmission in the substantia nigra pars reticulata. *Neuroscience, 97*(1), 89–97.

Szallasi, A., & Blumberg, P. M. (1999). Vanilloid (Capsaicin) receptors and mechanisms. *Pharmacological Reviews, 51*(2), 159–212.

Tanda, G., & Goldberg, S. R. (2003). Cannabinoids: Reward, dependence, and underlying neurochemical mechanisms–A review of recent preclinical data. *Psychopharmacology, 169*(2), 115–134.

Tanda, G., Munzar, P., & Goldberg, S. R. (2000). Self-administration behavior is maintained by the psychoactive ingredient of marijuana in squirrel monkeys. *Nature Neuroscience, 3*(11), 1073–1074.

Tang, K., Low, M. J., Grandy, D. K., & Lovinger, D. M. (2001). Dopamine-dependent synaptic plasticity in striatum during in vivo development. *Proceedings of the National Academy of Sciences of the United States of America, 98*(3), 1255–1260.

Tozzi, A., de Iure, A., Di Filippo, M., Tantucci, M., Costa, C., Borsini, F., ... Calabresi, P. (2011). The distinct role of medium spiny neurons and cholinergic interneurons in the D(2)/A(2)A receptor interaction in the striatum: implications for Parkinson's disease. *The Journal of Neuroscience: The Official Journal of the Society for Neuroscience, 31*(5), 1850–1862.

Tzavara, E. T., Li, D. L., Moutsimilli, L., Bisogno, T., Di Marzo, V., Phebus, L. A., ... Giros, B. (2006). Endocannabinoids activate transient receptor potential vanilloid 1 receptors to reduce hyperdopaminergia-related hyperactivity: Therapeutic implications. *Biological Psychiatry, 59*(6), 508–515.

Van Laere, K., Casteels, C., Lunskens, S., Goffin, K., Grachev, I. D., Bormans, G., & Vandenberghe, W. (2012). Regional changes in type 1 cannabinoid receptor availability in Parkinson's disease in vivo. *Neurobiology of Aging, 33*(3), 620.e1–620.e8.

Vila, M., Jackson-Lewis, V., Guégan, C., Wu, D. C., Teismann, P., Choi, D. K., ... Przedborski, S. (2001). The role of glial cells in Parkinson's disease. *Current Opinion in Neurology, 14*(4), 483–489.

Walsh, S., Mnich, K., Mackie, K., Gorman, A. M., Finn, D. P., & Dowd, E. (2010). Loss of cannabinoid CB1 receptor expression in the 6-hydroxydopamine-induced nigrostriatal terminal lesion model of Parkinson's disease in the rat. *Brain Research Bulletin, 81*(6), 543–548.

Walter, L., Franklin, A., Witting, A., Wade, C., Xie, Y., Kunos, G., ... Stella, N. (2003). Nonpsychotropic cannabinoid receptors regulate microglial cell migration. *The Journal of Neuroscience: The Official Journal of the Society for Neuroscience, 23*(4), 1398–1405.

Walter, L., & Stella, N. (2004). Cannabinoids and neuroinflammation. *British Journal of Pharmacology, 141*(5), 775–785.

Wang, Z., Kai, L., Day, M., Ronesi, J., Yin, H. H., Ding, J., ... Surmeier, D. J. (2006). Dopaminergic control of corticostriatal long-term synaptic depression in medium spiny neurons is mediated by cholinergic interneurons. *Neuron, 50*(3), 443–452.

Yin, H., Chu, A., Li, W., Wang, B., Shelton, F., Otero, F., ... Chen, Y. A. (2009). Lipid G protein-coupled receptor ligand identification using beta-arrestin PathHunter assay. *Journal of Biological Chemistry, 284*(18), 12328–12338.

Zeng, B. Y., Dass, B., Owen, A., Rose, S., Cannizzaro, C., Tel, B. C., & Jenner, P. (1999). Chronic L-DOPA treatment increases striatal cannabinoid CB1 receptor mRNA expression in 6-hydroxydopamine-lesioned rats. *Neuroscience Letters, 276*(2), 71–74.

Zhong, P., Pan, B., Gao, X. P., Blankman, J. L., Cravatt, B. F., & Liu, Q. S. (2011). Genetic deletion of monoacylglycerol lipase alters endocannabinoid-mediated retrograde synaptic depression in the cerebellum. *Journal of Physiology, 589*(Pt 20), 4847–4855.

Zygmunt, P. M., Petersson, J., Andersson, D. A., Chuang, H., Sørgård, M., Di Marzo, V., … Högestätt, E. D. (1999). Vanilloid receptors on sensory nerves mediate the vasodilator action of anandamide. *Nature, 400*(6743), 452–457.

CHAPTER 4

Putative Role of Endocannabinoids in Schizophrenia

F. Markus Leweke, Cathrin Rohleder
Heidelberg University, Mannheim, Germany

BACKGROUND

The endocannabinoid system (ECS) is a multifaceted modulating system, exerting activity in diverse peripheral tissues as well as acting on a variety of neurotransmitter systems in the central nervous system (CNS) (Maccarrone et al., 2010). As a consequence, the ECS influences various physiological processes, for example, cardiovascular, metabolic, reproductive, or immune functions, and controls, for instance, pain perception, wake/sleep cycles, appetite and food intake, as well as energy balance (Maccarrone et al., 2015, 2010; Mechoulam & Parker, 2013). Due to its comprehensive modulating nature, the ECS represents a major homeostatic system that seems to be in particular active if processes fall out of balance. The ECS appears to be involved in several disorders, including a number of psychiatric conditions (Hauer et al., 2013; Parolaro, Realini, Vigano, Guidali, & Rubino, 2010; Schaefer et al., 2013) and among them schizophrenia.

Schizophrenia affects about 1% of the population worldwide, but certain regional differences of prevalence rates have been observed (Kahn et al., 2015). The chronic psychiatric disorder is characterized by heterogeneous symptoms that can be divided into three main categories: (1) positive symptoms (delusions, thought disorder, hallucinations), (2) negative symptoms (anhedonia, blunted affect, social withdrawal), and (3) cognitive impairment (sensory information processing, attention, working memory, executive functions) (Freedman, 2003; Wong & Van Tol, 2003). These symptoms are often accompanied by more unspecific symptoms like anxiety (Freedman, 2003).

Initially it has been assumed that the underlying neurochemical pathology of schizophrenia involves only dopaminergic disturbances. This

The Endocannabinoid System
ISBN 978-0-12-809666-6
http://dx.doi.org/10.1016/B978-0-12-809666-6.00004-6

dopamine hypothesis was based on the preclinical observation that anti-psychotic drugs influenced catecholamine metabolism and inhibited dopamine receptors (Carlsson & Lindqvist, 1963). A first refinement of this basic theory led to the hypothesis of a general hyperactive dopamine system in patients with schizophrenia (van Rossum, 1966). Later, this hypothesis developed into the assumption that hyperactivation of dopamine D_2 receptors in mesolimbic pathways is primarily associated with positive symptoms, whereas negative symptoms result from hypoactivity of the dopamine system in mesocortical projections (Davis, Kahn, Ko, & Davidson, 1991). Nowadays, a much more complex pathophysiology is assumed, including dysfunctions of not only the dopaminergic but also of the glutamatergic (Kim, Kornhuber, Schmid-Burgk, & Holzmuller, 1980), GABAergic (Garbutt & van Kammen, 1983), serotonergic (Meltzer, 1989), and nicotinergic systems (Olincy et al., 2006) as well as immuno-logical, inflammatory (Kirkpatrick & Miller, 2013; Leza et al., 2015; Rajasekaran, Venkatasubramanian, Berk, & Debnath, 2015), and develop-mental hypotheses of the disorder (Debnath, Venkatasubramanian, & Berk, 2015; Laurens et al., 2015) and suggestions on the role of oxidative stress (Leza et al., 2015; Rajasekaran et al., 2015) and disturbed energy metabo-lism (Holmes et al., 2006; Rajasekaran et al., 2015) in its pathophysiology. Furthermore, there is markedly evidence that the ECS plays an important role in schizophrenia.

The endocannabinoid hypothesis of schizophrenia has been proposed by Emrich, Leweke, and Schneider (1997) about one decade after the discov-ery of the ECS in the late 1980s (Devane, Dysarz, Johnson, Melvin, & Howlett, 1988) and the early 1990s (Devane et al., 1992; Matsuda, Lolait, Brownstein, Young, & Bonner, 1990; Mechoulam et al., 1995; Sugiura et al., 1995). Based on the finding that the psychoactive ingredient of *Cannabis sativa*—Δ^9-tetrahydrocannabinol (Δ^9-THC)—induced neuropsychological and psychopathological alterations in healthy volunteers, which exhibited strong similarities with data from schizophrenic patients, and studies report-ing that cannabis abuse might be a stressor for psychotic relapse and schizo-phrenic symptom exacerbation, Emrich et al. (1997) assumed that a dysfunctional ECS is involved in causing psychoses.

In the following years, schizophrenia research has been more focused on the connection of endocannabinoids [anandamide, 2-arachidonoyl-*sn*-glycerol (2-AG)] and related fatty acid ethanolamides [palmitoylethanolamide (PEA), oleoylethanolamide (OEA)], cannabinoid receptors [mainly type 1 (CB_1R) and type 2 (CB_2R)], as well as enzymes synthesizing [in particular *N*-acyl phosphatidylethanolamine phospholipase (NAPE) and diacylglycerol lipase (DAGL)] and degrading endocannabinoids [mainly fatty acid amide hydrolase (FAAH) and monoacylglycerol lipase (MAGL)] to schizophrenia. (More detailed general information on the ECS compounds can be found in "Chapter 2: Basic mechanisms of synthesis and hydrolysis of major endocannabinoids" and "Chapter 3: Receptors for the endocannabinoids.")

One research focus has been the impact of cannabis consumption on schizophrenia. To date, cannabis use is considered as a significant environmental risk factor for the development of schizophrenia in vulnerable individuals as substantiated by epidemiological studies (for review see Gage et al., 2016).

Analysis of large cohorts revealed that already moderate cannabis consumption (more than 20 times in lifetime) is associated with an increased risk of developing psychotic symptoms (Andreasson, Allebeck, Engstrom, & Rydberg, 1987; Arseneault et al., 2002; Fergusson, Horwood, & Swain-Campbell, 2003; Henquet, Murray, Linszen, & van Os, 2005; van Os et al., 2002; Zammit, Allebeck, Andreasson, Lundberg, & Lewis, 2002) and an earlier onset of schizophrenia (Di Forti et al., 2009; Donoghue et al., 2014; Large, Sharma, Compton, Slade, & Nielssen, 2011). Thereby the risk seems to rise with the amount (Moore et al., 2007; Zammit et al., 2002) and frequency of cannabis consumption (Di Forti et al., 2009), the potency of cannabis preparations used (Di Forti et al., 2009, 2014), and the age of onset (Arseneault et al., 2002). Cannabis preparations or single cannabinoids, e.g., the main psychotomimetic ingredient of the *C. sativa* plant—Δ^9-THC—are able to induce psychopathological, perceptional, and cognitive changes in healthy volunteers that resemble those observed in at-risk mental states and schizophrenic patients (Bhattacharyya et al., 2009; D'Souza et al., 2004; Emrich et al., 1997; Fusar-Poli et al., 2009; Koethe et al., 2006; Koethe, Hoyer, & Leweke, 2009; Leweke, Schneider, Radwan, Schmidt, & Emrich, 2000; Leweke, Schneider, Thies, Münte, & Emrich, 1999; Mason et al., 2009; Sherif, Radhakrishnan, D'Souza, & Ranganathan, 2016). Furthermore, it has been shown that acute cannabis use leads to more pronounced psychotic symptoms in highly psychosis-prone individuals (Mason et al., 2009) and transiently exacerbates positive, negative, and general symptoms as well as cognitive deficits in schizophrenic patients (D'Souza et al., 2005).

As the exogenous cannabinoid Δ^9-THC and the endogenous cannabinoids are CB_1R agonists, several studies investigated the CB_1R density in patients with schizophrenia to elucidate how the ECS is involved in the etiology of psychoses.

The majority of studies reported alterations of CB_1R density in the brain of schizophrenic patients. Postmortem autoradiography using radiolabeled CB_1R ligands and found increased CB_1R densities in the anterior and posterior cingulate cortex as well as dorsolateral prefrontal cortex (PFC) of schizophrenic patients (Dalton, Long, Weickert, & Zavitsanou, 2011; Dean, Sundram, Bradbury, Scarr, & Copolov, 2001; Jenko et al., 2012; Newell, Deng, & Huang, 2006; Zavitsanou, Garrick, & Huang, 2004). In contrast, unchanged (Koethe et al., 2007) or decreased CB_1R protein immunoreactivity (Eggan, Hashimoto, & Lewis, 2008; Eggan, Stoyak, Verrico, & Lewis, 2010; Urigüen et al., 2009) as well as reduced (Eggan et al., 2008) or unaltered (Dalton et al., 2011; Urigüen et al., 2009; Volk, Eggan, & Lewis, 2010) CB_1R messenger RNA (mRNA) levels were observed. Interestingly, Volk, Eggan, Horti, Wong, and Lewis (2014) found that CB_1R ligand binding was negatively correlated to CB_1R mRNA and protein immunoreactivity levels and thus concluded that patients possess a lower total amount of CB_1R, but higher levels of available membrane-bound CB_1R due to altered receptor trafficking. Another explanation might be that the negative correlation reflects greater CB_1R affinity, which may lead to a reduced need for CB_1R receptors resulting in lower CB_1R transcription and translation (Volk et al., 2014).

Unfortunately, one disadvantage of postmortem studies is that only the end of a chronic psychiatric disease influenced by various personal histories (e.g., environmental influences, life circumstances, kind and duration of treatment) can be captured. Positron emission tomography (PET) opens up the valuable opportunity to study CB_1R availabilities at different time points of the disease. So far, two PET studies in schizophrenic patients using the CB_1R radioligand $[^{18}F]MK$-9470 or $[^{11}C]OMAR$ showed increased CB_1R availabilities across several brain regions, interpretable as elevated CB_1R density (Ceccarini et al., 2013; Wong et al., 2010). Interestingly, CB_1R availability seems to be influenced by antipsychotic treatment. First, the increase of CB_1R binding was more pronounced in first-onset antipsychotic-naïve compared to treated patients (Ceccarini et al., 2013). Second, antipsychotic monotherapies affected CB_1R binding patterns differently. For instance, the mesocortical upregulation of CB_1R in antipsychotic-naïve patients was treatment-specifically reduced (Ceccarini et al., 2013). In addition, it has been suggested that elevated CB_1R binding levels in specific brain regions may be associated with symptom severity. Although in treated patients no (Ceccarini et al., 2013) or just a tendentious correlation (Wong et al., 2010) between PET data and psychiatric rating scales was found, antipsychotic-naïve patients revealed a negative correlation between CB_1R binding and negative symptoms and depression scores, in particular in the nucleus accumbens (NAc) (Ceccarini et al., 2013). These data indicate that an upregulation of CB_1R may serve an adaptive role toward psychotic symptoms. However, another PET study using [^{11}C]OMAR reported opposite results (Ranganathan et al., 2015) with a CB_1R availability that was reduced in schizophrenic patients compared to healthy controls. Although no difference was found between antipsychotic-free and antipsychotic-treated patients, the reduction of CB_1R availability was more pronounced in antipsychotic-free patients compared to healthy volunteers. Nevertheless, these results should be regarded with prudence, since only one of seven antipsychotic-free patients was antipsychotic naive, while for all other antipsychotic-free patients 2–60 months had passed since the last intake of antipsychotics. In addition, patients exhibited more severe psychotic symptoms compared to the patients enrolled in the two previous studies. This may also contribute to opposing data. Methodological differences or varieties in the analyzed patient cohorts that lead to the diverse results should be addressed in further studies. However, the results of postmortem and in vivo PET studies indicate that the ECS shows disturbances at receptor level in schizophrenic patients.

In the following sections, a closer look is taken on preclinical and clinical studies addressing the putative role of the endogenous ligands—anandamide and 2-AG—of the cannabinoid receptors. Beside the analysis of endocannabinoid levels in animal models or patients, some studies were focused on the synthesizing and in particular the metabolizing enzymes, as their activity finally determines the endocannabinoid levels. In addition, randomized, controlled clinical trials have been conducted studying the therapeutic benefit of pharmacological compounds affecting the ECS.

INSIGHTS FROM PRECLINICAL STUDIES

Animal models can be used to study the putative role of endocannabinoids on schizophrenia at least to a certain degree. Although several disease characteristics, in particular positive symptoms, are human specific, negative symptoms (e.g., inadequate social behavior/social withdrawal or deficits in sensorimotor gating) and cognitive impairments as well as more unspecific aspects like anxiety can be studied in rodents.

Due to the complex nature of schizophrenia, several animal models based on different etiological factors of schizophrenia are available. For example, animal models of early cannabis exposure are used to characterize long-lasting behavioral effects of early and frequent cannabis use and to clarify the underlying cellular mechanisms (Rubino & Parolaro, 2016).

However, during the past years, most of the studies analyzing the role of endocannabinoids in schizophrenialike animal models were based on the phencyclidine (PCP) model. This model is founded on the hypothesis of hypofunctional N-methyl-D-aspartate (NMDA) receptors in schizophrenia (for review see Snyder & Gao, 2013). In humans, the NMDA antagonist PCP has psychotomimetic effects. It induces psychotic states that closely resemble schizophrenia symptoms and exacerbates psychotic symptoms in schizophrenic patients (Javitt & Zukin, 1991; Steinpreis, 1996). In animals, PCP is able to mimic various aspects of schizophrenia on a behavioral and neurochemical level (Seillier & Giuffrida, 2009; Steinpreis, 1996).

So far, four studies have assessed the endocannabinoid levels in two slightly different PCP animal models of schizophrenia per se (Guidali et al., 2011; Seillier, Advani, Cassano, Hensler, & Giuffrida, 2010; Seillier, Martinez, & Giuffrida, 2013; Vigano et al., 2009). The procedures of both models are compared in Fig. 4.1.

In addition to the assessment of endocannabinoid levels and behavioral aspects, different ECS modulators were used in these studies, to elucidate the underlying neurochemical principles. An overview of these compounds is given in Fig. 4.2.

On the behavioral level, chronic-intermittent PCP treatment of juvenile rats over 4 weeks followed by a 3-day washout period resulted in recognition memory impairment (Guidali et al., 2011; Vigano et al., 2009), as well as increased avolition, a negative symptom of schizophrenia (measured as immobility in the forced swim test) (Guidali et al., 2011). Quantification of endocannabinoid levels directly after the behavioral tasks revealed that these effects were accompanied by increased 2-AG levels in the PFC, whereas anandamide concentration remained unaltered (Guidali et al., 2011; Vigano et al., 2009). However, in one of the two studies, anandamide levels were reduced by trend (Vigano et al., 2009). In the hippocampus, both endocannabinoid levels were unaffected (Guidali et al., 2011; Vigano et al., 2009). Unfortunately, other brain regions were not assessed. The increase in 2-AG was interpreted as a compensatory reaction that may contribute to a local enhancement of glutamatergic transmission (Vigano et al., 2009).

Interestingly, chronic cotreatment with Δ^9-THC worsened the PCP-induced cognitive deficits and triggered a significant anandamide level reduction, but reversed the PCP-induced increase in 2-AG level in the PFC

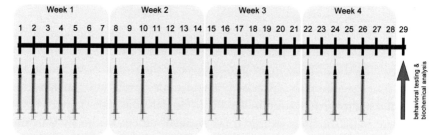

Chronic-intermittent PCP model

Sub-chronic PCP model

Figure 4.1 Schematic comparison of the two primarily used phencyclidine (PCP) animal models. In the chronic-intermittent PCP model, animals receive 14 intraperitoneal PCP injections over 4 weeks, according to the depicted schedule. After a washout period of 3 days, behavioral or biochemical testing is accomplished. The same number of injections is applied in the subchronic PCP model, but animals are treated twice a day for 7 days. Depending on the study, the washout period ranges between 36 h and 10 days.

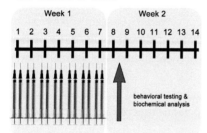

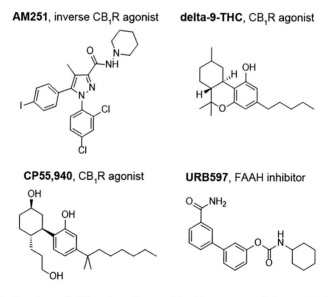

AM251, inverse CB$_1$R agonist   **delta-9-THC**, CB$_1$R agonist

CP55,940, CB$_1$R agonist   **URB597**, FAAH inhibitor

Figure 4.2 Structure of different endocannabinoid system modulators. While AM251 blocks CB$_1$R, both Δ^9-THC and CP55,940 activate this receptor. On the other hand, URB597 inhibits the anandamide-degrading enzyme fatty acid amide hydrolase (FAAH) and thus increases anandamide levels.

(Vigano et al., 2009). On the other hand, chronic cotreatment with the inverse CB_1R agonist AM251 reversed PCP-induced recognition memory impairments, the increase in avolition as well as the rise of 2-AG and at the same time, significantly increased anandamide levels and antagonized increased neuronal activation in the PFC (Guidali et al., 2011).

These initial findings indicated (1) an involvement of the ECS in this pharmacological schizophrenia-like animal model, (2) pointed to a different role of anandamide and 2-AG, and (3) revealed that probably CB_1R activation contributes to the behavioral impairments. Furthermore, the authors suggested an antipsychotic-like profile of inverse CB_1R agonist AM251 by restoring ECS functions and thus probably normalizing some of the neurochemical maladaptations present in this schizophrenia-like animal model (Guidali et al., 2011).

Also, results from studies using 7-day subchronic PCP treatment of adult rats, followed by a 5- or 7-day washout period indicated a contribution of the ECS. On the behavioral level, subchronic PCP treatment caused working memory deficits (Seillier et al., 2010) as well as social withdrawal (Seillier et al., 2010, 2013).

Under resting conditions (i.e., measurements were carried out in a separate group of rats that did not undergo behavioral testing), significantly increased anandamide levels were observed in the NAc, and by trend in the medial PFC. This may be related to an increased dopaminergic transmission in this brain area. In addition, subchronic PCP treatment led to a slow increase of 2-AG levels over several days, as in the ventral tegmental area (VTA) a significant increase was observed 10 days but not 5 days after the last PCP injection. The authors suggested that the slowed elevation was probably triggered by increased dopamine levels and turnover in the amygdala and PFC. In other brain areas or in cerebrospinal fluid (CSF), no changes in anandamide or 2-AG levels were found (Seillier et al., 2010).

In contrast, quantification of endocannabinoids immediately after the social interaction task revealed elevated 2-AG levels in the NAc and caudate putamen (and by trend in the medial PFC) of PCP-treated rats. Furthermore, PCP treatment resulted again in increased anandamide levels in the NAc, and also in significantly reduced anandamide levels in both medial PFC and amygdala compared to saline-treated rats. In other regions of the brain, the endocannabinoid concentrations remained unaltered. Results of nonlinear regression analysis led the authors to the conclusion that an association between anandamide levels and optimal social interaction (as in control animals) exists. Further experiments showed that anandamide-mediated activation of CB_1R is important for social interaction and that deficient endocannabinoid signaling contributes to social withdrawal in animals treated with PCP (Seillier et al., 2013). Interestingly, the blockade of CB_1R by acute AM251 treatment ameliorated the working memory deficit (Seillier et al., 2010) but had no effect on the social behavior in PCP-treated rats (Seillier et al., 2010, 2013). In addition, augmentation of anandamide levels via inhibition of the main degrading enzyme FAAH by URB597, reversed PCP-induced social withdrawal (Seillier et al., 2010, 2013) without affecting the working memory deficit (Seillier et al., 2010). As AM251 reversed the effects of URB597 on social withdrawal in PCP-treated rats, the beneficial effect of URB597 seems to be CB_1R dependent. This was further supported by the observation that the synthetic CB_1R agonist CP55,940 was also able to reverse social withdrawal in PCP-treated rats (Seillier et al., 2013).

Despite some minor discrepancies, likely based on methodological differences, all preclinical results indicate an involvement of endocannabinoids in processes underlying psychosis. It appears that deficient functionality of the ECS contributes to cognitive deficits and negative symptoms, but at the same time, elevation of anandamide levels seems to have beneficial effects. As already mentioned, the ECS is a homeostatic, modulating system, and therefore the optimal balance of endocannabinoid tone might be important.

During the past years, the effects of modulators of the ECS have been analyzed more deeply, using the subchronic PCP model without studying the endocannabinoid levels. Analysis of neuronal activation via c-Fos expression revealed that PCP treatment led to altered neuronal activation pattern induced by social interaction (Matricon, Seillier, & Giuffrida, 2016). For instance, it has been observed that subchronic PCP administration prevented neuronal activation in the infralimbic and orbitofrontal cortex. Interestingly, the reversal of PCP-induced social withdrawal by FAAH inhibition, and thus increased anandamide and related fatty acid ethanolamide (PEA and OEA) levels, was accompanied by restoring a similar neuronal activation pattern as observed in saline-treated rats (Matricon et al., 2016). The c-Fos expression data indicate that the orbitofrontal cortex, posterior central amygdala, and the dorsomedial part of the bed nucleus of the stria terminalis play a critical role in the reversal of PCP-induced social withdrawal by URB597. As already mentioned, one study, using the chronic-intermittent PCP model reported that the blockade of CB_1R by AM251 resulted not only in increased anandamide levels but also antagonized the

PCP-induced increased c-Fos expression in the PFC directly after a working memory task (Guidali et al., 2011). Thus alterations of neuronal activation pattern induced by PCP may depend on the behavioral task, but it seems that anandamide augmentation is able to normalize the activation pattern.

The influence of FAAH inhibition on neuronal activation has been also analyzed in an electrophysiological study. Chloral-hydrate-anesthetized PCP-treated rats displayed an increased dopamine neuron population activity in the VTA that was normalized by acute and chronic systemic URB597 treatment (Aguilar, Chen, & Lodge, 2015). The authors demonstrated that this elevation of dopamine neuron activity in PCP-treated rats was related to an aberrant increased activity in the ventral hippocampus (vHipp). Interestingly, initial studies of the group showed that the vHipp activity was not influenced by FAAH inhibition, so that URB597 seems to affect regions downstream of the vHipp, probably involving the NAc or ventral pallidum (VP). Since FAAH expression is higher in the VP compared to NAc, Aguilar et al. (2015) decided to analyze the role of VP more deeply. Indeed, they found that the vHipp-induced inhibition of the VP activity was reduced by URB597. In addition, intra-VP injections of the FAAH-inhibitor resulted in a reduction of the increased dopamine neuron activity, indicating that VP may be the site of action of anandamide level upregulation. Since endocannabinoids are produced on demand (Di Marzo et al., 1994; Di Tomaso et al., 1997; Piomelli, 2003), the same group decided to investigate the neuronal activity in conscious, freely moving rats and compared the effects of URB597 and Δ^9-THC on PCP-treated and control rats (Aguilar, Giuffrida, & Lodge, 2016). Interestingly, subchronic PCP treatment had no effect on the neuronal firing rates, spontaneous oscillatory activity, or baseline synchrony between vHipp and medial PFC. However, FAAH inhibition increased medial PFC firing rates and gamma oscillatory activity, while in vHipp, a reduced delta band activity, but a tendentiously increased firing rate was observed. Moreover, URB597 decreased vHipp-medial PFC delta synchrony. In contrast, differential effects were observed after acute Δ^9-THC treatment. Δ^9-THC did not affect the firing rate (beside a slight increase in vHipp) and oscillatory activity in the medial PFC and vHipp, but it increased vHipp–medial PFC delta synchrony.

The exact mechanism of restoring neuronal activation and normalizing behavior by anandamide augmentation still needs to be elucidated. As aforementioned, there is some evidence that the reversal of PCP-induced social withdrawal is CB_1R dependent, since the effect of URB597 treatment can be blocked by acute AM251 injection. In addition, CP55,940 was also able to counteract the PCP-induced deficient social behavior. However, the involvement of CB_1R activation may only apply for certain behavioral aspects. For instance, a CB_1R blockade by AM251 had no effect on social withdrawal but was able to reverse cognitive deficits as well as avolition in PCP-treated rats. Furthermore, CB_1R activation by systemic Δ^9-THC administration worsened the PCP-induced cognitive deficits in the chronic-intermittent PCP model instead. However, we have to keep in mind that endocannabinoids and exogenous cannabinoids like Δ^9-THC or CP55,940 probably differentially regulate neuronal activity and thus behavioral outcome, as demonstrated by the recent study of Aguilar et al. (2016). While exogenous cannabinoids activate CB_1R throughout the entire brain, URB597 administration results in region-specific CB_1R activation, as elevation of anandamide levels depends on the occurrence and relative abundance of the FAAH enzyme in neurons (Egertova, Cravatt, & Elphick, 2003). In addition, upregulation of anandamide levels may not only result in region-specific CB_1R activation but also in activation of other receptor types. At high concentrations, anandamide may also activate the transient receptor potential vanilloid type 1 (Di Marzo, De Petrocellis, Fezza, Ligresti, & Bisogno, 2002; Zygmunt et al., 1999) or peroxisome proliferator-activated receptor (PPAR) (for review see O'Sullivan, 2016).

Since the ECS is acting as a modulating system in the brain, it is conceivable that anandamide counteracts neurotransmitter imbalances. As already shortly discussed, some of the described alterations of endocannabinoid levels may be related to changes in the dopaminergic system.

It has been hypothesized that the ECS (in particular anandamide) acts as inhibitory feedback mechanism counteracting dopamine stimulation of motor activity (Giuffrida et al., 1999). This theory was based on the finding that neuronal activity stimulated anandamide release in the dorsal striatum and that administration of a D_2-like dopamine receptor agonist further increased anandamide release. In addition, this increase was prevented by a D_2-like dopamine receptor antagonist (Giuffrida et al., 1999). In the PCP model, elevated dopaminergic transmission seemed to induce anandamide elevation in the NAc, as elevated levels of the dopamine metabolite homovanillic acid were found in the NAc (Seillier et al., 2010). In addition, it has been suggested that the 2-AG increase in the VTA may be linked to the observed increases of dopamine levels and turnover (measured as ratio of the metabolite DOPAC and dopamine) in the amygdala and PFC, respectively (Seillier et al., 2010).

Another hypothesis is related to the GABAergic system. As described, PCP-treated rats are characterized by deficient anandamide mobilization and social behavior deficits. Augmentation of anandamide levels by FAAH or direct CB_1R activation reverses PCP-induced social withdrawal. Interestingly, blockade of a cholecystokinin (CCK, a neuropeptide expressed by GABAergic interneurons) receptor (CCK2) had the same effect (Seillier et al., 2013). As in normal rats, social interaction seems to be accompanied by increased anandamide and decreased CCK levels, the authors suggested that the endocannabinoid–induced inhibition of GABAergic interneurons expressing CCK is lost or reduced in PCP-treated rats and triggers social withdrawal. The inverse conclusion is that increasing anandamide levels by FAAH inhibition counteracts reduced inhibition of GABAergic interneurons in PCP-treated rats.

INVESTIGATIONS IN HUMANS

The first human study analyzing endocannabinoids and related fatty acid ethanolamides was already published in 1999 (Leweke, Giuffrida, Wurster, Emrich, & Piomelli, 1999). Analysis of CSF samples of 10 patients with acute schizophreniform psychotic symptoms and 11 healthy volunteers revealed significantly increased anandamide and PEA levels in schizophrenic patients. Thus Leweke, Giuffrida, et al. (1999) suggested that functional abnormalities in endogenous cannabinoid signaling might contribute to the pathogenesis of schizophrenia.

Some years later, De Marchi et al. (2003) compared peripheral blood samples of 12 patients suffering from schizophrenia for several years and 20 healthy volunteers. Anandamide levels were significantly elevated in patients, indicating that changes in the ECS may not only be restricted to the CNS. After treatment with olanzapine, 5 of the 12 patients achieved clinical remission, which was associated not only with a reduction of peripheral anandamide concentrations but also with decreased CB_2R mRNA transcripts and FAAH levels. Due to the alterations in CB_2R mRNA transcript levels, the authors suggested that the observed effects might be related to immunological changes in schizophrenia.

The data of these two studies supported the hypothesis that dysfunctional endocannabinoid signaling might contribute to schizophrenia. However, owing to the small number of investigated subjects and possible influences of concomitant medication, the interpretation of these results was not straightforward. However, in the following years, further studies shed light on the nature of the role of endocannabinoids in schizophrenia.

In 2004, Giuffrida et al. investigated endocannabinoid and related fatty acid ethanolamide levels in a larger cohort (237 subjects). In this study, CSF and serum samples of 47 antipsychotic-naïve as well as 71 patients with first-episode schizophrenia briefly treated with antipsychotics, 13 patients with dementia, 22 patients with affective disorder, and 84 healthy volunteers were collected and analyzed. CSF levels of anandamide were significantly increased in antipsychotic-naïve patients, whereas no alterations of anandamide concentration were found in patients with dementia or affective disorder (Fig. 4.3). Thus augmentation of anandamide levels in CSF seems to be disease specific. In contrast to De Marchi et al. (2003), serum levels did not differ, indicating that the measured alterations originate from CNS, most likely from temporoparietal structures and basal ganglia as these areas primarily drain to into the spinal CSF, whereas CSF from frontal brain areas may be cleaved through the bridge veins to the peripheral blood. Beside these different flow aspects of CSF, discrepant results may be attributed to differences in the investigated patient cohort (different disease duration) or experimental procedure (whole blood vs. serum). It has been suggested that the increase of anandamide levels in schizophrenic patients is not caused by a general disturbance of lipid signaling, since the concentrations of the related lipid OEA were comparable in all groups and only slightly but not significantly altered PEA levels were found in antipsychotic-naïve patients.

Interestingly, antipsychotic treatment seems to affect anandamide levels, whereby the extent of the effect depends on the class of antipsychotics. Patients with first episode treated with dopamine $D_{2/3}$ receptor antagonists, such as first-generation antipsychotics or amisulpride, had anandamide levels comparable to those of healthy volunteers, whereas patients treated with second-generation antipsychotics with additional antagonistic effects at serotonin $5HT_{2A}$ receptors still showed significantly increased anandamide concentrations. Nevertheless, these levels did not reach the concentrations of antipsychotic-naïve patients (Fig. 4.3).

Moreover, CSF anandamide levels were negatively correlated with psychotic symptoms including positive symptoms (e.g., delusions and hallucinations) as well as negative symptoms (apathy, paucity of speech, blunting, or incongruity of emotional responses) and more general psychopathological features (e.g., depressive mood or aggression) assessed with the Positive and Negative Syndrome Scale (PANSS). This finding led the authors to the conclusion that anandamide elevation in acute schizophrenia reflects a compensatory adaptation to the disease state counteracting exaggerated dopaminergic neurotransmission. If this applies, it is conceivable that the adaptive enhancement already slowly develops in early disease stages.

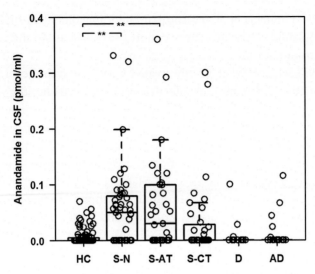

Figure 4.3 Cerebrospinal fluid (CSF) anandamide in healthy volunteers (C); antipsychotic-free schizophrenic patients with paranoid schizophrenia (S–N); acute schizophrenic patients (paranoid type) treated with $5HT_{2A}/D_2$ receptor antagonistic antipsychotics (S-AT) or D_2/D_3 receptor antagonistic antipsychotics (S-CT) antipsychotic drugs; and patients affected by dementia (D) or affective disorders (AD) without psychotic symptoms. Single values are given with mean ± standard error of mean as well as corresponding *box-plots* illustrating median, range, and quartiles for each group. Statistically significant differences between groups are shown (*$P \leq .01$; **$P \leq .001$; Kruskal–Wallis test followed by Mann–Whitney test; a Bonferroni correction (five tests) of the α-level was applied: $\alpha = 0.05/5 = 0.01$). *Modified from Giuffrida, A., Leweke, F.M., Gerth, C.W., Schreiber, D., Koethe, D., Faulhaber, J., ... Piomelli, D. (2004). Cerebrospinal anandamide levels are elevated in acute schizophrenia and are inversely correlated with psychotic symptoms.* Neuropsychopharmacology, 29(11), 2108–2114.

Koethe, Giuffrida, et al. (2009) actually found increased cerebrospinal anandamide in individuals of at-risk mental states (initial prodromal syndrome). These changes were CNS specific, since serum anandamide levels of patients (n = 27) were comparable to those of healthy controls (n = 81). Cerebrospinal and serum OEA concentrations were also comparable between both groups, strengthening the hypothesis that anandamide alterations are not caused by a generalized change in lipid signaling. Anandamide levels were additionally negatively correlated with cognitive aspects measured by PANSS (factor 3 "cognitive syndrome"). More importantly, anandamide concentrations in CSF were associated with transition time into frank psychosis by trend. The lower the anandamide levels, the higher the risk for early transition to psychosis. These data may suggest a protective role of anandamide rather than a participation in pathophysiology of schizophrenia.

At first glance this might be counterintuitive since as mentioned earlier, cannabis consumption is considered as one risk factor for the development of schizophrenia in vulnerable individuals and both the main psychotomimetic ingredient of cannabis—Δ^9-THC—and anandamide activate CB_1R. However, as discussed earlier, endo- and exogenous cannabinoids (i.e., anandamide and Δ^9-THC, respectively) probably differentially affect neuronal activity and behavioral outcome, even though both act on the same receptor type. Thus the suggested protective role of anandamide and the adverse effects of excessive cannabis use are not mutually exclusive. In fact, cannabis consumption may facilitate the development of psychotic episodes in susceptible individuals by reducing anandamide levels, as Leweke et al. (2007) found decreased anandamide concentrations in schizophrenic high-frequent cannabis users (lifetime use >20) compared to schizophrenic low-frequency users (lifetime use <5) as well as a negative correlation between anandamide levels in CSF and schizophrenic symptoms. Hence, the putative protective properties of anandamide can only take effect to a lesser extent in individuals frequently using cannabis. Downregulation of cerebrospinal anandamide levels was also observed in healthy volunteers, whereby it seems that higher amounts of cannabis are tolerated compared to individuals prone to psychosis. While no differences were found between low- versus high-frequency users (Leweke et al., 2007), lower CSF anandamide levels were observed in heavy (>10 times in a month) compared with light cannabis users (<10 times in a month) (Morgan et al., 2013). In addition, anandamide levels seems to influence the impact of cannabis consumption in subjects as they were negatively correlated with persisting psychotic symptoms when nonintoxicated (Morgan et al., 2013).

The underlying processes resulting in a downregulation of anandamide signaling in the CNS remain elusive, but one possible interpretation is that cannabis use, exceeding a certain threshold, decreases anandamide biosynthesis or increases anandamide degradation (Leweke et al., 2007).

Potvin et al. (2008) proposed that disturbances of ECS may be associated with the high prevalence of substance use disorders in patients with schizophrenia. Based on this hypothesis, they investigated whether baseline serum endocannabinoid and related eicosanoid (PEA, OEA) levels predict substance abuse progression during treatment with the second-generation antipsychotic quetiapine, preferentially antagonizing $5-HT_{2A}$- and D_2 receptors. Peripheral baseline levels of anandamide as well as PEA and OEA were significantly elevated in patients (n = 29) with a dual diagnosis of schizophrenia and substance use disorders compared with controls (n = 17), whereas 2-AG level were equal in both groups. Substance use disorder severity parameter improved during treatment with quetiapine for 12 weeks, but anandamide, PEA, and OEA levels remained unaltered. Peripheral anandamide baseline levels predicted the endpoints of psychoactive substance use (e.g., cannabis or alcohol), whereby higher anandamide levels were associated with worse substance abuse outcomes. This finding needs to be carefully interpreted because anandamide levels were also found to be elevated in a subgroup of seven patients with schizophrenia who showed a marked reduction in substance use at study endpoint. In addition, it should be considered that at baseline, patients were mostly on second-generation antipsychotics (clozapine or quetiapine), and thus baseline anandamide levels were already influenced by medication.

In contrast to the observed increased anandamide levels in CSF (Giuffrida et al., 2004; Koethe, Giuffrida, et al., 2009; Leweke et al., 2007; Leweke, Giuffrida, et al., 1999) and blood (De Marchi et al., 2003; Potvin et al., 2008), as well as unaltered 2-AG serum levels (Potvin et al., 2008), one postmortem study found decreased anandamide in the cerebellum and hippocampus but significantly increased 2-AG levels in the hippocampus and PFC (Muguruza et al., 2013). OEA levels were comparable to those found in brains of control subjects, whereas PEA levels were reduced in the cerebellum of schizophrenic patients. Importantly, a positive toxicological finding for cannabis was considered as exclusion criterion. Therefore the results were presumably not influenced by heavy cannabis consumption, although the influence of lifetime cannabis use remains elusive. For further analysis, the brains of schizophrenic subjects were divided into two groups according to the results of a toxicological screening: antipsychotic-free (n = 11) and antipsychotic-treated patients (n = 8). As observed in the overall analysis, patients without traceable antipsychotic levels showed equal anandamide levels compared to controls in PFC but decreased concentrations in cerebellum and hippocampus, whereas in antipsychotic-treated subjects, lower levels of anandamide were additionally observed in the PFC. On the other hand, 2-AG levels of antipsychotic-free patients with schizophrenic remained significantly higher in the hippocampus and cerebellum, whereas 2-AG levels of antipsychotic-treated patients with schizophrenia did not differ from those of control subjects. These data emphasize that pharmacological treatment is a confounding factor that we have to be aware of, when investigating the ECS in schizophrenic patients. Due to the much longer disease history of the patients studied in this postmortem study, the results cannot be adequately compared to those obtained in the CSF studies (Giuffrida et al., 2004; Koethe, Giuffrida, et al., 2009; Leweke et al., 2007;

Leweke, Giuffrida, et al., 1999). One possible explanation for the opposing results might be that in the initial states of the disease elevated anandamide levels in the CSF reflect the adaptive, even protective role of anandamide, whereas in the course of the disease, the years of illness and antipsychotic treatment, as well as life circumstances and social environment, influence anandamide signaling, finally resulting in decreased anandamide levels in several brain regions. In addition, we have to keep in mind, that one half of the patients (10 of 19 patients: 6 antipsychotic-free and 4 antipsychotic-treated patients) committed suicide. Hence, especially patients with a high psychological strain and probably marked symptoms were included in the study. As it was shown that anandamide levels in CSF are inversely correlated with symptom severity (i.e., individuals with higher anandamide levels suffered from less severe psychotic symptoms), the observed low anandamide levels may also reflect a malfunction of the endogenous, protective mechanism based on endocannabinoid signaling.

Beside studies analyzing endocannabinoid and related eicosanoids, some studies were published focusing on endocannabinoid-synthesizing and endocannabinoid-degrading enzymes. Bioque et al. (2013) reported reduced synthesizing enzyme levels (NAPE and DAGL) but increased expression of degrading enzymes (FAAH and MAGL) in peripheral blood mononuclear cells (PBMCs) in patients with first episode of psychosis (87 antipsychotic-treated, 8 nontreated patients), after controlling for possible confounders [age, gender, body mass index (BMI), cannabis consumption]. This means that a dysregulation of ECS compounds that control peripheral endocannabinoid levels can be shown in treated patients with a first-episode psychosis. Furthermore, a negative correlation between FAAH expression and total PANSS as well as Montgomery-Asberg Depression Rating Scale was observed. At first glance, this is contrary to the negative correlation of anandamide levels in CSF observed in antipsychotic-naïve schizophrenic patients (Giuffrida et al., 2004) as well as in individuals of at-risk mental states (Koethe, Giuffrida, et al., 2009), because higher PANSS scores were associated with lower FAAH expression and thus higher peripheral anandamide levels. However, in the analyzed cohort, the mean PANSS score was 53.75, indicating that the majority of patients were under remission. The authors argued that increased FAAH levels may depend on the state of the psychotic disorder, and that acute patients with higher PANSS scores would have higher anandamide and decreased FAAH levels. In addition, the peripheral and central ECS may behave independently, as no correlation of peripheral and central endocannabinoid levels were found in former studies (Giuffrida et al., 2004; Koethe, Giuffrida, et al., 2009).

In a subgroup, enzyme expression of patients who had fulfilled *Diagnostic and Statistical Manual of Mental Disorders-Fourth Edition* criteria for dependence or abuse (at least during 12 consecutive months) throughout their lives was compared with those of healthy controls (Bioque et al., 2013). NAPE and DAGL expression levels in PBMCs were significantly reduced, whereas FAAH and MAGL levels were comparable. This may point to reduction of peripheral endocannabinoid signaling due to heavy cannabis consumption and is in line with the aforementioned finding that already moderate cannabis consumption reduced CSF anandamide levels (Leweke et al., 2007). Interestingly, there were no differences between patients with a history of heavy cannabis use and patients who used cannabis sporadically or not at all in their lifetime (Bioque et al., 2013).

In 2016, the same group reported that the peripheral expression of ECS components was associated with some neuropsychological domains in the same patient cohort (Bioque et al., 2016). After controlling for potential confounding factors, working memory performance correlated significantly to the NAPE expression, whereby the short-term verbal memory was associated with DAGL and FAAH expression in PBMCs. Moreover, attention measures correlated to the MAGL expression. Importantly, in healthy controls, the expression of enzymes of the ECS did not correlate with any of the cognitive subtests. These results indicate that the ECS differently affects distinct cognitive processes in patients with a first episode psychosis. While 2-AG signaling might be important for attentional tasks, anandamide seems to modulate working memory performance. Other neuropsychological domains, for instance, short-term verbal memory, seems to be differentially influenced by anandamide and 2-AG.

In addition to these two studies analyzing ECS enzyme expression in peripheral PBMCs, two postmortem studies have been published. The first study reported comparable mRNA levels for endocannabinoid-synthesizing (DAGL) and endocannabinoid-metabolizing (MAGL) enzymes in the PFC of schizophrenic patients and healthy controls (Volk et al., 2010). This is not in line with the observed reduction in peripheral DAGL mRNA levels (Bioque et al., 2013). Although only one brain area has been analyzed, it might be suggested that central and peripheral alterations in ECS compounds are not directly comparable, as it was also suggested for peripheral and central endocannabinoid levels. On the other hand, methodological differences might also contribute to the different findings.

Some years later, the same group analyzed the serine hydrolase α-β-hydrolase domain 6 (ABHD6) mRNA levels in the PFC of schizophrenic patients and healthy controls. ABHD6 has been shown to degrade 2-AG (Blankman, Simon, & Cravatt, 2007), and thus alterations may influence 2-AG signaling in schizophrenic patients. The mean ABHD6 mRNA levels were comparable between both groups, but interestingly, ABHD6 mRNA levels were increased in schizophrenic patients who were younger (<40 years) and possessed shorter disease duration (<15 years). ABHD6 mRNA levels were not affected by chronic exposure to either antipsychotic medication or Δ^9-THC. These results indicate that 2-AG levels may be slightly reduced in the first years of the illness, in particular with regard to the previous study of this group, showing unaltered DAGL and MAGL levels.

THERAPEUTIC APPROACHES

Both preclinical and clinical data indicate that the ECS is involved in schizophrenia. It has been controversial for a while whether the ECS plays a protective or harmful role in the pathogenesis of schizophrenia (Leweke, Mueller, Lange, & Rohleder, 2016), but an increasing body of evidence indicates that especially anandamide has a protective role in schizophrenia and at-risk mental states, while a frequent regional unspecific CB_1R activation during adolescence by, e.g., Δ^9-THC might trigger the development of schizophrenia in vulnerable persons.

As reviewed by Leweke et al. (2016), two main therapeutic approaches have been investigated so far. First, the selective blockade/inverse agonism of CB_1R and second, the modulation of endocannabinoid levels by use of the phytocannabinoid cannabidiol.

Both molecular and preclinical data from rodents led to the suggestion that the CB_1R antagonist/inverse agonist SR141716 (rimonabant) (Alonso et al., 1999; Poncelet, Barnouin, Breliere, Le Fur, & Soubrie, 1999; Rinaldi-Carmona et al., 1996; Scatton & Sanger, 2000) might be a promising antipsychotic compound with a mechanism of action different from first- and second-generation antipsychotics. Unfortunately, in schizophrenic patients, rimonabant had no significant effects on psychopathology and cognition (Boggs et al., 2012; Meltzer, Arvanitis, Bauer, Rein, & Group M-TS, 2004). Preclinical data suggesting that the beneficial effects of the FAAH inhibitor URB597 on PCP-induced social withdrawal is CB_1R dependent (Seillier et al., 2013), as well as the inverse correlation between CSF anandamide levels and psychotic symptoms (Giuffrida et al., 2004), may explain why this treatment approach has failed.

Administration of cannabidiol represents the other new therapeutic approach. The use of cannabidiol is based on the hypothesis that it increases anandamide levels probably by acting as a moderate FAAH inhibitor (Bisogno et al., 2001; Leweke et al., 2012; Watanabe, Kayano, Matsunaga, Yamamoto, & Yoshimura, 1996), and thus taking advantage of the endogenous protective mechanism. However, the actual mechanism of action is still controversial (Bisogno et al., 2001; Pertwee, Ross, Craib, & Thomas, 2002; Thomas et al., 2007), and it might be that also other mechanisms contribute to the antipsychotic effect of this exogenous cannabinoid, but to date, studies directly linking other proposed mechanisms to the antipsychotic properties of cannabidiol are lacking.

The results of the first individual treatment attempt were reported in 1995 (Zuardi, Morais, Guimarães, & Mechoulam, 1995). Daily administration of up to 1500 mg/day over 4 weeks decreased the Brief Psychiatric Rating Scale (BPRS) and Interactive Observation Scale for Psychiatric Inpatients scores. The clinical improvement could not be increased by additional treatment with the conventional antipsychotic haloperidol. Nearly 10 years later, three treatment-resistant patients were treated with cannabidiol (Zuardi et al., 2006). One patient showed a slight improvement of both positive and negative symptoms, whereas cannabidiol treatment was not effective in the two other patients. However, none of the patients suffered from side effects.

The first controlled, randomized, double-blind clinical trial was published in 2012 (Leweke et al., 2012). In this 4-week trial, the efficacy of cannabidiol (600–800 mg/day) was compared to the second-generation antipsychotic amisulpride (600–800 mg/day), antagonizing highly effectively, dopamine $D_{2/3}$ receptors. Both drugs significantly reduced positive, negative, and general symptoms as assessed by PANSS and BPRS. The efficacy of cannabidiol was comparable to that of amisulpride, as depicted in Fig. 4.4, but cannabidiol possessed a superior side effect profile and did not induce weight gain, extrapyramidal symptoms, and increased prolactin levels. Importantly, the reduction of psychotic symptoms by cannabidiol but not by amisulpride treatment was significantly associated with an increase of serum anandamide levels. Interestingly, the levels of the two related fatty acid ethanolamides OEA and PEA were also increased in schizophrenic patients treated with cannabidiol, although their levels were not associated with the reduction of psychotic symptoms. This supports the aforementioned hypothesis that cannabidiol develops its antipsychotic properties by increasing anandamide levels via, e.g., inhibition of FAAH or anandamide transport among potential other mechanisms.

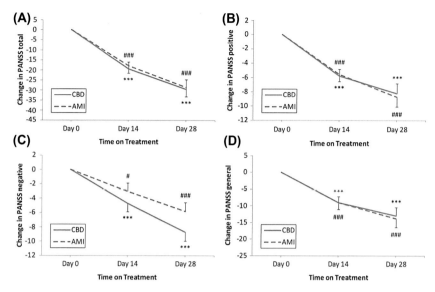

Figure 4.4 Changes from baseline in Positive and Negative Syndrome Scale (PANSS) scores determined using mixed effects repeated measures model analysis (adjusted for baseline). (A) PANSS total score. (B) PANSS-positive score. (C) PANSS-negative score. (D) PANSS general score. Data show predicted means ±standard error of mean at each week. Statistical significance is calculated between groups ($^{†}P \leq .05$, $^{††}P \leq .01$, and $^{†††}P \leq .001$) and versus baseline (i.e., 0; *CBD, #AMI; ***/###$P \leq .05$, **/##$P \leq .01$, */#$P \leq .001$). *From Leweke, F.M., Piomelli, D., Pahlisch, F., Muhl, D., Gerth, C.W., Hoyer, C., ..., Koethe, D. (2012). Cannabidiol enhances anandamide signaling and alleviates psychotic symptoms of schizophrenia.* Translational Psychiatry [Electronic Resource], 2, e94 *with permission.*

As summarized in Leweke et al. (2016), five further randomized clinical trials have been conducted or are currently ongoing. Hence, new data on the acute antipsychotic effects of cannabidiol will hopefully be available soon. However, current data already indicate promising antipsychotic properties and a preferable side effect profile. Nevertheless, large-scale trials are still needed to gain information on efficacy, long-term tolerability, and safety.

BEYOND ANANDAMIDE: OLEOYLETHANOLAMIDE AS PROMISING TARGET FOR TREATMENT OF DISTURBED EATING BEHAVIOR AND WEIGHT GAIN IN PSYCHOSIS

As mentioned earlier, Leweke et al. (2012) found not only elevated anandamide but also increased PEA and OEA levels in schizophrenic patients treated with cannabidiol. While anandamide seems to be relevant for the antipsychotic properties of cannabidiol, elevated OEA levels may have additional beneficial effects with respect to appetite disturbances often observed in psychotic patients. Results of several animal studies indicate that OEA triggers satiety and seems to be a biosensor for dietary fat (for review see DiPatrizio & Piomelli, 2015; Piomelli, 2013). In free-feeding rodents, OEA administration reduced food intake (Astarita et al., 2006; Fu et al., 2003; Oveisi, Gaetani, Eng, & Piomelli, 2004; Rodriguez de Fonseca et al., 2001) and body weight (Rodriguez de Fonseca et al., 2001). OEA exhibits characteristics of a satiety signal (DiPatrizio & Piomelli, 2015), as it decreases meal frequency (i.e., increases postmeal intervals; Astarita et al., 2006) without influencing meal size (Gaetani, Oveisi, & Piomelli, 2003), which possibly constitutes enhanced across-meal satiety (Fu, Kim, Oveisi, Astarita, & Piomelli, 2008). To date it is suggested that OEA is synthetized by small intestinal enterocytes during the digestion of fat-containing diets (DiPatrizio & Piomelli, 2015; Fu et al., 2007; Schwartz et al., 2008). To be more precise, food-derived oleic acid is internalized by enterocytes and seems to serve as substrate for biosynthesis of OEA (Schwartz et al., 2008). Finally, OEA activates PPAR-α (Astarita et al., 2006; Fu et al., 2003; Schwartz et al., 2008) and causes satiety by the recruitment of afferent sensory fibers as destruction of peripheral vagal and nonvagal fibers prevented hypophagic responses to OEA (Rodriguez de Fonseca et al., 2001).

These preclinical findings seem to be transferable to humans. Recently, it has been shown in humans that plasma OEA levels were increased after an experimental meal containing higher amounts of oleic acid (bread with high oleic sunflower oil or virgin olive oil) compared to a meal with same caloric content but lower oleic acid concentration (bread with sunflower seed oil) (Mennella, Savarese, Ferracane, Sacchi, & Vitaglione, 2015). Interestingly, the two meals with higher oleic acid amounts reduced energy intake at subsequent lunch (180 min after the experimental meal). In addition, increased fullness and satiety ratings were observed 30 and 60 min after the meals with high oleic acid content, but not after the control meal.

Thus augmentation of OEA or targeting PPAR-α might be a new pharmacological possibility to treat appetite disturbances in schizophrenic patients. However, it has also been reported that OEA signaling is disturbed in obese individuals (Grosshans et al., 2014). In controls, BMI was negatively associated with OEA levels by trend, whereas obese patients showed a positive correlation. Furthermore, food-cue-induced brain activity was differentially associated with OEA levels in obese and nonobese participants. The authors speculated that OEA is involved in the suppression of food-liking reactions in healthy controls, whereas obesity is accompanied by alterations in such processes (Grosshans et al., 2014).

Although in particular human studies are currently rare and thus certainly more research is needed in that area, these observations indicate that targeting OEA signaling might be a promising target for treatment of appetite disturbances if the individual's BMI is taken into account. As OEA seems to act via PPAR-α, the psychiatric adverse effects provoked by the inverse agonist rimonabant, which had to be withdrawn from the market (Cheung, Cheung, & Samaranayake, 2013), can hopefully be avoided.

CONCLUSIONS

Being characterized by heterogeneous symptoms, schizophrenia also has a quite complex pathophysiology. This psychiatric disease appears to be a result of various genetic and environmental risk factors that influence brain development (Kahn et al., 2015). Although disturbances of the dopaminergic system may account for a large portion of the neurobiological processes underlying schizophrenia, substantial contributions of other neurotransmitter imbalances, such as deficient glutamatergic and GABAergic signaling, can be assumed. Since Emrich et al. (1997) proposed that malfunctions of the ECS also contribute to the pathology of the schizophrenia, a body of studies elucidated this theory. It has been obscure for a while if the observed changes in the ECS system need to be counted among deleterious or protective effects. The individual role of ECS compounds still remains conjectural, but several preclinical and clinical studies indicate that anandamide counteracts neurotransmitter imbalances and has a protective effect in the progression of schizophrenia, whereas a deficient ECS contributes to the pathophysiology of schizophrenia.

Arguments for the deleterious role of ECS abnormalities are mainly related to the observed effects of cannabis or Δ^9-THC as well as CB_1R densities. (1) Early and excessive cannabis use seems to be one risk factor that favors the development of schizophrenia in vulnerable persons (for review see Gage et al., 2016). (2) In healthy volunteers, cannabis or single cannabinoids such as Δ^9-THC have psychotomimetic effects and induce a range of different symptoms that resemble those of schizophrenic patients (for review see Koethe, Hoyer, et al., 2009; Sherif et al., 2016). (3) Highly psychosis-prone individuals experienced more pronounced psychoticlike symptoms after acute cannabis use than healthy controls (Mason et al., 2009). (4) Δ^9-THC transiently exacerbated positive, negative, and general symptoms as well as cognitive deficits in schizophrenic patients (D'Souza et al., 2005). (5) Although the introductory described alterations of CB_1R density in the brain of schizophrenic patients are somehow contradictory, reduced CB_1R densities have been attributed to the pathology of schizophrenia (Ranganathan et al., 2015).

On the other hand, in particular, the inverse correlation of symptom severity with CSF anandamide levels in schizophrenic patients (Giuffrida et al., 2004) and individuals of at-risk mental states (Koethe, Giuffrida, et al., 2009) as well as with CB_1R density measured by PET imaging in antipsychotic-naïve patients (Ceccarini et al., 2013), point to a protective role of anandamide. This is substantiated by the finding that individuals of at risk mental states with lower CSF anandamide levels had a higher risk for transiting to psychosis earlier (Koethe, Giuffrida, et al., 2009). Further evidence is given by the observation that the reduction of psychotic symptoms by cannabidiol but not by amisulpride treatment was significantly associated with an increase of serum anandamide levels.

However, there is substantially more preclinical and clinical work needed to resolve contradicting findings and to improve our understanding of the complex neurobiological fundamentals of the ECS and its role in schizophrenia. The discovery of the ECS undoubtedly stimulated schizophrenia research and resulted in new promising therapeutic approaches that should be further investigated during the next years.

REFERENCES

Aguilar, D. D., Chen, L., & Lodge, D. J. (2015). Increasing endocannabinoid levels in the ventral pallidum restore aberrant dopamine neuron activity in the subchronic PCP rodent model of schizophrenia. *International Journal of Neuropsychopharmacology, 18*(1).

Aguilar, D. D., Giuffrida, A., & Lodge, D. J. (2016). THC and endocannabinoids differentially regulate neuronal activity in the prefrontal cortex and hippocampus in the subchronic PCP model of schizophrenia. *Journal of Psychopharmacology, 30*(2), 169–181.

Alonso, R., Voutsinos, B., Fournier, M., Labie, C., Stienberg, R., Souilhac, J., ... Soubrié, P. (1999). Blockade of cannabinoid receptors by SR141716 selectively increases Fos expression in rat mesocorticolimbic areas via reduced dopamine D2 function. *Neuroscience, 91*(2), 607–620.

Andreasson, S., Allebeck, P., Engstrom, A., & Rydberg, U. (1987). Cannabis and schizophrenia. A longitudinal study of Swedish conscripts. *Lancet, 2*(8574), 1483–1486.

Arseneault, L., Cannon, M., Poulton, R., Murray, R., Caspi, A., & Moffitt, T. E. (2002). Cannabis use in adolescence and risk for adult psychosis: Longitudinal prospective study. *British Medical Journal, 325*(7374), 1212–1213.

Astarita, G., Di Giacomo, B., Gaetani, S., Oveisi, F., Compton, T. R., Rivara, S., ... Piomelli, D. (2006). Pharmacological characterization of hydrolysis-resistant analogs of oleoylethanolamide with potent anorexiant properties. *The Journal of Pharmacology and Experimental Therapeutics, 318*(2), 563–570.

Bhattacharyya, S., Fusar-Poli, P., Borgwardt, S., Martin-Santos, R., Nosarti, C., O'Carroll, C., ... McGuire, P. (2009). Modulation of mediotemporal and ventrostriatal function in humans by Delta9-tetrahydrocannabinol: A neural basis for the effects of *Cannabis sativa* on learning and psychosis. *Archives of General Psychiatry, 66*(4), 442–451.

Bioque, M., Cabrera, B., Garcia-Bueno, B., Mac-Dowell, K. S., Torrent, C., Saiz, P. A., ... Bernardo, M. (2016). Dysregulated peripheral endocannabinoid system signaling is associated with cognitive deficits in first-episode psychosis. *Journal of Psychiatric Research, 75,* 14–21.

Bioque, M., Garcia-Bueno, B., Macdowell, K. S., Meseguer, A., Saiz, P. A., Parellada, M., ... Bernardo, M. (2013). Peripheral endocannabinoid system dysregulation in first-episode psychosis. *Neuropsychopharmacology, 38*(13), 2568–2577.

Bisogno, T., Hanus, L., De Petrocellis, L., Tchilibon, S., Ponde, D. E., Brandi, I., ... Di Marzo, V. (2001). Molecular targets for cannabidiol and its synthetic analogues: Effect on vanilloid VR1 receptors and on the cellular uptake and enzymatic hydrolysis of anandamide. *British Journal of Pharmacology, 134*(4), 845–852.

Blankman, J. L., Simon, G. M., & Cravatt, B. F. (2007). A comprehensive profile of brain enzymes that hydrolyze the endocannabinoid 2-arachidonoylglycerol. *Chemistry & Biology, 14*(12), 1347–1356.

Boggs, D. L., Kelly, D. L., McMahon, R. P., Gold, J. M., Gorelick, D. A., Linthicum, J., ... Buchanan, R. W. (2012). Rimonabant for neurocognition in schizophrenia: A 16-week double blind randomized placebo controlled trial. *Schizophrenia Research, 134*(2–3), 207–210.

Carlsson, A., & Lindqvist, M. (1963). Effect of chlorpromazine or haloperidol on formation of 3-methoxytyramine and normetanephrine in mouse brain. *Acta Pharmacologica et Toxicologica, 20,* 140–144.

Ceccarini, J., De Hert, M., Van Winkel, R., Peuskens, J., Bormans, G., Kranaster, L., ... Van Laere, K. (2013). Increased ventral striatal CB1 receptor binding is related to negative symptoms in drug-free patients with schizophrenia. *Neuroimage, 79,* 304–312.

Cheung, B. M., Cheung, T. T., & Samaranayake, N. R. (2013). Safety of antiobesity drugs. *Therapeutic Advances in Drug Safety, 4*(4), 171–181.

D'Souza, D. C., Abi-Saab, W. M., Madonick, S., Forselius-Bielen, K., Doersch, A., Braley, G., ... Krystal, J. H. (2005). Delta-9-tetrahydrocannabinol effects in schizophrenia: Implications for cognition, psychosis, and addiction. *Biological Psychiatry, 57*(6), 594–608.

D'Souza, D. C., Perry, E., MacDougall, L., Ammerman, Y., Cooper, T., Wu, Y. T., ... Krystal, J. H. (2004). The psychotomimetic effects of intravenous delta-9-tetrahydrocannabinol in healthy individuals: Implications for psychosis. *Neuropsychopharmacology, 29*(8), 1558–1572.

Dalton, V. S., Long, L. E., Weickert, C. S., & Zavitsanou, K. (2011). Paranoid schizophrenia is characterized by increased CB1 receptor binding in the dorsolateral prefrontal cortex. *Neuropsychopharmacology, 36*(8), 1620–1630.

Davis, K. L., Kahn, R. S., Ko, G., & Davidson, M. (1991). Dopamine in schizophrenia: A review and reconceptualization. *The American Journal of Psychiatry, 148*(11), 1474–1486.

De Marchi, N., De Petrocellis, L., Orlando, P., Daniele, F., Fezza, F., & Di Marzo, V. (2003). Endocannabinoid signalling in the blood of patients with schizophrenia. *Lipids in Health and Disease, 2,* 5.

Dean, B., Sundram, S., Bradbury, R., Scarr, E., & Copolov, D. (2001). Studies on [3H] CP-55940 binding in the human central nervous system: Regional specific changes in density of cannabinoid-1 receptors associated with schizophrenia and cannabis use. *Neuroscience, 103*(1), 9–15.

Debnath, M., Venkatasubramanian, G., & Berk, M. (2015). Fetal programming of schizophrenia: Select mechanisms. *Neuroscience and Biobehavioral Reviews, 49,* 90–104.

Devane, W. A., Dysarz, F. A., Johnson, M. R., Melvin, L. S., & Howlett, A. C. (1988). Determination and characterization of a cannabinoid receptor in rat brain. *Molecular Pharmacology, 34*, 605–613.

Devane, W. A., Hanus, L., Breuer, A., Pertwee, R. G., Stevenson, L. A., Griffin, G., ... Mechoulam, R. (1992). Isolation and structure of a brain constituent that binds to the cannabinoid receptor. *Science (New York, N.Y.), 258*(5090), 1946–1949.

Di Forti, M., Morgan, C., Dazzan, P., Pariante, C., Mondelli, V., Marques, T. R., ... Murray, R. M. (2009). High-potency cannabis and the risk of psychosis. *The British Journal of Psychiatry, 195*(6), 488–491.

Di Forti, M., Sallis, H., Allegri, F., Trotta, A., Ferraro, L., Stilo, S. A., ... Murray, R. M. (2014). Daily use, especially of high-potency cannabis, drives the earlier onset of psychosis in cannabis users. *Schizophrenia Bulletin, 40*.

Di Marzo, V., De Petrocellis, L., Fezza, F., Ligresti, A., & Bisogno, T. (2002). Anandamide receptors. *Prostaglandins, Leukotrienes, and Essential Fatty Acids, 66*(2–3), 377–391.

Di Marzo, V., Fontana, A., Cadas, H., Schinelli, S., Cimino, G., Schwartz, J. C., & Piomelli, D. (1994). Formation and inactivation of endogenous cannabinoid anandamide in central neurons. *Nature, 372*(6507), 686–691.

Di Tomaso, E., Cadas, H., Gaillet, S., Beltramo, M., Desarnaud, F., Venance, L., & Piomelli, D. (1997). Endogenous lipids that activate cannabinoid receptors. Formation and inactivation. *Advances in Experimental Medicine and Biology, 407*, 335–340.

DiPatrizio, N. V., & Piomelli, D. (2015). Intestinal lipid-derived signals that sense dietary fat. *Journal of Clinical Investigation, 125*(3), 891–898.

Donoghue, K., Doody, G. A., Murray, R. M., Jones, P. B., Morgan, C., Dazzan, P., ... Maccabe, J. H. (2014). Cannabis use, gender and age of onset of schizophrenia: Data from the ÆSOP study. *Psychiatry Research, 215*(3), 528–532.

Egertova, M., Cravatt, B. F., & Elphick, M. R. (2003). Comparative analysis of fatty acid amide hydrolase and cb(1) cannabinoid receptor expression in the mouse brain: Evidence of a widespread role for fatty acid amide hydrolase in regulation of endocannabinoid signaling. *Neuroscience, 119*(2), 481–496.

Eggan, S. M., Hashimoto, T., & Lewis, D. A. (2008). Reduced cortical cannabinoid 1 receptor messenger RNA and protein expression in schizophrenia. *Archives of General Psychiatry, 65*(7), 772–784.

Eggan, S. M., Stoyak, S. R., Verrico, C. D., & Lewis, D. A. (2010). Cannabinoid CB1 receptor immunoreactivity in the prefrontal cortex: Comparison of schizophrenia and major depressive disorder. *Neuropsychopharmacology, 35*(10), 2060–2071.

Emrich, H. M., Leweke, F. M., & Schneider, U. (1997). Towards a cannabinoid hypothesis of schizophrenia: Cognitive impairments due to dysregulation of the endogenous cannabinoid system. *Pharmacology, Biochemistry, and Behavior, 56*(4), 803–807.

Fergusson, D. M., Horwood, L. J., & Swain-Campbell, N. R. (2003). Cannabis dependence and psychotic symptoms in young people. *Psychological Medicine, 33*(1), 15–21.

Freedman, R. (2003). Schizophrenia. *The New England Journal of Medicine, 349*(18), 1738–1749.

Fu, J., Astarita, G., Gaetani, S., Kim, J., Cravatt, B. F., Mackie, K., & Piomelli, D. (2007). Food intake regulates oleoylethanolamide formation and degradation in the proximal small intestine. *The Journal of Biological Chemistry, 282*(2), 1518–1528.

Fu, J., Gaetani, S., Oveisi, F., Lo Verme, J., Serrano, A., Rodríguez De Fonseca, F., ... Piomelli, D. (2003). Oleylethanolamide regulates feeding and body weight through activation of the nuclear receptor PPAR-alpha. *Nature, 425*(6953), 90–93.

Fu, J., Kim, J., Oveisi, F., Astarita, G., & Piomelli, D. (2008). Targeted enhancement of oleoylethanolamide production in proximal small intestine induces across-meal satiety in rats. *American Journal of Physiology. Regulatory, Integrative and Comparative Physiology, 295*(1), R45–R50.

Fusar-Poli, P., Crippa, J. A., Bhattacharyya, S., Borgwardt, S. J., Allen, P., Martin-Santos, R., … McGuire, P. K. (2009). Distinct effects of {Delta}9-tetrahydrocannabinol and cannabidiol on neural activation during emotional processing. *Archives of General Psychiatry, 66*(1), 95–105.

Gaetani, S., Oveisi, F., & Piomelli, D. (2003). Modulation of meal pattern in the rat by the anorexic lipid mediator oleoylethanolamide. *Neuropsychopharmacology, 28*(7), 1311–1316.

Gage, S. H., Hickman, M., & Zammit, S. (2016). Association between cannabis and psychosis: Epidemiologic evidence. *Biological Psychiatry, 79*(7), 549–556.

Garbutt, J. C., & van Kammen, D. P. (1983). The interaction between GABA and dopamine: Implications for schizophrenia. *Schizophrenia Bulletin, 9*(3), 336–353.

Giuffrida, A., Leweke, F. M., Gerth, C. W., Schreiber, D., Koethe, D., Faulhaber, J., … Piomelli, D. (2004). Cerebrospinal anandamide levels are elevated in acute schizophrenia and are inversely correlated with psychotic symptoms. *Neuropsychopharmacology, 29*(11), 2108–2114.

Giuffrida, A., Parsons, L. H., Kerr, T. M., Rodriguez de Fonseca, F., Navarro, M., & Piomelli, D. (1999). Dopamine activation of endogenous cannabinoid signaling in dorsal striatum. *Nature Neuroscience, 2*(4), 358–363.

Grosshans, M., Schwarz, E., Bumb, J. M., Schaefer, C., Rohleder, C., Vollmert, C., … Leweke, F. M. (2014). Oleoylethanolamide and human neural responses to food stimuli in obesity. *JAMA Psychiatry, 71*(11), 1254–1261.

Guidali, C., Vigano, D., Petrosino, S., Zamberletti, E., Realini, N., Binelli, G., … Parolaro, D. (2011). Cannabinoid CB1 receptor antagonism prevents neurochemical and behavioural deficits induced by chronic phencyclidine. *International Journal of Neuropsychopharmacology, 14*(1), 17–28.

Hauer, D., Schelling, G., Gola, H., Campolongo, P., Morath, J., Roozendaal, B., … Kolassa, I. T. (2013). Plasma concentrations of endocannabinoids and related primary fatty acid amides in patients with post-traumatic stress disorder. *PLoS One, 8*(5), e62741.

Henquet, C., Murray, R., Linszen, D., & van Os, J. (2005). The environment and schizophrenia: The role of cannabis use. *Schizophrenia Bulletin, 31*(3), 608–612.

Holmes, E., Tsang, T. M., Huang, J. T., Leweke, F. M., Koethe, D., Gerth, C. W., … Bahn, S. (2006). Metabolic profiling of CSF: Evidence that early intervention may impact on disease progression and outcome in schizophrenia. *PLoS Medicine, 3*(8), e327 1421.

Javitt, D. C., & Zukin, S. R. (1991). Recent advances in the phencyclidine model of schizophrenia. *The American Journal of Psychiatry, 148*(10), 1301–1308.

Jenko, K. J., Hirvonen, J., Henter, I. D., et al. (2012). Binding of a tritiated inverse agonist to cannabinoid CB1 receptors is increased in patients with schizophrenia. *Schizophrenia Research, 141*(2–3), 185–188.

Kahn, R. S., Sommer, I. E., Murray, R. M., Meyer-Lindenberg, A., Weinberger, D. R., Cannon, T. D., … Insel, T. R. (2015). Schizophrenia. *Nature Reviews. Disease Primers, 1,* 15067.

Kim, J. S., Kornhuber, H. H., Schmid-Burgk, W., & Holzmuller, B. (1980). Low cerebrospinal fluid glutamate in schizophrenic patients and a new hypothesis on schizophrenia. *Neuroscience Letters, 20*(3), 379–382.

Kirkpatrick, B., & Miller, B. J. (2013). Inflammation and schizophrenia. *Schizophrenia Bulletin, 39*(6), 1174–1179.

Koethe, D., Gerth, C. W., Neatby, M. A., Haensel, A., Thies, M., Schneider, U., … Leweke, F. M. (2006). Disturbances of visual information processing in early states of psychosis and experimental delta-9-tetrahydrocannabinol altered states of consciousness. *Schizophrenia Research, 88,* 142–150.

Koethe, D., Giuffrida, A., Schreiber, D., Hellmich, M., Schultze-Lutter, F., Ruhrmann, S., … Leweke, F. M. (2009). Anandamide elevation in cerebrospinal fluid in initial prodromal states of psychosis. *The British Journal of Psychiatry: The Journal of Mental Science, 194*(4), 371–372.

Koethe, D., Hoyer, C., & Leweke, F. M. (2009). The endocannabinoid system as a target for modelling psychosis. *Psychopharmacology, 206*(4), 551–561.

Koethe, D., Llenos, I. C., Dulay, J. R., Hoyer, C., Torrey, E. F., Leweke, F. M., & Weis, S. (2007). Expression of CB1 cannabinoid receptor in the anterior cingulate cortex in schizophrenia, bipolar disorder, and major depression. *Journal of Neural Transmission, 114*(8), 1055–1063.

Large, M., Sharma, S., Compton, M. T., Slade, T., & Nielssen, O. (2011). Cannabis use and earlier onset of psychosis: A systematic meta-analysis. *Archives of General Psychiatry, 68*(6), 555–561.

Laurens, K. R., Luo, L., Matheson, S. L., Carr, V. J., Raudino, A., Harris, F., & Green, M. J. (2015). Common or distinct pathways to psychosis? A systematic review of evidence from prospective studies for developmental risk factors and antecedents of the schizophrenia spectrum disorders and affective psychoses. *BMC Psychiatry, 15*, 205.

Leweke, F. M., Giuffrida, A., Koethe, D., Schreiber, D., Nolden, B. M., Kranaster, L., ... Piomelli, D. (2007). Anandamide levels in cerebrospinal fluid of first-episode schizophrenic patients: Impact of cannabis use. *Schizophrenia Research, 94*(1–3), 29–36.

Leweke, F. M., Giuffrida, A., Wurster, U., Emrich, H. M., & Piomelli, D. (1999). Elevated endogenous cannabinoids in schizophrenia. *Neuroreport, 10*(8), 1665–1669.

Leweke, F. M., Mueller, J. K., Lange, B., & Rohleder, C. (2016). Therapeutic potential of cannabinoids in psychosis. *Biological Psychiatry, 79*(7), 604–612.

Leweke, F. M., Piomelli, D., Pahlisch, F., Muhl, D., Gerth, C. W., Hoyer, C., ... Koethe, D. (2012). Cannabidiol enhances anandamide signaling and alleviates psychotic symptoms of schizophrenia. *Translational Psychiatry [Electronic Resource], 2*, e94.

Leweke, F. M., Schneider, U., Radwan, M., Schmidt, E., & Emrich, H. M. (2000). Different effects of nabilone and cannabidiol on binocular depth inversion in man. *Pharmacology, Biochemistry, and Behavior, 66*(1), 175–181.

Leweke, F. M., Schneider, U., Thies, M., Münte, T. F., & Emrich, H. M. (1999). Effects of synthetic Δ9-tetrahydrocannabinol on binocular depth inversion of natural and artificial objects in man. *Psychopharmacology, 142*, 230–235.

Leza, J. C., Garcia-Bueno, B., Bioque, M., Arango, C., Parellada, M., Do, K., ... Bernardo, M. (2015). Inflammation in schizophrenia: A question of balance. *Neuroscience and Biobehavioral Reviews, 55*, 612–626.

Maccarrone, M., Bab, I., Biro, T., Cabral, G. A., Dey, S. K., Di Marzo, V., ... Zimmer, A. (2015). Endocannabinoid signaling at the periphery: 50 Years after THC. *Trends in Pharmacological Sciences, 36*(5), 277–296.

Maccarrone, M., Gasperi, V., Catani, M. V., Diep, T. A., Dainese, E., Hansen, H. S., & Avigliano, L. (2010). The endocannabinoid system and its relevance for nutrition. *Annual Review of Nutrition, 30*, 423–440.

Mason, O., Morgan, C. J., Dhiman, S. K., Patel, A., Parti, N., Patel, A., & Curran, H. V. (2009). Acute cannabis use causes increased psychotomimetic experiences in individuals prone to psychosis. *Psychological Medicine, 39*(6), 951–956.

Matricon, J., Seillier, A., & Giuffrida, A. (2016). Distinct neuronal activation patterns are associated with PCP-induced social withdrawal and its reversal by the endocannabinoid-enhancing drug URB597. *Neuroscience Research, 110*.

Matsuda, L. A., Lolait, S. J., Brownstein, M. J., Young, A. C., & Bonner, T. I. (1990). Structure of a cannabinoid receptor and functional expression of the cloned cDNA. *Nature, 346*(6284), 561–564.

Mechoulam, R., Ben-Shabat, S., Hanus, L., Ligumsky, M., Kaminski, N. E., Schatz, A. R., ... Vogel, Z. (1995). Identification of an endogenous 2-monoglyceride, present in canine gut, that binds to cannabinoid receptors. *Biochemical Pharmacology, 50*(1), 83–90.

Mechoulam, R., & Parker, L. A. (2013). The endocannabinoid system and the brain. *Annual Review of Psychology, 64*, 21–47.

Meltzer, H. Y. (1989). Clinical studies on the mechanism of action of clozapine: The dopamine-serotonin hypothesis of schizophrenia. *Psychopharmacology, 99*(Suppl), S18–S27.

Meltzer, H.Y., Arvanitis, L., Bauer, D., Rein, W., & Group M-TS (2004). Placebo-controlled evaluation of four novel compounds for the treatment of schizophrenia and schizoaffective disorder. *American Journal of Psychiatry, 161*(6), 975–984.

Mennella, I., Savarese, M., Ferracane, R., Sacchi, R., & Vitaglione, P. (2015). Oleic acid content of a meal promotes oleoylethanolamide response and reduces subsequent energy intake in humans. *Food and Function, 6*(1), 204–210.

Moore, T. H., Zammit, S., Lingford-Hughes, A., Barnes, T. R., Jones, P. B., Burke, M., & Lewis, G. (2007). Cannabis use and risk of psychotic or affective mental health outcomes: A systematic review. *Lancet, 370*(9584), 319–328.

Morgan, C. J., Page, E., Schaefer, C., Chatten, K., Manocha, A., Gulati, S., … Leweke, F. M. (2013). Cerebrospinal fluid anandamide levels, cannabis use and psychotic like symptoms. *The British Journal of Psychiatry, 202*(5), 381–382.

Muguruza, C., Lehtonen, M., Aaltonen, N., Morentin, B., Meana, J. J., & Callado, L. F. (2013). Quantification of endocannabinoids in postmortem brain of schizophrenic subjects. *Schizophrenia Research, 148*(1–3), 145–150.

Newell, K. A., Deng, C., & Huang, X.-F. (2006). Increased cannabinoid receptor density in the posterior cingulate cortex in schizophrenia. *Experimental Brain Research, 172*(4), 556–560.

O'Sullivan, S. E. (2016). An update on PPAR activation by cannabinoids. *British Journal of Pharmacology, 173*(12), 1899–1910.

Olincy, A., Harris, J. G., Johnson, L. L., Pender, V., Kongs, S., Allensworth, D., … Freedman, R. (2006). Proof-of-concept trial of an alpha7 nicotinic agonist in schizophrenia. *Archives of General Psychiatry, 63*(6), 630–638.

van Os, J., Bak, M., Hanssen, M., Bijl, R.V., de Graaf, R., & Verdoux, H. (2002). Cannabis use and psychosis: A longitudinal population-based study. *American Journal of Epidemiology, 156*(4), 319–327.

Oveisi, F., Gaetani, S., Eng, K. T., & Piomelli, D. (2004). Oleoylethanolamide inhibits food intake in free-feeding rats after oral administration. *Pharmacological Research, 49*(5), 461–466.

Parolaro, D., Realini, N., Vigano, D., Guidali, C., & Rubino, T. (2010). The endocannabinoid system and psychiatric disorders. *Experimental Neurology, 224*(1), 3–14.

Pertwee, R. G., Ross, R. A., Craib, S. J., & Thomas, A. (2002). (-)-Cannabidiol antagonizes cannabinoid receptor agonists and noradrenaline in the mouse vas deferens. *European Journal of Pharmacology, 456*(1–3), 99–106.

Piomelli, D. (2003). The molecular logic of endocannabinoid signalling. *Nature Reviews, 4*(11), 873–884.

Piomelli, D. (2013). A fatty gut feeling. *Trends in Endocrinology and Metabolism: TEM, 24*(7), 332–341.

Poncelet, M., Barnouin, M. C., Breliere, J. C., Le Fur, G., & Soubrie, P. (1999). Blockade of cannabinoid (CB1) receptors by 141716 selectively antagonizes drug-induced reinstatement of exploratory behaviour in gerbils. *Psychopharmacology, 144*(2), 144–150.

Potvin, S., Kouassi, E., Lipp, O., Bouchard, R. H., Roy, M. A., Demers, M. F., … Stip, E. (2008). Endogenous cannabinoids in patients with schizophrenia and substance use disorder during quetiapine therapy. *Journal of Psychopharmacology, 22*(3), 262–269.

Rajasekaran, A., Venkatasubramanian, G., Berk, M., & Debnath, M. (2015). Mitochondrial dysfunction in schizophrenia: Pathways, mechanisms and implications. *Neuroscience and Biobehavioral Reviews, 48*, 10–21.

Ranganathan, M., Cortes-Briones, J., Radhakrishnan, R., Thurnauer, H., Planeta, B., Skosnik, P., … D'Souza, D. C. (2015). Reduced brain cannabinoid receptor availability in schizophrenia. *Biological Psychiatry, 79*.

Rinaldi-Carmona, M., Pialot, F., Congy, C., Redon, E., Barth, F., Bachy, A., … Le Fur, G. (1996). Characterization and distribution of binding sites for [3H]-SR 141716A, a selective brain (CB1) cannabinoid receptor antagonist, in rodent brain. *Life Sciences, 58*(15), 1239–1247.

Rodriguez de Fonseca, F., Navarro, M., Gomez, R., Escuredo, L., Nava, F., Fu, J., … Piomelli, D. (2001). An anorexic lipid mediator regulated by feeding. *Nature, 414*(6860), 209–212.

van Rossum, J. M. (1966). The significance of dopamine-receptor blockade for the mechanism of action of neuroleptic drugs. *Archives Internationales de Pharmacodynamie et de Thérapie, 160*(2), 492–494.

Rubino, T., & Parolaro, D. (2016). The impact of exposure to cannabinoids in adolescence: Insights from animal models. *Biological Psychiatry, 79*(7), 578–585.

Scatton, B., & Sanger, D. J. (2000). Pharmacological and molecular targets in the search for novel antipsychotics. *Behavioural Pharmacology, 11*(3–4), 243–256.

Schaefer, C., Enning, F., Mueller, J., Bumb, J. M., Rohleder, C., Odorfer, T. M., … Leweke, F. M. (2013). Fatty acid ethanolamide levels are altered in borderline personality and complex posttraumatic stress disorders. *European Archives of Psychiatry and Clinical Neuroscience*, 1–5.

Schwartz, G. J., Fu, J., Astarita, G., Li, X., Gaetani, S., Campolongo, P., … Piomelli, D. (2008). The lipid messenger OEA links dietary fat intake to satiety. *Cell Metabolism, 8*(4), 281–288.

Seillier, A., Advani, T., Cassano, T., Hensler, J. G., & Giuffrida, A. (2010). Inhibition of fatty-acid amide hydrolase and CB1 receptor antagonism differentially affect behavioural responses in normal and PCP-treated rats. *International Journal of Neuropsychopharmacology, 13*(3), 373–386.

Seillier, A., & Giuffrida, A. (2009). Evaluation of NMDA receptor models of schizophrenia: Divergences in the behavioral effects of sub-chronic PCP and MK-801. *Behavioural Brain Research, 204*(2), 410–415.

Seillier, A., Martinez, A. A., & Giuffrida, A. (2013). Phencyclidine-induced social withdrawal results from deficient stimulation of cannabinoid CB(1) receptors: Implications for schizophrenia. *Neuropsychopharmacology, 38*(9), 1816–1824.

Sherif, M., Radhakrishnan, R., D'Souza, D. C., & Ranganathan, M. (2016). Human laboratory studies on cannabinoids and psychosis. *Biological Psychiatry, 79*(7), 526–538.

Snyder, M. A., & Gao, W. J. (2013). NMDA hypofunction as a convergence point for progression and symptoms of schizophrenia. *Frontiers in Cellular Neuroscience, 7*, 31.

Steinpreis, R. E. (1996). The behavioral and neurochemical effects of phencyclidine in humans and animals: Some implications for modeling psychosis. *Behavioural Brain Research, 74*(1–2), 45–55.

Sugiura, T., Kondo, S., Sukagawa, A., Nakane, S., Shinoda, A., Itoh, K., … Waku, K. (1995). 2-Arachidonoylglycerol: A possible endogenous cannabinoid receptor ligand in brain. *Biochemical and Biophysical Research Communications, 215*(1), 89–97.

Thomas, A., Baillie, G. L., Phillips, A. M., Razdan, R. K., Ross, R. A., & Pertwee, R. G. (2007). Cannabidiol displays unexpectedly high potency as an antagonist of CB1 and CB2 receptor agonists in vitro. *British Journal of Pharmacology, 150*(5), 613–623.

Urigüen, L., García-Fuster, M. J., Callado, L. F., Morentin, B., La Harpe, R., Casadó, V., … Meana, J. J. (2009). Immunodensity and mRNA expression of A2A adenosine, D2 dopamine, and CB1 cannabinoid receptors in postmortem frontal cortex of subjects with schizophrenia: Effect of antipsychotic treatment. *Psychopharmacology, 206*(2), 313–324.

Vigano, D., Guidali, C., Petrosino, S., Realini, N., Rubino, T., Di Marzo, V., & Parolaro, D. (2009). Involvement of the endocannabinoid system in phencyclidine-induced cognitive deficits modelling schizophrenia. *International Journal of Neuropsychopharmacology, 12*(5), 599–614.

Volk, D. W., Eggan, S. M., Horti, A. G., Wong, D. F., & Lewis, D. A. (2014). Reciprocal alterations in cortical cannabinoid receptor 1 binding relative to protein immunoreactivity and transcript levels in schizophrenia. *Schizophrenia Research, 159*(1), 124–129.

Volk, D. W., Eggan, S. M., & Lewis, D. A. (2010). Alterations in metabotropic glutamate receptor 1α and Regulator of G protein signaling 4 in the prefrontal cortex in schizophrenia. *American Journal of Psychiatry, 167*(12), 1489–1498.

Watanabe, K., Kayano, Y., Matsunaga, T., Yamamoto, I., & Yoshimura, H. (1996). Inhibition of anandamide amidase activity in mouse brain microsomes by cannabinoids. *Biological & Pharmaceutical Bulletin, 19*(8), 1109–1111.

Wong, D. F., Kuwabara, H., Horti, A. G., Raymont, V., Brasic, J., Guevara, M., ... Cascella, N. (2010). Quantification of cerebral cannabinoid receptors subtype 1 (CB1) in healthy subjects and schizophrenia by the novel PET radioligand [11C]OMAR. *Neuroimage, 52*(4), 1505–1513.

Wong, A. H. C., & Van Tol, H. H. M. (2003). Schizophrenia: From phenomenology to neurobiology. *Neuroscience and Biobehavioral Reviews, 27*(3), 269–306.

Zammit, S., Allebeck, P., Andreasson, S., Lundberg, I., & Lewis, G. (2002). Self reported cannabis use as a risk factor for schizophrenia in Swedish conscripts of 1969: Historical cohort study. *BMJ: British Medical Journal, 325*(7374), 1199.

Zavitsanou, K., Garrick, T., & Huang, X. F. (2004). Selective antagonist [3H]SR141716A binding to cannabinoid CB1 receptors is increased in the anterior cingulate cortex in schizophrenia. *Progress in Neuro-psychopharmacology & Biological Psychiatry, 28*(2), 355–360.

Zuardi, A. W., Hallak, J. E., Dursun, S. M., Morais, S. L., Sanches, R. F., Musty, R. E., & Crippa, J. A. (2006). Cannabidiol monotherapy for treatment-resistant schizophrenia. *Journal of Psychopharmacology, 20*(5), 683–686.

Zuardi, A. W., Morais, S. L., Guimarães, F. S., & Mechoulam, R. (1995). Antipsychotic effect of cannabidiol. *Journal of Clinical Psychiatry, 56*(10), 485–486.

Zygmunt, P. M., Petersson, J., Andersson, D. A., Chuang, H., Sørgård, M., Di Marzo, V., ... Högestätt, E. D. (1999). Vanilloid receptors on sensory nerves mediate the vasodilator action of anandamide. *Nature, 400*(6743), 452–457.

CHAPTER 5

The Endocannabinoid System and Human Brain Functions: Insight From Memory, Motor, and Mood Pathologies

John C. Ashton[1], Megan J. Dowie[2], Michelle Glass[2]

[1]University of Otago, Dunedin, New Zealand; [2]University of Auckland, Auckland, New Zealand

The endogenous cannabinoid system is integral to normal physiology, including the control of movement and appetite, pain modulation, and the creation of memories. Due to its abundance and extensive distribution throughout the mammalian brain, and its ability to modulate neurotransmitter action, the system has been implicated in the pathology of a number of neurological conditions. As cannabinoid CB1 receptors are localized in areas important to memory (e.g., hippocampus), motor (e.g., basal ganglia), and mood (e.g., prefrontal cortex) (Fig. 5.1), it is not surprising that the endocannabinoid system is postulated to be involved in diseases in which these processes are dysfunctional. The study of endocannabinoid function (or dysfunction) in disease states has also provided greater insight into the role of this system in normal physiological brain function.

The endocannabinoid system consists of at least two cannabinoid receptors, CB1 and CB2, and two key endocannabinoids, anandamide and 2-arachidonylglycerol (2-AG), as well as the enzymes involved in the synthesis and metabolism of these compounds. CB1 receptors are localized on neurons and widely distributed in the mammalian central nervous system (CNS), whereas the CB2 receptor is found in the immune system and to date is convincingly described only on a small number of neurons of the brainstem (Van Sickle et al., 2005) and cerebellum (Ashton, Friberg, Darlington, & Smith, 2006), and in activated glial cells in the brain (Nunez et al., 2004) (for a review see Atwood and Mackie (2010)). The endocannabinoids and their related enzymes display widespread neuronal distribution.

The Endocannabinoid System
ISBN 978-0-12-809666-6
http://dx.doi.org/10.1016/B978-0-12-809666-6.00005-8

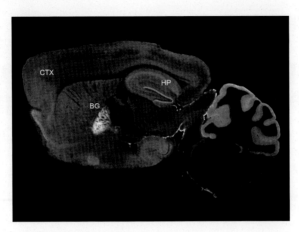

Figure 5.1 The broad distribution of the cannabinoid receptor 1 (CB1) in a sagittal section through the mouse brain is illustrated here by immunohistochemistry. High levels are apparent in regions involved in memory [e.g., hippocampus (HP)], motor function [e.g., basal ganglia (BG)], and mood [e.g., cortex CTX)]. *Reproduced from Dowie, M. J., & Glass, M. (2011). The endocannabinoid system and human brain functions: Insight from memory, motor and mood pathologies. In E. Murillo-Rodríguez (Ed)., Endocannabinoids: Molecular, pharmacological, behavioral and clinical features. Mérida, Yucatán, México: Universidad Anáhuac Mayab.*

ENDOCANNABINOID ACTIVITY IN THE BRAIN

The endocannabinoid system has explicitly been implicated in a number of neurological conditions. Some of the processes by which endocannabinoids can have impact are conserved across several diseases; these are discussed later. Excitotoxicity and inflammation are both mechanisms that have been postulated to be responsible for at least some of the dysfunction and degeneration in Huntington's disease (HD), Alzheimer's disease (AD), and other disorders of the CNS. The ability of endocannabinoids to modulate these acute processes, coupled with enhancement of neurogenesis, suggest a potential beneficial role for the modulation of the endocannabinoid system in the treatment of neurological disease. In short, it appears that the cannabinoid system has the potential to reduce neuroinflammation and associated detrimental effects while in fact supporting natural repair and protection mechanisms.

Neurogenesis

An exciting area in which cannabinoids have shown promise is in the regulation of neurogenesis (Galve-Roperh, Aguado, Palazuelos, & Guzman, 2007).

Embryonic and adult hippocampal progenitor cells produce endocannabinoids and express the CB1 receptor and the endocannabinoid-inactivating enzyme fatty acid amide hydrolase (FAAH) (Aguado et al., 2005). CB1 receptor activation promotes cell proliferation and neurosphere generation (Aguado et al., 2005; Jiang et al., 2005). Furthermore, proliferation of hippocampal neural progenitors is increased in FAAH-deficient mice (Aguado et al., 2005) and decreased in CB1 knockout mice (Dubreucq, Koehl, Abrous, Marsicano, & Chaouloff, 2010; Wolf et al., 2010). The precise role of CB1 in the neurogenic pathway has yet to be elucidated, but studies have indicated that in the absence of CB1 receptors, cell proliferation was increased and neuronal differentiation was reduced (Wolf et al., 2010).

It is postulated that the anxiolytic and antidepressant effects of chronic HU210 treatment (a potent CB1 agonist) in rats is a functional consequence of hippocampal neurogenesis (Jiang et al., 2005). Furthermore, the neurogenic effect of environmental enrichment and voluntary wheel running depends on the presence of the CB1 receptor (Wolf et al., 2010) and can be blocked by CB1 antagonists (Hill et al., 2010). Interpretation of these studies could, however, be complicated by the finding that CB1 knockout mice engage in 40% less voluntary wheel running than their wild-type counterparts (Dubreucq et al., 2010). CB2 activation has also been demonstrated to stimulate proliferation of neural progenitors (Palazuelos et al., 2006).

Although most neurogenic studies have focused on rodent brains, a subpopulation of proliferating cells have been found in the subependymal layer in adult human HD brains, which are CB1 receptor positive (Curtis, Faull, & Glass, 2006). Although it is known that CB1 receptors are lost from basal ganglia areas early in the disease, the presence of CB1 on these proliferating cells indicates a cell population not immediately susceptible to disease processes and therefore with potential to act in endogenous cell replacement, across a number of neurodegenerative conditions (Curtis et al., 2006).

Considerable progress has been made in understanding the molecular pathways through which cannabinoids and cannabinoid receptors act to influence neurogenesis (Prenderville, Kelly, & Downer, 2015). In particular, evidence has accumulated that cannabinoid receptors drive cell division through mitogen-activated protein kinase (MAPK/ERK) and phosphoinositide 3-kinase (PI3K/Akt) pathways largely via the CB2 receptor (Molina-Holgado et al., 2007; Palazuelos et al., 2006; Palazuelos, Ortega, Diaz-Alonso, Guzman, & Galve-Roperh, 2012). For instance, the CB1 agonist arachidonyl-2′-chloroethylamide (ACEA) inhibits ERK phosphorylation in neural progenitor cells (Compagnucci et al., 2013), and the CB2 agonist HU-308 increases neural progenitor cell proliferation through both the ERK and PI3K/Akt pathways, driving neural progenitor proliferation (Palazuelos et al., 2006, 2012). The various pathways that are activated by cannabinoid receptors in neural progenitor cells converge at transcription factors such as cyclic adenosine monophosphate (cAMP) response element–binding protein (Soltys, Yushak, & Mao-Draayer, 2010), T-box transcription factor (Tbr) (Saez, Aronne, Caltana, & Brusco, 2014), and the Sox2 family, the last of these via Notch signaling through CB1 receptor activation (Xapelli et al., 2013).

The endocannabinoid system interacts with other receptor systems in regulating neural stem cell growth and division, with both CB1 and CB2 receptors interacting with epidermal growth factor receptor (Sutterlin et al., 2013). Moreover, growth factors such as brain-derived neurotrophic factor (BDNF) interact with CB1 receptors in the brain (De Chiara et al., 2010) and CB2 receptor activation has been suggested to stimulate BDNF expression (Choi et al., 2013). Other interacting growth factors include protein kinase C (PKC), adenosine, and interleukin-1 (IL-1) (Shinjyo & Di Marzo, 2013). IL signaling has also been demonstrated to interact with cannabinoid receptors through coexpression with IL-1 types I and II (Garcia-Ovejero et al., 2013).

Beyond neurodegenerative diseases such as HD, recent investigations have identified a range of human conditions in which the effects of cannabinoids on neurogenesis may be important, including the effect of synthetic cannabinoids as drugs of abuse (Xu, Loh, & Law, 2015), human immunodeficiency virus (HIV)/HIV-associated dementia (Avraham et al., 2014), and a possible effect on neural progenitor cell cycle progression from use of cannabidiol (CBD) (Campos et al., 2013); CBD at the time of writing is of particular interest, due to its popularity as an alternative treatment for children with refractory epilepsy (Porter & Jacobson, 2013; Reddy & Golub, 2016).

Neuroprotection: Excitotoxicity and Inflammation

Cannabinoids have been extensively linked to excitatory and inflammatory neuroprotection through in vitro and in vivo studies, utilizing both receptor- and nonreceptor-mediated mechanisms (Bisogno & Di Marzo, 2010; Centonze, Finazzi-Agro, Bernardi, & Maccarrone, 2007). Cannabinoids have been shown to provide protection against acute hypoxic, oxidative, excitotoxic, and traumatic insults (Shen & Thayer, 1998; Nagayama et al., 1999; Panikashvili et al., 2001). For example, following a closed head injury in mice, levels of the endocannabinoid 2-AG are enhanced dramatically, with further exogenous administration of 2-AG reducing brain edema, infarct volume, and hippocampal cell death, and facilitating better clinical recovery (Panikashvili et al., 2001). In traumatic injury using rat models of global and focal cerebral ischemia, the cannabinoid agonist WIN55,212-2 decreased hippocampal cell loss and infarct volume (Nagayama et al., 1999). In similar injury models the cannabinoid agonist BAY 398-7271 also exhibited potent neuroprotective properties (Mauler, Mittendorf, Horvath, & De Vry, 2002). In perhaps the most dramatic protection observed yet, it was reported that the combination of a CB1 antagonist with a CB2 agonist elevated the cerebral blood flow during ischemia and reduced infarction by 75% after 1h of middle cerebral artery occlusion followed by 23h reperfusion in mice (Zhang et al., 2008). Some endocannabinoid mimics are now being tested in clinical trials for neuroprotection in brain trauma and ischemia (Osuna-Zazuetal, Ponce-Gomez, & Perez-Neri, 2015).

One CB1-receptor-dependent mechanism that has been described is through inhibition of calcium entry and subsequent suppression of glutamate release (Gerdeman & Lovinger, 2001). The production of endocannabinoids has been suggested to be calcium dependent, implying a feedback inhibition for excitotoxicity (Di Marzo et al., 1994). Activation of the CB1 receptor results in the inhibition of the release of neurotransmitters such as noradrenaline, serotonin, glutamate, γ-aminobutyric acid (GABA), and acetylcholine, depending on the neuron in which the receptors are expressed (Kim & Thayer, 2000; Petschner et al., 2013; Steffens et al., 2003). Retrograde inhibition can thereby suppress activity rates and energy expenditure of the postsynaptic neuron (Brenowitz, Best, & Regehr, 2006). Endocannabinoids released from the postsynaptic neuron diffuse across the synapse to bind to presynaptic CB1 receptors, decreasing vesicle fusion with the membrane, and reducing levels of cAMP (Demuth & Molleman, 2006) thereby altering protein kinase A (PKA)–mediated phosphorylation and MAPK activation (Borner, Smida, Hollt, Schraven, & Kraus, 2009). Effects on A-type K^+ channel phosphorylation further reduce excitability, along with cAMP-independent inactivation of Ca^{2+} channels (Demuth & Molleman, 2006).

Glutamate is the major excitatory neurotransmitter in the CNS with excitatory stimulation important to multiple neurophysiological processes; however, excessive activation is thought to be responsible for neuronal injury in both acute (e.g,. seizures) and chronic (e.g., AD and HD) neurological disorders (Lewerenz & Maher, 2015; Skaper et al., 1996). Cannabinoids have been described as displaying neuroprotective and antioxidant properties against glutamate toxicity in a range of model systems. These include cannabinoids Δ^9-tetrahydrocannabinol (THC), CBD, and WIN55,212-2 in cortical neuron cultures (Hampson, Grimaldi, Axelrod, & Wink, 1998; Nagayama et al., 1999), and WIN55,212-2 and CP55,940 in hippocampal neuronal cultures (Shen & Thayer, 1998).

Whether suppression of glutamate release by cannabinoids is protective will depend on the precise location of the neuronal insult. For example, Schaffer collateral/commissural fiber–CA1 pyramidal cell synaptic terminals in the hippocampus express CB1 receptors that regulate glutamate release. However, CB1 receptors are also expressed on GABAergic neurons (Elphick & Egertova, 2001), and suppression of GABA release in these cells by the administration of an exogenous cannabinoid could have a disinhibiting and hence proexcitotoxic effect. In the forebrain, glutamatergic neurons only weakly express CB1, in contrast to GABAergic neurons, which strongly express CB1, a fact vividly demonstrated in comparisons of CB1 immuno-labeling in the hippocampi of mice with CB1 deficiency in glutamatergic knock-out and GABAergic knock-out neurons (Lutz, Marsicano, Maldo-nado, & Hillard, 2015).

Several studies have also suggested that cannabinoids can provide pro-tection against kainic acid–mediated excitotoxicity. Abood, Rizvi, Sallapudi, and McAllister (2001) observed that the addition of THC to primary neuronal cultures from mouse spinal cord reduced kainate toxicity. Furthermore, CB1 receptors on glutamatergic hippocampal neurons provide protection against kainic acid–induced seizures in mice (Monory et al., 2006). It has been shown that both genetic ablation and pharmacological blockade of CB1 receptors increases the susceptibility of organotypic hippocampal slice cultures to kainic acid–induced excitotoxicity, leading to cell death. Kainic acid application results in an increase in BDNF, which is prevented by CB1 antagonist treatment. In the presence of CB1 antagonist, cell death could then be prevented by exogenous application of BDNF, suggesting BDNF to be a key mediator in CB1 receptor–dependent protection against excito-toxicity (De March et al., 2008; Khaspekov et al., 2004). This finding is consistent with studies that have investigated the relationship that BDNF and CB1 have in neurological diseases including HD (De March et al., 2008; Khaspekov et al., 2004; Spires et al., 2004), AD (Tapia-Arancibia, Aliaga, Silhol, & Arancibia, 2008), Parkinson's disease (PD) (Evans & Barker, 2008) and schizophrenia (Pillai, 2008). More recently, cannabinoids have been shown to be protective against excitoxicity via the suppression of glu-tamate activation of N-methyl-D-aspartic acid (NMDA) receptors, particularly where CB1 receptors are coexpressed with adenosine receptors (Serpa, Correia, Ribeiro, Sebastiao, & Cascalheira, 2015; Serpa, Pinto, Bernardino, & Cascalheira, 2015).

Some cannabinoid neuroprotective effects do not require the CB1 receptor, although results of experiments on the CB2 receptor have been mixed (Rivers-Auty, Smith, & Ashton, 2014; Yu et al., 2015). Several cannabinoids provide protection from oxidative neuronal cell death in cultured cerebellar granule cells from both normal and CB1 knockout mice, through a nonreceptor-mediated antioxidant pathway (Marsicano, Moosmann, Hermann, Lutz, & Behl, 2002). Panikashvili et al. (2005) found that 2-AG can increase levels of endogenous antioxidants and other cannabinoids can upregulate the expression of one such mediator, superoxide dismutase (Panikashvili, Mechoulam, Beni, Alexandrovich, & Shohami 2005; Garcia-Arencibia et al., 2007). Evidence is accumulating that the cellular location of antioxidant actions of cannabinoids is the mitochondria (Ma et al., 2015). Intriguingly, the transient receptor potential cation channel subfamily V member 1 (TRPV1) has recently been implicated in mitochondrial function in tissue damage (Lang et al., 2015). TRPV1 receptors are pivotal in the regulation of the ischemic cascade (Marinelli et al., 2003), and Pegorini, Zani, Braida, Guerini-Rocco, and Sala (2006) found that the neuroprotective properties of rimonabant (the CB1 inverse agonist SR141716) are TRPVR1 dependent, possibly indirectly through effects on endocannabinoid displacement from CB1 receptors. Direct effects on TRPV1 may also explain the neuroprotective effects of WIN-55,212-2 (Carletti, Gambino, Rizzo, Ferraro, & Sardo, 2016).

Inflammation is a defensive response to injury or insult, designed to limit detrimental effects through inhibition or inactivation of the insult. Although this primary role is protective, inflammation can also cause damage and disease. Neuroinflammation occurs in degenerating disorders such as multiple sclerosis (MS), AD, ischemia and traumatic brain injury, and treatment of inflammation is a major therapeutic target in these diseases (van Dijk et al., 2015). The cannabinoid signaling system is functionally expressed in immune cells, and evidence has shown that cannabinoids may be able to modulate neuroinflammation (Rivers & Ashton, 2010; Walter & Stella, 2004). The localization of cannabinoid receptors on glial cells has seen many conflicting results, but it is currently accepted that CB2 is found on rodent microglia and infiltrating immune cells (Atwood & Mackie, 2010), and that CB1 is expressed on rodent astrocytes (Navarrete & Araque, 2008, 2010; Oliveira da Cruz, Robin, Drago, Marsicano, & Metna-Laurent, 2015). Data suggest that cannabinoid agonists may decrease glial neurotoxicity and modulate glial cell function. Some of the beneficial effects of cannabinoids may therefore be due to inhibition of glial release of proinflammatory mediators, inhibition of microglial migration and enhancement of oligodendrocyte survival (Franco & Fernandez-Suarez, 2015). Release of endocannabinoids in CNS injury may then represent a means of communication between neurological and immune components, to mediate potential damage (Eljaschewitsch et al., 2006). Recently, the neuroprotection effect of the powerful antiinflammatory agent minocycline has been shown to be blocked by CB1 and CB2 antagonists (Lopez-Rodriguez et al., 2015). Finally, the regulation of the blood–brain barrier by cannabinoids is an emerging area of research as a mechanism of cannabinoid neuroprotection (Vendel & de Lange, 2014).

ENDOCANNABINOIDS AND MEMORY DYSFUNCTION

Dysfunction in memory, learning, and cognition are characteristics of many neurodegenerative diseases. The regions of the brain important to these functions are primarily the hippocampus, amygdala, entorhinal cortex, and temporal cortex. Damage occurring here can result in impaired thought, memory, and language, and over time can affect a person's ability to undertake daily tasks. Memory deficits and dementia become more common as people age, but are not a normal part of the aging process. One in 100 people have dementia in their 60s, while this increases to 25% of people 85 years and older (Alzheimer's Association, 2015).

Activation of CB1 receptors in the hippocampus leads to well-described memory changes. Upon recreational consumption of cannabis, people experience impairments in short-term memory (Schwartz, Gruenewald, Klitzner, & Fedio, 1989). This is likely mediated by activation of CB1 receptors reducing glutamate release from hippocampal neurons, inhibiting long-term potentiation and impairing memory formation (Sullivan, 2000). Considerable research has investigated the potential of CB1 antagonists, such as rimonabant, to enhance memory formation, and this has been reviewed (Varvel, Wise, & Lichtman, 2009). In general, the results have been highly dependent on the task studied and occur in some, but certainly not all, models of learning and memory.

Alzheimer's Disease Postmortem Studies

An age-related chronic neurodegenerative disease, AD is the most common form of dementia in old age and is characterized by progressive cognitive decline. AD is typically sporadic, with a genetic basis only accounting for a small proportion of cases (Bekris, Yu, Bird, & Tsuang, 2010). Key neuropathic features of the disease include extracellular neuritic plaques made from a toxic form of β-amyloid protein (Aβ) and intracellular neurofibrillary tangles resulting from hyperphosphorylation of tau protein (Hebert, Scherr, Bienias, Bennett, & Evans, 2003). Further degeneration occurs in the form of synaptic dysfunction and neuronal loss, gliosis, and a chronic inflammatory response including activated microglia clustering at the plaques (McGeer, Itagaki, Tago, & McGeer, 1987). Cellular energy deficits have been suggested to be a major contributing factor to pathology (Ferreira, Resende, Ferreiro, Rego, & Pereira, 2010, Struble, Ala, Patrylo, Brewer, & Yan, 2010).

There have been conflicting reports on cannabinoid receptor levels in the postmortem AD brain. An early, autoradiographic study of postmortem tissue showed significantly decreased CB1 receptor binding in hippocampal regions (37%–45%) and the caudate nucleus (49%) (Westlake, Howlett, Bonner, Matsuda, & Herkenham, 1994). However, no significant changes in CB1 messenger RNA (mRNA) expression were measured (Westlake et al., 1994), and a subsequent autoradiographic study found unchanged CB1 binding (Lee et al., 2010).

Several immunohistochemical studies of the hippocampus and entorhinal cortex in AD postmortem tissue found no change in CB1 receptor protein levels associated with plaques (Benito et al., 2003; Mulder et al., 2011); however, decreased levels of CB1 in cortical brain regions were recently reported by Western blotting, but these demonstrated no correlation with plaques or cognitive status (Solas, Francis, Franco, & Ramirez, 2013). Importantly, recent in vivo positron emission tomographic (PET) analysis of human subjects found no evidence of altered CB1 availability in AD compared to age-matched controls (Ahmad et al., 2014).

High levels of CB2 receptors have been identified to be immunohistochemically selectively localized in neuritic plaque-associated microglia (Benito et al., 2003; Ramirez, Blazquez, Gomez del Pulgar, Guzman, & de Ceballos, 2005). CB2 receptors were also localized to AD neurofibrillary tangles and dystrophic neurites (Ramirez et al., 2005). An additional feature of AD is increased nitration of proteins in the brain as a result of reaction between nitric oxide–forming and superoxide-forming toxic free radicals. Ramirez et al. (2005) described increased nitration of both CB1 and CB2, which may contribute to impairments in coupling to downstream signaling pathways. More recently, the same authors have described increased levels of CB2 and astrocytes (40%) in the cortex in AD; although these changes did not correlate with cognitive status, they did correlate with Aβ(42) levels and senile plaque score (Solas et al., 2013).

Benito et al. (2003) also observed high levels of the endocannabinoid-metabolizing enzyme FAAH selectively in neuritic plaque–associated astrocytes and increased enzyme activity in plaques, which would contribute to an increase in endocannabinoid metabolites such as arachidonic acid. These compounds and related pathways have been implicated in increases in prostaglandin production and other proinflammatory substances (Diez-Dacal & Perez-Sala, 2010). An elegant study by Mulder et al. (2011) used subcellular fractionation to demonstrate impaired monoacylglycerol lipase recruitment to biological membranes in postmortem AD tissues, suggesting that disease progression slows the termination of 2-AG signaling. Thus, differential endocannabinoid regulation, with low anandamide but high 2-AG may be occurring.

In Vitro Studies

Immunohistochemical and biochemical studies in C6 cells and in hippocampal homogenates found that Aβ treatment significantly reduced CB1 receptors and anandamide concentration, while increasing CB2 receptors and 2-AG concentration (Esposito et al., 2007). Selective CB1 agonism and CB2 antagonism reduced Aβ-induced astrogliosis, demonstrating an unusual dual regulation of Aβ (Esposito et al., 2007). In vitro studies of the endocannabinoids noladin ether and anandamide have shown concentration-dependent inhibition of Aβ-induced toxicity in cultured neurons (Milton, 2002). This neuroprotective effect was prevented by the CB1 receptor antagonist AM251, as well as by a MAPK pathway inhibitor.

A number of synthetic cannabinoid agonists, including HU210, WIN55,212-2, and the CB2 selective agonist JWH-133, have been found to inhibit Aβ-induced activation of cultured microglia (Ramirez et al., 2005). CB1 receptor stimulation by CP55,940 also dose dependently inhibits nitric oxide release from cultured microglia, of relevance to AD since nitric oxide plays a significant role in gliosis initiation and formation of plaques and tangles (Ramirez et al., 2005; Waksman, Olson, Carlisle, & Cabral, 1999).

The potential role of non-CB1/non-CB2 cannabinoid effects also needs to be further investigated. CBD, a nonpsychoactive cannabis constituent with potent antioxidant properties, has been shown to be protective against Aβ toxicity, reducing neuronal cell death in PC12 cells (Iuvone et al., 2004). CBD also reduces the activation of inducible nitric oxide synthase, nitric oxide production, and activation of inflammatory mediators (Esposito, De Filippis, Maiuri, et al., 2006) and reverses tau hyperphosphorylation (Esposito, De Filippis, Carnuccio, Izzo, & Iuvone, 2006). Recently, CBD was demonstrated to act through peroxisome proliferator-activated receptor-γ receptors to induce the ubiquitination of amyloid precursor protein, which led to a substantial decrease in the full-length protein levels in SHSY5Y cells with the consequent decrease in Aβ production (Scuderi, Steardo, & Esposito, 2014).

In Vivo Studies

In vivo rodent models of AD involve the induction of cognitive deficits, microglial activation, and neuronal loss by administering or overexpressing Aβ (Van Dam & De Deyn, 2011). Aβ treatment has been reported to increase the production of the endocannabinoid 2-AG (van der Stelt et al., 2006). Early administration of the endocannabinoid uptake inhibitor VDM11 significantly increased endocannabinoid levels further and resulted in a reversal of hippocampal damage and memory retention deficits caused by Aβ treatment (van der Stelt et al., 2006), while late administration of VDM11 in fact worsened the disease pathology and memory retention of rodents, suggesting that there may be a specific window for therapeutic intervention (van der Stelt et al., 2006). A possible mechanism by which the endocannabinoids may be protective in early stages of this disease model is through the reduction of microglial activation. Ramirez et al. (2005) have demonstrated a significant reduction in microglial activation in the cortex and improvement in cognitive performance following WIN55,212-2 administration in Aβ-treated rats. However, in the APP23/PS45–double transgenic mouse model of AD, HU210 treatment had no effect on amyloid precursor protein processing, Aβ generation, and neuritic plaque formation (Chen et al., 2010). Indeed, HU210 treatment worsened memory performance and decreased the survival rate of the mice by 80% (Chen et al., 2010). CBD and WIN55,212-2 were able to prevent learning of a spatial navigation task and cytokine gene expression in Aβ-injected mice (Shakespeare, Boggild, & Young, 2003).

A recent study demonstrated that THC and CBD in combination (Sativex) preserved memory and reduced learning impairment in AβPP/PS1 transgenic mice when chronically administered during the early symptomatic stage (Aso, Sanchez-Pla, Vegas-Lozano, Maldonado, & Ferrer, 2015). Interestingly, these investigators looked to understand the role of CB2 in these processes, by crossing AβPP/PS1 transgenic mice with CB2 knockout mice (Aso, Andres-Benito, Carmona, Maldonado, & Ferrer, 2016). This study showed that lack of CB2 exacerbates cortical Aβ deposition and increases the levels of soluble Aβ40 but did not affect the viability of AβPP/PS1 mice, accelerate their memory impairment, or modify tau hyperphosphorylation in dystrophic neurites associated to Aβ plaques, suggesting that CB2 may not be important in these processes as previously considered. Finally, knockout of CB2 did not attenuate the positive cognitive effect induced by the cannabis-based medicine in these animals (Aso et al., 2016).

An interaction may occur between THC and the acetylcholinesterase enzyme (AChE), which indirectly induces Aβ aggregation and is the principal target of most currently used clinical treatments for AD (Eubanks et al., 2006; Martorana, Esposito, & Koch, 2010). THC competitively inhibits AChE by binding to a site on the enzyme that is critical to amyloidogenesis. Interestingly, anandamide and arachidonic acid have been found to be inhibitors of muscarinic acetylcholine receptors by preventing agonist binding, which is integral to memory function (Lagalwar, Bordayo, Hoffmann, Fawcett, & Frey, 1999). Further investigation is necessary to determine the effects of exogenous and endogenous cannabinoids in the disease state and to understand the discrepancies between these different models. There remains a clear need to understand which effects are actually cannabinoid receptor mediated.

Human Studies

A recent study has reported no difference in circulating or cerebrospinal fluid (CSF) endocannabinoid levels between patients with AD and healthy controls (Koppel et al., 2009). There are limited data investigating the clinical use of cannabinoids in the treatment of dementia; indeed, a Cochrane review (Krishnan, Cairns, & Howard, 2009) found only one study meeting their criteria of a randomized double-blind placebo-controlled study (Volicer, Stelly, Morris, McLaughlin, & Volicer, 1997), and it had insufficient quantitative data to validate the results.

Studies have suggested that cannabinoids may produce some improvement in AD symptoms other than dementia. Volicer et al. (1997) described improvements in appetite and disturbed behavior in patients with AD treated orally with synthetic THC (Dronabinol). Similarly, nighttime agitation and nocturnal motor activity was reduced with Dronabinol treatment in patients with late-stage AD (Walther, Mahlberg, Eichmann, & Kunz, 2006). A case study with Nabilone, another THC analog, demonstrated similar beneficial effects on persistent dementia–related agitation and aggression (Passmore, 2008), a finding that was confirmed in a retrospective chart review study of patients with dementia (Karst,

Wippermann, & Ahrens, 2010). However, a recent randomized, double-blind, placebo-controlled study (Wade, Collin, Stott, & Duncombe, 2010) tested THC (1.5 mg 3 times per day for 3 weeks) in 24 patients with dementia and found that although this was well tolerated, no significant changes were observed in neuropsychiatric symptoms, agitation, or quality of life. Notably, none of these studies appear to have explicitly addressed cannabinoid effects on learning or memory or disease progression.

In summary, as CB1 receptors appear, in the majority of studies, to be unaltered in AD; they remain available as a potential therapeutic target. The few human studies carried out to date suggest that cannabinoids may have use in treating dementia-related symptoms such as aggression and agitation, although this was not supported in the most recent study, thus more adequately powered randomized and appropriately controlled clinical trials are needed. As prolonged use of cannabis in the general population may be associated with memory deficits and cognitive impairment (Rubino, Zamberletti, & Parolaro, 2012), it is critical that studies investigate longer term treatment and consider these conditions.

ENDOCANNABINOIDS AND MOTOR DYSFUNCTION

Cannabinoids have been investigated in multiple conditions exhibiting movement disorders. Cannabis consumption results in recognized psychomotor effects, with rodent studies demonstrating that plant-derived and synthetic cannabinoid compounds produce motor impairments, which are linked to neurotransmitter alterations in basal ganglia circuitry (Sanudo-Pena, Tsou, & Walker, 1999).

CB1 receptors have particularly high expression in the output nuclei of the basal ganglia, the globus pallidus, and substantia nigra, supporting a role for cannabinoids in the control of movement. The striatum itself exhibits only moderate expression; however, both striatopallidal and striatonigral projection neurons have high densities of CB1 receptors located presynaptically (Herkenham et al., 1991). CB1 is expressed on both D1 and D2 dopamine receptors expressing medium spiny GABAergic neurons (Hohmann & Herkenham, 2000). Matching receptor densities, the highest levels in the brain of the endocannabinoids anandamide and 2-AG are in the globus pallidus and substantia nigra (Di Marzo, Hill, Bisogno, Crossman, & Brotchie, 2000).

Motor abnormalities are characteristic of a number of neurological conditions, most notably PD, HD, and MS, therefore these disorders will be discussed further.

Huntington's Disease

HD is an autosomal dominant inherited neurodegenerative disease with onset usually in the fifth or sixth decade, increased symptom severity over time, and ultimately premature death. An expanded CAG repeat sequence in the *huntingtin* gene leads to a mutated polyglutamine expansion in the expressed protein, resulting in complex dysfunctions including cellular excitotoxicity and transcriptional dysregulation (Huntington's Disease Collaborative Research Group, 1993). Individuals with >39 CAG repeats usually develop HD, with symptoms including cognitive deficits, psychiatric changes, and a movement disorder often referred to as Huntington's chorea, which involves characteristic involuntary dance-like writhing movements. Neuropathologically HD is characterized by neuronal dysfunction and death in the striatum and cortex with an overall decrease in cerebral volume (Ho et al., 2001).

In HD the endocannabinoid system is described as being hypofunctional. CB1 receptors are found at high densities in areas of basal ganglia circuitry specifically affected by cellular dysfunction and death in HD. Extensive studies in human HD postmortem tissue as well as HD animal models have characterized loss of receptors, suggesting deterioration in the endocannabinoid system. Loss of CB1 receptors from basal ganglia nuclei is one of the earliest neurological changes observed, with a loss in CB1 densities correlating with severity of neuropathological grade (Glass, Faull, & Dragunow, 1993; Richfield & Herkenham, 1994). An autoradiographic comparative study of CB1, D1 and D2, adenosine A_{2A}, and $GABA_A$ receptors in the basal ganglia of graded HD revealed a complex pattern of degeneration, where CB1 receptors are reduced in all regions prior to other receptor changes (Glass, Dragunow, & Faull, 2000). These cannabinoid system changes have subsequently been confirmed by immunohistochemical characterization (Allen, Waldvogel, Glass, & Faull, 2009) and a human HD in vivo PET study (Van Laere et al., 2010). This early cannabinoid change indicates that dysfunction is occurring before cell death and suggests a potential role for the endocannabinoid system in neurodegenerative progression in HD.

Decreased CB1 receptor binding has been seen in transgenic animal models of HD (Dowie et al., 2009; Glass, van Dellen, Blakemore, Hannan, & Faull,

2004; Jacobsen et al., 2010; Lastres-Becker, Berrendero, et al., 2002), corresponding to changes seen in the human disease. The 3-nitropropionic acid (3-NP)–induced lesion model of HD also exhibits decreased CB1 receptor binding (Page et al., 2000) and reduced G-protein activation in the basal ganglia as well as changes in endocannabinoid transmission in the form of decreased levels of anandamide and 2-AG in the striatum and increased anandamide in the substantia nigra (Lastres-Becker, Fezza, et al., 2001). It can be speculated that this hypofunctionality may contribute to the course of HD degeneration, considering the known cannabinoid-mediated control of striatal function. Indeed, two studies using CB1-deficient mice have demonstrated that when either treated with 3-NP neurotoxin or when crossed with transgenic HD models, there is a worsening of pathophysiological disease development, strongly implicating CB1 function in HD (Blazquez et al., 2011; Mievis, Blum, & Ledent, 2011). Intriguingly a recent elegant study has suggested that this detrimental effect may be due to loss of CB1 on corticostriatal glutamatergic neurons (Chiarlone et al., 2014.). This finding is particularly interesting as it implies that targeting these corticostriatal CB1 receptors, which may be preserved in the disease, could provide therapeutic benefit, despite the loss of the medium spiny neuron CB1 receptors.

Due to the neuromodulatory nature of the endocannabinoid system, its dysfunction in HD will inevitably result in consequences for other neurotransmitter systems. CB1 receptors are expressed presynaptically on GABAergic striatal terminals and exert modulatory control upon inhibitory synaptic neurotransmission (Wilson & Nicoll, 2002). Spontaneous GABAergic synaptic currents demonstrated increased frequency in striatal medium spiny neurons of R6/2 mice (Cepeda et al., 2004). Altered cannabinoid-mediated control of striatal GABAergic transmission has also been observed in R6/2 mice, correlating with early cannabinoid system dysfunction (Centonze et al., 2005). The R6/2 model therefore suggests enhanced GABA release to be associated with HD, potentially as a result of cannabinoid dysfunction. However, it is likely that complex alterations occur to multiple neurotransmitter systems; for example, electrophysiology studies have demonstrated that corticostriatal glutamate-mediated excitatory transmission is also severely impaired in the R6/2 model (Cepeda et al., 2003). $GABA_A$ receptor binding in presymptomatic and symptomatic transgenic HD mice is also increased in the striatum (Cepeda et al., 2004; Glass et al., 2004). Finally, it has also been found that in the striatum of R6/1 transgenic HD mice, glutamate release is reduced and GABA release is enhanced, even prior to the development of an obvious phenotype

(Nicniocaill, Haraldsson, Hansson, O'Connor, & Brundin, 2001). The mechanism in HD for GABA synapse sensitivity to be altered in response to CB1 stimulation is inherently complex. One hypothesis is that the early dysfunction of the cannabinoid system in HD might represent a compensatory mechanism aimed to counteract the initial excitotoxic damage, by increasing GABA release (Glass et al., 2000). Preserving function in the cannabinoid system may therefore prove a beneficial therapeutic pathway.

An investigation of the endocannabinoid system in patients with HD that focused on the endocannabinoid degradative enzyme FAAH found peripheral activity to be dramatically decreased in circulating lymphocytes. A corresponding sixfold elevation in levels of the endocannabinoid anandamide was recorded. Patients known to have the HD gene but who were presymptomatic already had defective FAAH activity, suggesting this enzyme to be a potentially useful early peripheral marker that associates with the HD mutation (Battista et al., 2007).

Effect of Environmental Enrichment

When exposed to a stimulating enriched environment from an early age, R6/1 mice show a delayed onset of motor disorders and decreased loss of cerebral volume (van Dellen, Blakemore, Deacon, York, & Hannan, 2000). Similarly, even low-level enrichment mediates significant improvements in both behavioral and molecular analyses in R6/2 mice (Hockly et al., 2002). The potential modifying role of genetics and environment on HD onset and progression was assessed in humans as part of the Venezuelan Kindred Study (Wexler et al., 2004), which involved the largest and most well-characterized HD population in the world. It was concluded that ~40% of the variation in onset age among patients with HD is due to genes other than the HD gene, with 60% due to environmental effects (Wexler et al., 2004). As CB1 loss from the basal ganglia is known to be an early neuropathological marker in several animal models as well as human HD model (Glass, 2001), CB1-binding sites were examined in R6/1 mice following environmental enrichment. The observed reduction in loss of cerebral volume and delayed disease progression seen with enrichment correlated with an upregulation of CB1 receptors in striatal output nuclei, when assessed at 20 weeks of age (Glass et al., 2004). Dopamine D1 and D2 receptors, also on striatal projection neurons, did not show altered expression with enrichment, suggesting that the upregulation of CB1 is selective, and not a function of a global increase in transcription (Glass et al., 2004).

CB1 Messenger RNA Changes in Huntington's Disease Models

Several studies have investigated CB1 mRNA in animal models of HD, revealing significant reductions early in disease progression prior to observation of behavioral deficits. Investigation of the endocannabinoid system in the HD94 conditional transgenic mouse model of HD revealed a decrease in overall activity in the basal ganglia, with reduced CB1 binding in striatal projection neurons accompanied by a loss in CB1 mRNA levels and decreased activation of GTP-binding proteins (Lastres-Becker, Berrendero, et al., 2002). Differences were observed between striatonigral and striatopallidal projections, indicating variation in vulnerability to the effects of mutant *huntingtin*. Neuronal cell death did not accompany these changes, suggesting changes as a result of neuronal plasticity and cellular dysfunction and implicating cannabinoids in early stages of the disease (Lastres-Becker, Berrendero, et al., 2002).

The R6/1 and R6/2 transgenic mouse lines have similarly been investigated for CB1 alterations at various stages of disease. Little neuronal degeneration is reported in these transgenic models, such that the observed HD-like phenotype is primarily the result of basal ganglia cellular dysfunction (Naver et al., 2003). Several studies have found that prior to the onset of motor deficits, CB1 mRNA levels were significantly downregulated in both the R6/1 and R6/2 lines, with differential onset and rate due to CAG repeat length (Denovan-Wright & Robertson, 2000; Dowie et al., 2009; McCaw et al., 2004; Naver et al., 2003). As the mechanism through which mutant *huntingtin* causes these changes was not clear, McCaw et al. (2004) sought to determine whether this decline was a result of decreased transcription, altered mRNA processing, or increased mRNA turnover, and demonstrated that CB1 transcription rate is reduced in the striatum of R6 mice, prior to the decrease in CB1 mRNA level to a new lowered steady state.

Cannabinoid Treatment in Animal Models

A number of in vivo cannabinoid drug studies have been carried out in lesion models of HD. Most in vivo studies to date have been in rodents with disease simulation a result of toxic stimulus. Lastres–Becker et al. investigated the effect of cannabinoid agonist THC and CB1 inverse agonist SR141716 in a model produced by unilateral malonate striatal injection; a toxin that mimics the striatal degeneration and mitochondrial complex II deficiency seen in patients with HD. Both compounds were found to increase the malonate-induced striatal lesion compared to vehicle (Lastres–Becker, Bizat, et al., 2003). Administration of UCM707, an endocannabinoid uptake inhibitor, to malonate-lesioned rats also did not protect against neurodegeneration (de Lago et al., 2006). UCM707 was similarly investigated in a bilateral striatal injection model using mitochondrial toxin 3-NP, which also replicates the characteristic complex II deficiency. Behavioral results showed a significant antihyperkinetic activity (de Lago et al., 2006).

Additional cannabinoid compounds have been investigated using the 3-NP model. The endocannabinoid uptake inhibitor and TRPV1 agonist AM404 and Arvanil, a hybrid CB1/vanilloid TRPV1 agonist and endocannabinoid transport inhibitor, both acted as antihyperkinetic agents. Arvanil increased glutamate content in the globus pallidus, and thus the behavioral response is thought to be due to this enhancement of excitatory transmission (de Lago, Urbani, Ramos, Di Marzo, & Fernandez-Ruiz, 2005). By increasing endocannabinoid activity as well as acting through the vanilloid system, AM404 facilitated recovery from 3-NP induced deficits in GABA and dopamine in the basal ganglia (Lastres-Becker, Hansen, et al., 2002; Lastres-Becker, de Miguel, et al., 2003). THC was found to be neuroprotective for striatal degeneration (Lastres-Becker et al., 2004), whereas SR141716, when administered to late-stage 3-NP treated rats, failed to reduce the observed akinesia (Lastres-Becker, Gomez, De Miguel, Ramos, & Fernandez-Ruiz, 2002). CBD, but not the CB1 agonist ACEA or CB2 agonist HU-308, was found to reverse the 3-NP-induced decreases in GABA levels and improve other in vitro measures, suggesting CBD to possess neuroprotective properties preferentially toward striatonigral projection neurons. CBD effects were not reversed by CB1, TRPV1, or adenosine A_{2A} antagonists, suggesting the beneficial effects to be a result of antioxidant properties (Sagredo, Ramos, Decio, Mechoulam, & Fernandez-Ruiz, 2007).

The CB1 agonist WIN55,212-2 has been found to attenuate the effects induced by the quinolinic acid (QA) lesion model of HD in rats. QA is an excitotoxin that, when injected into the striatum, stimulates glutamate outflow to mimic HD. Although WIN55,212-2 prevented the QA-induced increase in glutamate and reduced striatal damage, no significant behavioral rescue was observed (Pintor et al., 2006).

No behavioral impact was found in the R6/1 transgenic HD model, following chronic treatment with the cannabinoid agonists THC and HU210, or FAAH inhibitor URB597 (Dowie et al., 2010). However, URB597 treatment preserved CB1 receptors in the striatum, suggesting that manipulating endocannabinoid levels could prove a valuable approach. Interestingly, HU210 treatment led to an increase in striatal huntingtin aggregates (Dowie et al., 2010).

This result correlates with work in an in vitro model of HD, also demonstrating increased aggregates with HU210 (Scotter, Goodfellow, Graham, Dragunow, & Glass, 2010). A recent further study described therapeutic benefits of THC in the R6/2 transgenic model, improving symptoms as well as reducing striatal atrophy and aggregate accumulation (Blazquez et al., 2011). The contrast in results between R6/1 and R6/2 mice may be due to differences between models, age at commencement of treatment, and drug dosing, and requires further investigation.

Collectively these results support the idea that manipulation of the endocannabinoid system may modify HD neurodegeneration.

CB2 Receptors

Recent studies have described an upregulation of CB2 receptors in microglia in the striatum following malonate lesion, in R6/1 and R6/2 transgenic mice, and in human postmortem HD tissue samples (Palazuelos et al., 2009; Sagredo et al., 2009), although other studies have failed to find upregulated CB2 in human HD striatum (Dowie, Grimsey, Hoffman, Faull, & Glass, 2014) or R6/2 mice (Bouchard et al., 2012). Treatment with CB2 agonists has demonstrated protection from malonate or QA toxicity (Palazuelos et al., 2009; Sagredo et al., 2009). Furthermore, CB2-null mice appear more sensitive to malonate or QA toxicity, whereas genetic ablation of CB2 in R6/2 mice or BACHD mice similarly worsened disease state (Bouchard et al., 2012; Palazuelos et al., 2009; Sagredo et al., 2009). Intriguingly, a subsequent study suggested that protection by CB2 agonists was due to activation of peripheral CB2 on immune cells regulating IL-6 levels, with the effect blocked by the peripherally restricted CB2 antagonist SR144528 (Bouchard et al., 2012).

Human Trials

A preliminary open trial suggested that CBD reduced choreiform movements in HD (Sandyk, Consroe, Stern, Snider, & Bliklen, 1988). This was followed by a double-blind randomized crossover clinical trial carried out to investigate CBD as a potential therapeutic agent for symptomatic treatment in HD (Consroe et al., 1991). All patients had a family history; however, this was prior to the gene test being available. Although patients were neuroleptic free, participation included subjects with mild or moderate HD progression and chorea, with a range of severities and disease stages. The study concluded that while no toxicity occurred, 6 weeks of daily oral CBD treatment was ineffective in reducing chorea severity, the major therapeutic outcome variable.

A case study evaluating the synthetic THC analog Nabilone as a single-dose treatment in one patient with symptomatic HD resulted in marked increases in choreatic movements (Muller-Vahl, Schneider, & Emrich, 1999). Motor exaggeration may be due to the early degeneration of striatal projection neurons and abnormal excitatory/inhibitory neurotransmission, such as through interaction with the dopaminergic system. A later case study evaluating Nabilone, prescribed as a consequence of previous patient self-medication with cannabis, found improvements in behavior and reduction in chorea, as reported by caregivers (Curtis & Rickards, 2006). Subsequently, the same group carried out a small pilot randomized controlled trial demonstrating improvements in two neuropsychiatric outcomes and no worsening of chorea (Curtis, Mitchell, Patel, Ives, & Rickards, 2009). Only one clinical trial has considered if cannabinoids could have disease-modifying properties in HD (https://clinicaltrials.gov/ct2/show/NCT01502046). Although the results of this trial are not yet published, it was reported in a review article by some of the authors that the trial successfully demonstrated that Sativex was safe and well tolerated in patients with HD, but did not slow down disease progression (Fernandez-Ruiz, Romero, & Ramos, 2015).

In summary, there is an overall hypofunctionality of the endocannabinoid system in HD, with decreases in CB1 receptor binding in human studies and mRNA expression in transgenic models early in the disease. A significant impact is apparent upon disease course when CB1-null mice are evaluated. Furthermore, corticostriatal CB1, which are preserved in the disease, appear to be able to mediate some neuroprotection. The early CB1 receptor changes occuring prior to changes in colocalized receptors points to the system being involved in dysfunction prior to cell death. Preservation of CB1 receptors following environmental enrichment further supports this putative functional role. Despite a number of in vivo studies, the therapeutic value of CB1-targeting cannabinoid compounds is still contentious. The CB2 receptor is only beginning to be investigated in HD, and may yet provide a further valuable target for therapeutic investigation.

Parkinson's Disease and the Cannabinoid System

PD is a chronic progressive neurodegenerative disease usually with late onset, characterized by motor symptoms of involuntary tremor, muscle rigidity, akinesia, postural instability, and gait abnormalities. Mood and cognitive symptoms may also be apparent including depression and dementia. PD is typically idiopathic with a small proportion being the result of genetic susceptibility. Neuropathologically, it is primarily the substantia nigra pars compacta that is affected in PD, where great atrophy occurs, and 70%–80% of dopamine-producing neurons are lost even prior to symptom onset and clinical diagnosis. These neurons are vital for dopamine signaling within the basal ganglia and projecting to the motor cortex. Dysfunctional mechanisms are postulated to include mitochondrial defects, oxidative stress, and activation of glia and glutamate toxicity. There are no treatments to slow the progression of nigral degeneration, with available therapies primarily targeting dopamine replacement for symptomatic control in early to mid- phases of the disease (Robinson, 2008). Contrasting with HD, research suggests that the pathology of PD includes a hyperfunctionality of the cannabinoid system. Consistent with this, it has been described that anandamide levels more than double in the CSF of untreated patients with PD compared to normal controls, independent of the stage of disease, symptom scoring, or drug washout (Pisani et al., 2005).

Parkinson's Disease: Postmortem Studies

Analysis of the endocannabinoid system in postmortem tissue from patients diagnosed with PD has revealed that WIN55,212-2-stimulated $[^{35}S]GTP\gamma S$ binding in areas of the basal ganglia is increased, indicating an increased efficacy of CB1 signaling. This was accompanied by CB1 binding increases in the striatum, but no changes in the globus pallidus or substantia nigra (Lastres-Becker, Cebeira, et al., 2001). In contrast, however, the expression of CB1 mRNA in PD postmortem tissue is reported to be decreased in the caudate nucleus, anterior dorsal putamen, and external globus pallidus but is unchanged in other basal ganglia regions (Hurley, Mash, & Jenner, 2003).

Insight From Animal Models

A range of different animal models for PD have been used to investigate potential alterations in the endocannabinoid system in the disease and the influence of cannabinoid-based therapies. Each of the models has different advantages and disadvantages, and often they model different aspects of the disease process, which may account for the wide-ranging and often conflicting data generated in this area.

Reserpine blocks transport of monoamine neurotransmitters to the synaptic cleft, reducing the activity of these compounds including serotonin and dopamine. Early cannabinoid research in a reserpine-induced rat model of PD found that THC potentiated the observed hypokinesia (Moss, McMaster, & Rogers, 1981). When evaluated in conjunction with a D2 agonist that alleviated the reserpine akinesia, it was found that the cannabinoid agonist WIN55,212-2 reduced this beneficial effect in a CB1-dependent manner (Maneuf, Crossman, & Brotchie, 1997). As these effects differed from those seen with D1 activation, it was suggested that cannabinoids modulate neurotransmission through the indirect pathway of the basal ganglia. It has been described that medium spiny neurons in this indirect pathway, projecting from the striatum to the lateral globus pallidus, selectively exhibit endocannabinoid-mediated long-term depression, requiring D2 receptor activation. In both reserpine and 6-hydroxy-dopamine (6-OHDA) models of PD, this long-term depression is lost but can be reinstated by D2 agonists or inhibitors of endocannabinoid degradation. When these drugs are coadministered, parkinsonian motor deficits are reduced, suggesting that both systems are critical to movement control (Kreitzer & Malenka, 2007).

A common toxin-induced rat model of PD involves unilateral injection of 6-OHDA, which leads to depletion of dopamine content, neurodegeneration in the lesioned striatum, and a reduction in mRNA levels of the dopamine-catalyzing enzyme tyrosine hydroxylase in the substantia nigra. Administration of THC, THC-V (a phytocannabinoid thought to possess CB1 antagonist and CB2 agonist properties), or CBD immediately following 6-OHDA lesion decreased these effects, with neuroprotective activity likely due to antioxidant or antiinflammatory properties of the compounds (Garcia et al., 2011; Lastres-Becker, Molina-Holgado, Ramos, Mechoulam, & Fernandez-Ruiz, 2005). AM404, an inhibitor of endocannabinoid inactivation, results in anti-parkinsonian effects and produced marked recovery in dopamine and tyrosine hydroxylase levels (Fernandez-Espejo et al., 2004; Garcia-Arencibia et al., 2007), although the effects may not be receptor mediated as no neuroprotection was observed with the CB1 agonist ACEA, the nonselective agonist WIN55,212-2, or UCM707, an inhibitor of endocannabinoid inactivation (Garcia-Arencibia et al., 2007; de Lago et al., 2006).

Several studies in rats lesioned with 6-OHDA show this to be associated with an increase in CB1 mRNA (Mailleux & Vanderhaeghen, 1993; Romero et al., 2000), whereas others suggested no change following lesion, but rather found that chronic dopamine replacement therapy resulted in increased CB1 mRNA expression in the striatum, suggesting CB1 activity to be altered by the action of the dopamine replacement (Zeng et al., 1999). In vivo PET imaging for CB1 receptors following 6-OHDA also suggests changes in endocannabinoid transmission in motor circuits including the striatum, cerebellum, and somatosensory cortex (Casteels et al., 2010).

Bilaterally, 6-OHDA-treated rats show similar degeneration, hypokinesia, and increases in CB1 density in the nigra as unilateral lesions. Treatment of bilaterally lesioned rats with CB1 inverse agonist SR141716 did partially attenuate the hypokinesia present, and also improved early emotional and cognitive alterations, with minor affects upon dopaminergic, GABAergic, or glutamatergic striatal transmission (Garcia-Arencibia, Ferraro, Tanganelli, & Fernandez-Ruiz, 2008; Gonzalez et al., 2006; Tadaiesky, Dombrowski, Da Cunha, & Takahashi, 2010). In the reserpine model, CB1 receptor mRNA expression is reduced in the striatum in an organized way with the greatest decreases observed dorsolaterally. This reduction is likely influenced by endocannabinoid level increases (Silverdale, McGuire, McInnes, Crossman, & Brotchie, 2001). Despite these differences, the majority of data suggests overactivity in the endocannabinoid system in PD.

A significant limitation of long-term use of dopamine replacement therapies for PD is the development of motor fluctuations and levodopa-induced dyskinesias. In the reserpine-treated rat model of PD, levodopa-induced dyskinesia was attenuated by CB1 inverse agonist SR141716 (Segovia, Mora, Crossman, & Brotchie, 2003). The potent CB1 agonist HU210 reduced the rotations and abnormal involuntary movements induced by levodopa in 6-OHDA lesioned rats, whereas CBD and its primary metabolite had no effect on rotational behavior (Gilgun-Sherki, Melamed, Mechoulam, & Offen, 2003; Walsh, Gorman, Finn, & Dowd, 2010). SR141716 also reduced motor asymmetry through a dopamine system interaction (El-Banoua, Caraballo, Flores, Galan-Rodriguez, & Fernandez-Espejo, 2004), and indeed when administered in combination with levodopa, SR141716 exhibited a synergistic beneficial effect on contralateral forepaw stepping (Kelsey, Harris, & Cassin, 2009). Intriguingly, when administered to non-dyskinetic levodopa-treated 6-OHDA-lesioned rats, SR141716 appeared to precipitate certain involuntary movements, suggesting that endocannabinoid tone may protect against levodopa-induced dyskinesia (Walsh et al., 2010). Subsequent studies have suggested that in addition to protecting against dyskinesia, SR141716 treatment may also preserve dopaminergic cells, delaying disease progression (Gutierrez-Valdez et al., 2013).

Somewhat paradoxically, agonist treatment also reduces levodopa-induced dyskinesia. Levodopa-treated 6-OHDA-lesioned rats also displayed decreases in abnormal involuntary movements on treatment with WIN55,212-2 (Segovia et al., 2003), correlating with changes in dopamine and glutamate outputs, and decreased striatal PKA activity (Martinez,

Macheda, Morgese, Trabace, & Giuffrida, 2012; Morgese, Cassano, Cuomo, & Giuffrida, 2007; Morgese, Cassano, Gaetani, et al., 2009). Furthermore, treatment with URB597, an inhibitor of endocannabinoid degradation, had no effect on involuntary movements alone; however, when administered with a TRPV1 antagonist, which prevented binding of anandamide to TRPV1 receptors, the ability of URB597 to enhance cannabinoid signaling via CB1 was unmasked and abnormal involuntary movements were reduced through this mechanism (Morgese et al., 2007).

When evaluated in early- and late-stage non-human primate 1-methyl-4-phenyl-1,2,3,6-tetrahydropyridine (MPTP)-induced parkinsonism, SR141716 failed to alleviate motor deficits (Meschler, Howlett, & Madras, 2001). However, when administered to animals exhibiting levodopa-induced dyskinesia, SR141716 potentiated the antiparkinsonian actions (van der Stelt et al., 2005). In MPTP-treated rhesus monkeys, a CB1 selective antagonist, 1-[7-(2-chlorophenyl)-8-(4-chlorophenyl)-2-methylpyrazolo[1,5-a]-[1,3,5] triazin-4-yl]-3-ethylaminoazetidine-3-carboxylic acid amide benzenesulfonate, similarly had no effect on parkinsonism alone but potentiated the response to levodopa, while not modifying levodopa-induced dyskinesias (Cao et al., 2007). In one of the few studies to consider the effect of cannabinoids on the pharmacokinetics of levodopa, URB597 was found to attenuate levodopa-induced hyperactivity in MPTP-lesioned marmosets without compromising the antiparkinsonian actions of levodopa or interfering with its metabolism (Johnston et al., 2011). Varying results for antagonist treatment may be explained by rodent data demonstrating that CB1 antagonist effects are only anti-parkinsonian in animals with very severe nigral lesions (>95% cell loss), possibly due to the progression of cell death in basal ganglia structures and alterations in dopamine signaling. This may suggest that antagonists have value as advanced stage PD therapeutics (Fernandez-Espejo et al., 2005).

A study of CB1 agonism in MPTP-lesioned animals found that THC improved locomotor activity in marmosets but induced significant adverse effects (van Vliet, Vanwersch, Jongsma, Olivier, & Philippens, 2008). The agonist WIN55,212-2 also protected against neurodegeneration and behavioral impairments, but its mechanism was found to be independent of CB1 activation and potentially via CB2 (Price et al., 2009). Upon administration of CB2 agonist JWH015, there was a reduction in MPTP-induced microglial activation, the putative site of the receptors, and genetic ablation of CB2 worsened the effects of MPTP, suggesting CB2 receptors to have therapeutic relevance in PD (Price et al., 2009).

Again in contrast to studies suggesting that antagonism of CB1 may be helpful in alleviating levodopa-induced dyskinesia, when MPTP-lesioned marmosets were treated with levodopa and the clinically available cannabinoid agonist Nabilone, there was a significant reduction in dyskinesia, compared to levodopa alone (Fox, Henry, Hill, Crossman, & Brotchie, 2002).

Further models, generated by deletion of various PARK genes associated with heritable PD, have shown significant biphasic changes with age in CB1 mRNA levels and receptor binding in the striatum, with decreases apparent in early phases and upregulation as the disease progresses (Garcia-Arencibia et al., 2009). The parkin-null mouse, a genetic model lacking the Park-2 gene, exhibits a progressive phenotype without cell death and is considered representative of early PD deficits. In these mice, cannabinoid agonists and antagonists lead to gender-dependent alterations in CB1 levels and motor responses (Gonzalez et al., 2005).

Endocannabinoids in Parkinson's disease

In the reserpine rat model of PD, levels of endocannabinoid 2-AG are increased sevenfold in the globus pallidus; 2-AG levels are reduced and locomotion is stimulated by treatment with either D1 or D2 agonists. Full locomotion is restored when a D2 agonist is coadministered with CB1 inverse agonist SR141716, suggesting that endocannabinoid system modulation may be useful in therapy development (Di Marzo et al., 2000). Consistent with these findings, MPTP-lesioned marmosets develop large increases in 2-AG and anandamide in the striatum, and 2-AG in the substantia nigra, aligning with the suggestions that endocannabinoid system changes may be compensating for dopamine loss (van der Stelt et al., 2005). Anandamide levels that were increased in the lateral globus pallidus were normalized by levodopa (van der Stelt et al., 2005). In the 6-OHDA model, one study showed striatal anandamide levels to be increased, correlating with decreased activity of FAAH, but with no change in receptor binding, or levels of 2-AG (Gubellini et al., 2002). Thus inhibition of endocannabinoid degradation could be exploited as a potential therapeutic target to help tackle abnormally high glutamatergic activity in PD (Gubellini et al., 2002). The finding that abnormalities in the endocannabinoid system are reversed with chronic levodopa in parkinsonian models suggests that dopamine replacement may be addressing endocannabinoid deficits.

Cannabinoid-Based Therapies in Humans

To date, there is very little information available on the effects of cannabinoid therapies in humans with PD. In an anonymous retrospective questionnaire-based study evaluating spontaneous use of cannabis in a population of patients with PD, it was found that 25% of responders reported cannabis use, most commonly for self-medication of bradykinesia, muscle rigidity, and tremor (Venderova, Ruzicka, Vorisek, & Visnovsky, 2004). In a relatively small (22 patients) open-label study, smoked cannabis resulted in improved sleep and pain scores (Lotan, Treves, Roditi, & Djaldetti, 2014).

A pilot randomized, double-blind, placebo-controlled, crossover trial demonstrated that Nabilone reduces GABA reuptake in the lateral globus pallidus and significantly reduces dyskinesias in patients with PD (Sieradzan et al., 2001). However, a similarly designed trial of oral cannabis extracts in patients with PD exhibiting dyskinesia recorded no pro- or anti-parkinsonian outcomes (Carroll et al., 2004). In a pilot randomized, double-blind, placebo-controlled study with 24 patients, SR141716 did not improve motor deficits (Mesnage et al., 2004). Clearly before any conclusions can be drawn, these studies would need to be extended. The nonpsychoactive CBD has also been tested in two studies and suggested some improvement in sleep (Chagas, Eckeli, et al., 2014) and quality-of-life scores (Chagas, Zuardi, et al., 2014), but again, these are preliminary studies (seven and four patients, respectively).

A different objective was investigated in an open-label pilot study of psychosis in PD. CBD treatment was given to six outpatients with PD and was found to significantly decrease psychotic symptoms, while not worsening motor function or inducing adverse effects. These preliminary results of safety and efficacy suggest that a larger study may be valid (Zuardi et al., 2009).

Overall, there is a complex interplay between neurotransmitter systems in PD. The overactivity in the endocannabinoid system, demonstrated by increases in endocannabinoid levels and receptor changes in the basal ganglia, suggests an insufficient compensatory mechanism in place. Enhancement of this effect may prove beneficial in PD, along with exploitation of the antioxidant properties of some cannabinoids. Concurrently, there is therapeutic potential of cannabinoids, in particular CB1 antagonists, in the treatment of dopamine replacement therapy adverse events. Further investigations are warranted to pursue potentially biphasic dynamic adaptive changes occurring in the system in PD.

Multiple Sclerosis and Cannabis Use

MS is an inflammatory autoimmune demyelinating disorder of the CNS where oligodendrocytes that form the myelin sheath of nerve fibers in the spinal cord and brain are damaged, along with the nerves themselves (Compston & Coles, 2002); MS is the most common inflammatory neuro-degenerative disease worldwide (Rizzo, Hadjimichael, Preiningerova, & Vollmer, 2004). A series of pathological events occur, including immune system activation, inflammatory injury of glia and axons, recovery of function and repair, gliosis, and neurodegeneration (Stadelmann, Wegner, & Bruck, 2011). These events may happen in phases, resulting in a variable clinical course with progression leading to severe persistent deficits interspersed with episodes of remission or recovery. There are a number of environmental and genetic factors linked to disease susceptibility, with a lifetime risk of 1 in 400 and onset common in younger adults. Current treatments are primarily antiimmunological, targeting the autoimmune response, with the related chronic irreversible neurodegeneration resulting in progressive disability (Dendrou, Fugger, & Friese, 2015). Treatments that could reduce the frequency, severity, or long-term disability associated with relapses, or those able to relieve symptoms such as spasticity and pain are sought.

The two distinct disease processes of neuroinflammation and neurodegeneration may both be differentially targeted by cannabinoids in patients and models of the disease (Baker et al., 2000; Pryce et al., 2003). Initial interest in the cannabinioid system came from the long-standing anecdotal association of cannabis use with MS. In surveys, patients have reported perceived improvements in symptoms from reduced spasticity, chronic pain, and tremor to improvements in emotional dysfunction, anorexia, and bladder control (Chong et al., 2006; Consroe, Musty, Rein, Tillery, & Pertwee, 1997; Ware, Adams, & Guy, 2005).

Medicinal cannabis has been approved for use to control symptoms of spasticity and chronic pain in a number of countries and several states in the United States. However, the availability of the plant product is a hotly debated topic, particularly due to adverse effects and societal pressures (Feinstein, Banwell, & Pavisian, 2015). It is often argued that the many phytocannabinoids and other plant constituents are important to the benefits seen with cannabis use such as antioxidant activity and rate of benefit onset (Grundy, 2002; Wilkinson et al., 2003). Many patients with MS are reported to prefer self-medication with cannabis to prescription cannabinoids, and research reports have supported further exploration of cannabis as an MS treatment (Mechoulam & Golan, 1998), while significant research has also investigated alternative cannabinoid compounds for therapeutic use in MS. This topic has been extensively reviewed elsewhere [for example, Zajicek and Apostu (2011)].

Cannabinoid-Based Drugs in Humans

Cannabinoid-based clinical trials have been hindered by potential mood-enhancing effects, which can create challenges in controlled studies, and difficulties can be encountered in determining relevant doses and route of administration (Zajicek, 2002). Early research in small controlled studies of oral THC in patients with MS revealed mixed outcomes. As research has progressed, clinical trials of medicinal cannabinoids are now consistently indicating effectiveness in the treatment of a range MS symptoms (Smith, 2007).

It is important to distinguish between the effects of cannabinoids on MS symptoms and on disease progression. Reduction in symptoms can be a result of remission, but it can also be the result of direct effects on the neurological outputs that cause spasticity and pain independently of disease suppression. CB1 receptors in the basal ganglia most likely explain the ability of cannabinoids to reduce spasticity and painful muscle spasm in MS. CB1 is highly expressed in the efferent neurons of the substantia nigra pars reticulata and globus pallidus. Activation of these receptors can reduce excessive motor activity and hence muscle spasm. In addition, cannabinoids reduce pain in MS through direct actions on neurons of the pain pathways. Symptom control is an important issue in MS, as the usual drugs prescribed to treat spasticity such as benzodiazepines, gabapentin, baclofen, tizanidine and dantrolene are only partially effective (Shakespeare et al., 2003).

Several early reports indicated oral THC to be beneficial for spasticity by subjective patient evaluation as well as clinical analysis (Meinck, Schonle, & Conrad, 1989; Petro & Ellenberger, 1981; Ungerleider, Andyrsiak, Fairbanks, Ellison, & Myers, 1987; Vaney et al., 2004). One small study of oral THC in severe tremor and ataxia reported motor benefits in 25% of patients with MS (Clifford, 1983). A thorough double-blind, placebo-controlled study with oral THC and cannabis extract in spasticity found that neither substance significantly reduced spasticity compared to placebo, with more adverse events associated with the plant extract (Killestein et al., 2002). While limited by small sample size, this study design was effective, and since then the large-scale UK National Study of Cannabinoids in MS (CAMS) has mimicked this design, including the same treatment groups, but involving recruitment of 667 patients. Despite no significant improvements found according to assessment using the Ashworth scale of muscle tone, subjective patient reports suggested cannabinoid treatment to be beneficial for spasticity and pain, indicating there may still be a clinical role for cannabinoids in MS (Zajicek et al., 2003). A substudy of the CAMS study described clinical benefit from active treatments over placebo in reducing incontinence episodes in patients with MS (Freeman et al., 2006). In the CAMS study, despite putative immunomodulatory properties, no evidence was recorded for cannabinoids influencing serum levels of several cytokines and immune mediators (Katona, Kaminski, Sanders, & Zajicek, 2005). A 12-month follow-up of patients in the CAMS study who chose to continue medication revealed that there was a small beneficial effect of treatment on muscle spasticity according to the Ashworth scale from baseline to 12 months (Zajicek et al., 2005).

The oral THC analog Nabilone has been investigated in blinded placebo-controlled studies, showing variable improvements most notably in pain associated with muscle spasms, nocturia, and general well-being (Martyn, Illis, & Thom, 1995; Wissel et al., 2006). The oral synthetic THC compound Dronabinol similarly decreased pain intensity in a larger controlled trial of patients with MS (Svendsen, Jensen, & Bach, 2004). A study of patients with MS treated with oral cannabis plant extracts exhibited increased tumor necrosis factor-α in lipopolysaccharide-stimulated whole blood, suggesting there is the potential for cannabinoids to in fact act via unexpected proinflammatory mechanisms (Killestein et al., 2003).

Cannabis-Based Medicinal Extracts

Research into cannabis as a treatment of MS spasticity has led to the approval of Sativex (GW Pharmaceuticals) in a number of countries. This is a cannabis-based pharmaceutical product administered by oromucosal (sublingual) spray, derived from cloned plants that yield specific concentrations of the primary components of cannabis, THC and CBD, as well as small amounts of naturally occurring terpenes and flavonoids. The combination of these cannabinoids has been deemed relevant for medicinal benefit (Russo & Guy, 2006), while the sublingual spray delivery method ensures less health risk than smoking and allows greater control over titration of dose than oral administration (Notcutt et al., 2004). Cannabis-based medicinal extracts became available for clinical study from GW Pharmaceuticals in 2000, prior to the first market approval in 2005 in Canada, enabling preliminary clinical trials to be undertaken. The initial development program focused on MS treatment, including determining a therapeutic window for the extracts when administered sublingually and evaluating the effects of medicinal extracts in three different compositions: THC alone, CBD alone, or a 1:1 mixture of both compounds (Sativex). Preliminary studies found that in patients experiencing chronic pain, THC-containing extracts were most effective in symptom management (Notcutt et al., 2004). Pain relief and improvements in spasticity and tremor, sleep disturbances, bladder dysfunction, and lower urinary tract symptoms were associated with THC and CBD in combination, or as an adjunctive therapy (Brady et al., 2004; Wade, Robson, House, Makela, & Aram, 2003). A number of studies have since evaluated Sativex in patients with MS (Collin, Davies, Mutiboko, & Ratcliffe, 2007; Conte et al., 2008; Rog, Nurmikko, Friede, & Young, 2005; Russo, Guy, & Robson, 2007; Wade, Makela, Robson, House, & Bateman, 2004), confirming symptom improvements (Karst et al., 2010). Meta-analysis of trials by Wade et al. (2010) calculated that the odds ratio for improvement in spasticity by Sativex over placebo is 1.67, while Collin et al. (2010) determined that 40% of patients achieved a greater than 30% benefit from using Sativex (21.9% of patients with placebo).

Evidence continues to accumulate that Sativex can help alleviate neuropathic pain in patients with MS (Russo et al., 2007, 2016) in addition to spasticity. Increasingly attention has turned to the ability of cannabinoids to help with other symptoms of MS (Notcutt, 2015). These include improvements in patients' sleeplessness and incontinence (Freeman et al., 2006; Kavia, De Ridder, Constantinescu, Stott, & Fowler, 2010) and function in daily living (Arroyo, Vila, & Dechant, 2014). In addition, a randomized clinical trial has now shown that another THC formulation, Namisol, which is a rapidly absorbed oral formulation, is safe to use in patients with MS. The results of effectiveness trials have yet to be published (Klumpers et al., 2012).

Long-term follow-up of Sativex use indicates that beneficial effects are maintained in patients who perceived benefit at treatment commencement, with dosage increases or tolerance not of great concern. Due to some undesired adverse events such as oral pain, dizziness, feelings of intoxication, and a low incidence of seizures, further large long-term safety studies should be undertaken, along with analysis of the potential for dependence and abuse (Perras, 2005; Rog, Nurmikko, & Young, 2007; Russo et al., 2007; Wade, Makela, House, Bateman, & Robson, 2006). Most evidence appears to suggest that cannabinoid treatment may be particularly helpful for patients with advanced MS that is refractory to conventional treatments as an "add-on" treatment while continuing other medications. In a randomized placebo-controlled trial over 19 weeks, 48% of patients with MS with treatment-refractory symptoms experienced greater than 20% symptom improvement (Novotna et al., 2011). Notwithstanding effects on symptom control, the ability of cannabinoid therapy to retard disease progression in MS has been disappointing (Ball et al., 2015; Novotna et al., 2011).

Cannabinoids in Animal Models of Multiple Sclerosis

Chronic relapsing experimental allergic (or autoimmune) encephalomyelitis (CREAE) is a commonly used model of MS, induced by inoculation with CNS extracts in Freund complete adjuvant, to cause sensitization to CNS myelin [see Baker, Gerritsen, Rundle, and Amor (2011) for a critical appraisal of model]. Most commonly used in rodents, this model develops a relapsing and remitting course, spasticity, and tremor and exhibits inflammatory and demyelinating characteristics similar to MS in humans (Kubajewska & Constantinescu, 2010), but with severe nerve inflammation and progressed time course (Baker et al., 2001). THC was first investigated in rat and guinea pig CREAE models of MS, anticipating that its immunosuppressive properties may prove beneficial for a primarily immune-mediated condition. THC-treated animals showed improved clinical signs, with delayed disease onset and improved survival, and decreased inflammation at a neuropathological level (Lyman, Sonett, Brosnan, Elkin, & Bornstein, 1989). Further investigation in a CREAE mouse model of disease showed that by intravenous administration, cannabinoid agonists WIN55,212-2, THC, methanandamide, and JWH-133 all decreased the frequency and extent of tremor and reduced hindlimb spasticity (Baker et al., 2000). These signs worsened with CB1 and CB2 antagonists, suggesting receptor-mediated activity and a tonic regulation of spasticity and tremor. Further investigations of WIN55,212-2 in experimental allergic encephalomyelitis (EAE) rodents have found that treatment involves immune regulatory actions by inhibiting an observed increase in leukocyte/endothelial interaction (Ni et al., 2004) and inducing apoptosis of encephalitogenic T cells (Sanchez, Gonzalez-Perez, Galve-Roperh, & Garcia-Merino, 2006), through a CB2-mediated pathway and likely other receptor-independent mechanisms [for example, Downer et al. (2011) showed that WIN55,212-2 protects in an EAE model via TL3/4-induced expression of IFN-β). When administered to CREAE rats, anandamide, 2-AG, and palmitoylethanolamide all significantly ameliorated spasticity to some degree (Baker et al., 2001). In EAE mice, 2-AG treatment acutely delayed disease onset and reduced chronic disability, with increased microglial activation, suggesting receptor- and immune system–mediated benefits (Lourbopoulos et al., 2011).

Compounds administered to EAE rodents to enhance the endogenous cannabinoid levels at their endogenous site of action by preventing uptake or degradation, including AM404, arvanil, AM374, and OMDM2, were similarly successful in decreasing the impact of MS deficits (Baker et al., 2001; Cabranes et al., 2005). The favorable activity of some of these compounds appears to be partially mediated through cannabinoid and vanilloid system mechanisms. Anandamide is known to also act as a full agonist at the TRPV1 vanilloid receptor, and TRPV1 agonists including capsaicin have been found to display a small but significant amelioration in CREAE mouse spasticity, blocked by TRPV1 antagonist capsazepine. Investigations using the hybrid cannabinoid–vanilloid compound arvanil have shown potent antispastic and analgesic effects, which are maintained in CB1 knockout mice and in the presence of antagonists for both receptors (Brooks et al., 2002). These results suggest that arvanil is working through alternative sites remote from CB1/CB2 and TRPV1 receptors (Brooks et al., 2002). Administration of the anandamide transport inhibitor VDM11 has produced conflicting data (Baker et al., 2001; Cabranes et al., 2005), whereas CBD (Baker et al., 2000) did not reduce spasticity.

When EAE is induced in CB1-deficient mice, inflammation and excitotoxic damage are exacerbated and significant neurodegeneration is observed (Pryce et al., 2003). There appears to be concomitant neuronal and axonal loss and demyelination in these mice, with CB1 loss implicated in caspase activation and dysregulation of myelin, supporting the notion that cannabinoid neuroprotective effects may be receptor mediated (Jackson, Pryce, Diemel, Cuzner, & Baker, 2005). This is further supported by studies of cannabinoid treatment in these EAE CB1-deficient mice in which the antispastic activities of nonselective CB1/CB2 agonists WIN55,212-2 and CP55,940 are lost, as well as those of CB2 agonist RWJ400065 (Pryce & Baker, 2007). The authors propose that CB1 controls spasticity, with some cross-reactivity accounting for CB2 agonist effects. EAE induction in CB2 knockout mice also led to worsening of behavioral scores compared to EAE-treated wild-type mice, along with increased axonal loss, activation of microglia, infiltration of T-lymphocytes, and recruitment of myeloid progenitor cells to the spinal cord (Palazuelos et al., 2008). Treatment of EAE wild-type mice with CB2 agonists reduced disease symptoms, axonal loss, and microglial activation, and influenced the expression of several chemokines, suggesting a CB2-mediated protection in the EAE condition (Palazuelos et al., 2008). Induction of a chronic

EAE model in FAAH null mice, which express elevated levels of endocannabinoids, causes initial inflammation similar to that seen in wild-type mice, but their remission phase was significantly improved, confirming the neuroprotective effect of potentiation of the endocannabinoid system (Webb, Luo, Ma, & Tham, 2008). The complexity of cannabinoid system signaling and of the disease pathogenesis makes it difficult to segregate individual effects, but determining the pathological origin of specific symptoms may help in precise treatment development. One development that demonstrates this principle is the discovery that activation of CB1 receptors in the peripheral nervous system is sufficient for the control of spasticity in mice (Pryce & Baker, 2007), which suggests that it may be possible to treat some of the symptoms of MS with cannabinoids without adverse psychoactive effects.

Another inflammatory model of MS utilizes infection with Theiler murine encephalomyelitis virus in susceptible mice to cause chronic immune-mediated demyelination similar to MS. Associated with this model are increases in proinflammatory cytokines, CB2 receptor expression, and levels of 2-AG and palmitoylethanolamide (Loria et al., 2008). Work with this model has shown that a range of cannabinoid agonists and inhibitors of endocannabinoid transport or uptake improve the disease course and progression, and exhibit immune regulatory properties (Arevalo-Martin, Vela, Molina-Holgado, Borrell, & Guaza, 2003; Croxford & Miller, 2003; Docagne et al., 2007; de Lago et al., 2006; Loria et al., 2008; Mestre et al., 2005; Ortega-Gutierrez et al., 2005).

Thus it appears that cannabinoids can act in anti-inflammatory symptom management as well as to slow the neurodegenerative processes that cause chronic deficits in MS, potentially with one pathway predominantly mediated through each cannabinoid receptor (Pryce et al., 2003). The challenge is now for insights that have been accumulating with studies using animal models to be translated to positive clinical outcomes in human patients (Fernandez-Ruiz, Moro, & Martinez-Orgado, 2015). In addition to clinical trials using phytocannabinoids, more human research is needed into the effects of more nuanced manipulations of the endocannabinoid system in MS, to fully capitalize on the advances made in preclinical animal models. To take three examples from animal research, clinical research may explore the use of combination cannabinoid treatments (Rahimi et al., 2015), selective CB2 receptor ligands (Kong, Li, Tuma, & Ganea, 2014), or inhibitors of FAAH (Pryce et al., 2013).

Endocannabinoid System Components in Multiple Sclerosis

Levels of endocannabinoids have been measured in spinal cord and brain tissue from spastic CREAE rats. Compared to normal rats, spastic animals had significantly elevated levels of endocannabinoids in the brain, where lesions are mainly observed only in the cerebellum. Anandamide was increased to around 200% of normal in the spinal cord, where dramatic pathological changes occur in this model (Baker et al., 2001). These results suggest that endocannabinoid levels are upregulated in specific areas of injury in the CREAE model, perhaps in a compensatory manner, or as a simple consequence of the spasticity. Non-spastic CREAE remission rats, which have some spinal cord lesions and two to three paralytic episodes, had unchanged endocannabinoid levels compared to normal rats (Baker et al., 2001). EAE mice also have high levels of anandamide in the acute phase, with increased synthesis and reduced degradation. Recordings from single neurons in EAE slices found that CB1 activation led to reduced excitatory neurotransmission, suggesting that anandamide increases act as a compensatory mechanism decreasing excitotoxicity (Centonze, Bari, et al., 2007).

Despite considerable preclinical and clinical data supporting cannabinoid therapeutic investigations in MS, data regarding cannabinoid receptors in human MS are only starting to become available. Intriguingly, studies have suggested that a polymorphism in the CB1 gene (long AAT repeats) leads to both a higher incidence of the disease (Ramil et al., 2010) and more severe disease and higher risk of progression (Rossi et al., 2011). In human MS spinal cord, there is an increased density of CB2-immunoreactive microglial/macrophage cells in affected regions, consistent with data from animal models, suggesting that CB2 agonists should be evaluated for potential therapeutic value (Yiangou et al., 2006). In human MS brain tissue samples, double immunohistochemistry revealed CB1 to be expressed in cortical neurons, oligodendrocytes, some astrocytes, and oligodendrocyte precursors (Benito et al., 2007; Zhang, Hilton, Hanemann, & Zajicek, 2011). CB1 and CB2 were both found on macrophages and infiltrated T-lymphocytes and endothelial cells, whereas CB2 receptors were also found on astrocytes and perivascular and reactive microglia in MS plaques (Benito et al., 2007; Zhang et al., 2011). Endothelial CB2 expression was increased in chronic inactive plaques, often in areas of blood–brain barrier damage (Zhang et al., 2011). In the CSF and peripheral lymphocytes of patients with relapsing MS, the concentration of anandamide was increased (Centonze, Bari, et al., 2007).

A comprehensive analysis of CB1 has been described in EAE rats that exhibited a high degree of neurological signs prior to death. CB1 receptor binding and mRNA levels showed marked decreases in the striatum, but not striatal output nuclei. Decreases were also observed in CB1 binding in the cerebral cortex. Despite receptor binding decreases, WIN55,212-2-stimulated [^{35}S]GTPγS binding was not different, suggesting CB1 receptors to be more efficiently coupled to signaling mechanisms in the striatum and cortex (Berrendero et al., 2001).

A similar characterization study was undertaken in the acute, remission, and chronic phases of CREAE mice. No changes were observed in any phase in CB1 mRNA levels in the brain regions examined, and no overall CB1 receptor changes were recorded in regions involved in cognition, including the cortex and hippocampus. Changes in motor structures differed as the disease phases progressed, with a decrease in CB1 density in the striatum during the acute phase, which was recovered during remission, but further exacerbation in the chronic phase. The cerebellum exhibited the same pattern, accompanied by progressive deficits in GTP binding, and the globus pallidus had decreased CB1 density in the acute phase, which was not recovered. The CB2 receptor has not been as extensively studied as the CB1 receptor in models of MS (Pryce & Baker, 2015). However, there is some evidence that CB2 may be a targetable receptor for treating MS (Di Iorio et al., 2013). CB2 activation has been shown to suppress T-cell response (Rom & Persidsky, 2013; Rom et al., 2013), thus reducing subsequent immune cell accumulation (Kong et al., 2014).

There appears to be a shift in the equilibrium of the endocannabinoid system in MS, particularly during spastic events, potentially as a response to signaling dysregulation or nerve degeneration. Treatment with cannabinoid-based therapeutics clearly holds promise, particularly for the clinical signs of spasticity and chronic pain, but the current activity must be enhanced for greatest benefit. This may be through manipulation and enhancement of endocannabinoids, continued investigation of the specific roles of CB1 and CB2 receptors, and delineation of the neuroinflammatory and neurodegenerative components of this demyelinating disease.

ENDOCANNABINOIDS IN MOOD AND THOUGHT DYSFUNCTION

Cannabis has been used recreationally for centuries, with its primary desired effects being its ability to alter perceptions and mood (Mechoulam & Hanu, 2001). It is perhaps not surprising then that the endocannabinoid system has been implicated in the pathophysiology and psychological symptoms of a number of mood and thought disorders. These symptoms include dysphoria, agitation, irritability, anxiety, and depression. Patients with AD, PD, and HD all display at least some cognitive, emotional, and personality alterations at different stages in the diseases (Koskderelioglu, Binbay, Arac, & Veznedaroglu, 2008; Muslimovic, Post, Speelman, Schmand, & de Haan, 2008; Walther et al., 2006). Moreover, cannabinoids have a controversial role in the causing psychiatric conditions such as schizophrenia, bipolar disorders, and depression; cannabis use has been associated with disease onset or triggering of symptoms in susceptible individuals; and although complicating matters, cannabis is sometimes used by such individuals for symptom management (Leweke & Koethe, 2008).

Putative Role of the Cannabinoid System in Schizophrenia

Schizophrenia is a complex neurological disease characterized by abnormal mental functions and disturbed behavior, with an incidence of 1% in the general population (Lewis & Lieberman, 2000). Diagnosis relies upon simultaneous presentation of a range of symptoms such as delusions and hallucinations, loss of motivation, and social withdrawal. Specific cognitive impairments, such as disturbances in attention, executive functions, and working memory, may contribute to the behavioral disability seen in schizophrenia, with disease-related mood alterations of depression and anxiety being linked to a high incidence of suicide (Lewis & Lieberman, 2000). The disease's pathophysiology is believed to be multifactorial in origin, with evidence for a genetic link as well as environmental conditions being implicated in the initial pathological vulnerability (Thaker & Carpenter, 2001). Onset is usually in late teens or early 20s, when maturational processes are thought to reveal a predisposition to the disease and symptoms can be triggered by environmental stressors. Schizophrenia is considered to be a neurodevelopmental disease due to the static nature of disease symptoms upon presentation, and the distinct absence of gliosis (Harrison, 1999). Structural abnormalities include enlargement of the ventricles and decreased cortical volume. Symptoms have been associated with frontal and temporal lobe dysfunction due to their roles in executive, cognitive, language and auditory functions (Shenton et al., 1992). Volume reductions in the hippocampus and amygdala have been associated with emotional behaviors and memory (Dwork, 1997).

The main neurochemical hypothesis of schizophrenia centers on dysfunction in the dopamine system, which is targeted by the majority of antipsychotic drugs available and closely interacts with the endocannabinoid system, as outlined earlier. Alterations are observed in a number of neurotransmitter systems, including glutamate, GABA, and serotonin, suggesting a complicated biochemical basis (Harrison, 1999). In 1992, Kaiya suggested a second messenger imbalance theory of schizophrenia; however, a related system could not be identified that could cause extensive mediation of neurotransmission. This was before the characterization of the endocannabinoid system, and the discovery of the endocannabinoid 2-AG, which may be Kaiya's missing second messenger, as several factors suggest it is likely to be overproduced by platelets in schizophrenia. As 2-AG is a neuromodulatory compound, Pryor hypothesized that there is potential for this to be a key mediator of disease symptomatology, and have a role in response to treatment and control of psychotic relapse (Pryor, 2000). Overall, it is proposed that the cannabinoid system is overactive in schizophrenia (Muller-Vahl & Emrich, 2008). Despite these findings, how the neurochemistry of the endocannabinoid system might influence the development of schizophrenia — as opposed to simply precipitating psychotic episodes — remains speculative. One area of potential relevance where evidence has been accumulating is in the study of early use of THC on neuroanatomical development. In a meta-analysis of 31 individual studies, neuroanatomic changes associated with THC use were associated with brain regions that highly express cannabinoid receptors, particularly in studies with greater doses and earlier onset of use (Lorenzetti, Solowij, & Yucel, 2016). These changes may be associated with deficits in memory, attention, and executive functions and can remain after cessation of use (Broyd, van Hell, Beale, Yucel, & Solowij, 2016). Overall, these studies support the theory that cannabis and cannabinoids affect the endocannabinoid system regulation of the inhibitory circuitry in the prefrontal cortex that is at the center of core cognitive processes (Volk & Lewis, 2016).

Cannabinoid-Induced Psychosis

It is important to clearly distinguish between the ability of cannabis and cannabinoids to cause acute psychotic episodes, and the association relating cannabis use and schizophrenia, a chronic disease that has psychosis as one of its symptoms. Complicating this distinction is strong evidence from clinical laboratory studies that cannabinoids can temporarily intensify psychotic symptoms in individuals with schizophrenia (Sherif, Radhakrishnan, D'Souza, & Ranganathan, 2016).

Hallucinations and psychotic symptoms such as paranoia are common in patients with cannabinoid toxicity (Bossong & Niesink, 2010; Simmons, Cookman, Kang, & Skinner, 2011; Simmons, Skinner, et al., 2011). Abuse of synthetic cannabinoids can also trigger both the onset and exacerbation of recurrent psychosis (Muller et al., 2010). Synthetic cannabinoid intoxication in particular is associated with acute psychosis (Rodgman, Kinzie, & Leimbach, 2011) particularly in those with other risk factors for psychosis (Fattore & Fratta, 2011). This is largely because of the great increase in potency that abused synthetic cannabinoids have over THC (Fattore, 2016). Randomized, blinded human laboratory studies have now shown conclusively that both cannabis and, especially potent synthetic cannabinoids, transiently induce psychotic behaviors (Sherif et al., 2016).

Cannabis Use and Schizophrenia

Multiple lines of evidence have converged to support the hypothesis that the cannabinoid system is involved in causing schizophrenia. The first strong evidence came from a longitudinal study in Sweden, which appeared to show an association between cannabis use and schizophrenia (Andreasson, Allebeck, Engstrom, & Rydberg, 1987). This was consistent with circumstantial evidence based around the high incidence and frequent use of cannabis by people with schizophrenia, to the extent of often meeting criteria for cannabis use disorder (Kovasznay et al., 1997). Schizophrenic drug abusers have been found to often have a significantly lower age of onset of psychotic symptoms and subsequent disease diagnosis, with the onset suggested to be precipitated by cannabis use in those predisposed to the disease. A comprehensive study systematically reviewed previous evidence of a link between cannabis use and occurrence of persistent psychotic or affective symptoms and found an average 41% increase in psychotic outcome in individuals who had used cannabis. The risk was greatly increased for people who used cannabis frequently and those with existing risk of psychotic disorder (Moore et al., 2007).

The role that the endocannabinoid system plays in disease pathology is often ambiguous as many psychiatric conditions have inherently complex characteristics and neurology; the cause or effect role that cannabis plays in schizophrenia is not entirely clear (Di Forti, Morrison, Butt, & Murray, 2007), with substance abuse and psychosis often beginning in the same month (Buhler, Hambrecht, Loffler, An der Heiden, & Hafner, 2002). A study that showed support for cannabis being an independent risk factor for schizophrenia found a six-fold greater incidence of disease development in heavy cannabis users compared to non-users (Andreasson et al., 1987). The prolonged and excessive use of cannabis is also shown to trigger relapses of symptoms in people with schizophrenia, with the intensity of cannabis abuse found to significantly affect the time until psychotic relapse (Linszen, Dingemans, & Lenior, 1994). People with comorbid schizophrenia and substance use disorders exhibit increased anandamide levels compared to controls (Potvin et al., 2008). Behavioral responses to cannabis use have been suggested to show some similarities to the various cognitive impairments seen in schizophrenic psychosis, including symptoms of paranoia, changes in working memory, and attentional disinhibition (Kuperberg & Heckers, 2000; Skosnik, Spatz-Glenn, & Park, 2001). Emrich, Leweke, and Schneider (1997) investigated psychotic disturbances by considering visual illusionary perception and observed similar disturbances of perception in cannabis-intoxicated healthy volunteers and people with schizophrenia, showing an example of comparable cognitive impairment and supporting the hypothesis of a cannabinoid imbalance in the condition.

In a comprehensive review of the many longitudinal and other studies to date Gage, Hickman, and Zammit (2016) concluded, after critically assessing various potential confounds such as bias and reverse causation, that epidemiologic studies provide strong enough evidence to warrant a public health message that "cannabis use can increase the risk of psychotic disorders." Despite this, puzzles remain; such as the lack of any detectable correlation between rising societal use of cannabis and the prevalence of schizophrenia (Degenhardt, Hall, & Lynskey, 2003). Whether there has been an increased prevalence of schizophrenia in the past decade as a result of the rise of synthetic cannabinoid abuse remains to be studied (Gururajan, Manning, Klug, & van den Buuse, 2012).

Despite its apparent contribution to the disease etiology, cannabis use is also proposed to have a therapeutic role in schizophrenia, with the suggestion that schizophrenic patients "self-medicate" for symptomatic relief. It has long been proposed that this self-medication can explain cannabis abuse, providing the patient with anxiety relief and a way to escape the devastating loss of control associated with the disease. Perhaps consistent with this, cannabis-using schizophrenics show an increase in the presentation of positive symptoms, including hallucinations and thought disorders, which can potentially be attributed to a direct dopamine neurotransmission effect, or reduced compliance to medications (Buhler et al., 2002; Skosnik, Park,

Dobbs, & Gardner, 2008). A unique in vivo study characterized dopaminergic function in a schizophrenia case study in which the patient secretly consumed cannabis during a pause in the imaging session for a single-photon emission computerized tomographic study. The drug had an immediate calming effect on the patient, but it worsened psychotic symptoms after several hours. The images demonstrated a 20% decrease in the striatal D2 receptor binding ratio, suggesting increased dopaminergic activity, following cannabis consumption. This activity may contribute to the psychotogenic effects of cannabis use in vulnerable people (Voruganti, Slomka, Zabel, Mattar, & Awad, 2001). Initial data from a study using a novel PET tracer for CB1 itself in patients with schizophrenia suggests elevated binding in some brain regions, with further studies warranted (Wong et al., 2010).

Cannabinoid Treatment for Schizophrenia

Aside from the anecdotal reports, cannabinoids have been administered to humans to evaluate the effects of these compounds, and in controlled attempts at symptomatic control in schizophrenia. Intravenous administration of THC to healthy individuals produces schizophrenia-like positive and negative symptoms as well as altered perception and increased anxiety and euphoria, indicating that cannabinoids lead to behaviors resembling psychoses and may contribute to the pathology of psychiatric conditions (D'Souza et al., 2004; Morrison et al., 2009). Chronic but not acute cannabis use also results in a schizophrenia like disruption in prepulse inhibition of acoustic startle response (PPI, a measure of sensorimotor gating) in healthy individuals (Kedzior & Martin-Iverson, 2006), and cannabis users tend to display neural synchronization deficits and increased schizotypal personality characteristics (Skosnik, Krishnan, Aydt, Kuhlenshmidt, & O'Donnell, 2006). In patients with antipsychotic-treated schizophrenia, intravenous THC transiently increased positive and negative symptoms, and increased cognitive deficits in learning, recall, and perception to a greater degree than control subjects (D'Souza et al., 2005). In contrast, a small compassionate use study of dronabinol (synthetic THC) described improvements in patients with treatment–refractory chronic schizophrenia who self-reported cannabis use, adding to the complex role of cannabinoids in psychosis (Schwarcz, Karajgi, & McCarthy, 2009).

CBD and the CB1 inverse agonist SR141716 have both been proposed as potentially valuable therapeutics (Roser, Vollenweider, & Kawohl, 2008); however, the evidence to date is inconsistent. Early preclinical and clinical evidence suggested that CBD demonstrated antipsychotic effects (Zuardi, Crippa, Hallak, Moreira, & Guimaraes, 2006). However, although well tolerated, a monotherapy clinical trial of CBD in three patients with treatment-resistant schizophrenia found only mild improvement in a single participant, suggesting the compounds may not be effective in this population (Zuardi, Hallak, et al., 2006). SR141716 was tested in adults with schizophrenia or schizoaffective disorder, and was found to produce no difference on any outcome measure compared to placebo, although the authors proposed further study of cannabinoid mechanisms to treat schizophrenia (Meltzer, Arvanitis, Bauer, & Rein, 2004). Despite the many clues linking schizophrenia and the endocannabinoid system, the search for cannabinoid-based treatments remains elusive. Leweke, Mueller, Lange, and Rohleder (2016) concluded that although there exists some evidence that modulating the endocannabinoid system may be of help in acute psychotic episodes in schizophrenia, to date sample sizes of patient groups have been insufficient to make a decisive judgment. Furthermore, there have been no longer term studies or adequate evaluations of efficacy and safety (Leweke et al., 2016).

Genetic Factors

Multiple chromosomes have been proposed to contain loci relevant to schizophrenia vulnerability. Catechol-O-methyltransferase (COMT) metabolizes dopamine in the brain and the V(108/158)M mutation of this gene reduces dopamine breakdown, possibly increasing the risk of psychosis. The $Val^{158}Met$ genotype reduces the effects of THC on psychosis, whereas the $Val^{158}Val$ genotype increases the effects of THC on cognition. It has been suggested that COMT mutational status increases the risk of developing schizophrenia following cannabis use. In a longitudinal study, carriers of the COMT $Val^{158}Val$ allele were most likely to develop schizophreniform disorder with cannabis use. This effect may be particularly important in people who begin using cannabis early in life (Caspi et al., 2005).

In addition, mutations in the DAT1 and AKT1 genes are associated with reductions in dopamine transporter and glycogen synthase kinase activity, respectively, both of which increase sensitivity to THC-induced psychosis (Sherif et al., 2016). The CNR1 gene encodes CB1 (Matsuda, Lolait, Brownstein, Young, & Bonner, 1990), and a locus on the chromosome containing the CB1-coding region was found to confer increased susceptibility to schizophrenia (Cao et al., 1997). An additional genetic connection has been explored in the first exon of the CB1 gene in patients with schizophrenia, with the non-substance-abusing subtype of schizophrenia exhibiting a lack of a specific genotype (Leroy et al., 2001). This provided the first support for CB1 genetic variants to be associated with a differential risk for substance abuse in schizophrenia (Leroy et al., 2001). Furthermore, an AAT triplet repeat polymorphism in the 3′ flanking region was found to be significantly associated with the hebephrenic subtype of schizophrenia in several distinct ethnic populations (Chavarria-Siles et al., 2008; Martinez-Gras et al., 2006; Ujike et al., 2002). This type of schizophrenia involves a symptom profile similar to chronic cannabinoid-induced psychosis. Several other studies have evaluated the CNR1 gene in normal and specific population samples and have found limited or no variations associated with schizophrenia (Fritzsche, 2001; Hamdani et al., 2008; Ho, Wassink, Ziebell, & Andreasen, 2011; Seifert, Ossege, Emrich, Schneider, & Stuhrmann, 2007; Zammit et al., 2007). Interestingly, CB1 knockout mice have been proposed as a potential model for schizophrenia, as they display a number of behavioral alterations that represent schizophrenia symptoms (Fritzsche, 2001).

An association of schizophrenia risk with variations in major histocompatibility complex (MHC) alleles has appeared to be explained by variations in expression of the C4A and C4B alleles in the brain (Sekar et al., 2016). Interaction between the MHC and endocannabinoid systems is poorly studied. However, there is evidence that THC may reduce the risk of graft-versus-host disease, which is caused by MHC antigen mismatches. Whether this is relevant to the cause and treatment of schizophrenia remains to be investigated.

Endocannabinoids and Schizophrenia

Further molecular evidence comes from studies reporting an elevation in anandamide levels in the CSF of people with schizophrenia as well as prodromal states of psychosis, in comparison to non-schizophrenic controls (Koethe et al., 2009; Leweke, Giuffrida, Wurster, Emrich, & Piomelli, 1999). A more in-depth study found anandamide to be eightfold higher in patients with antipsychotic-naive first-episode paranoid schizophrenia, and these levels negatively correlated with psychotic symptoms (Giuffrida et al., 2004). This change in anandamide levels was not seen in patients with schizophrenia treated with typical antipsychotics (Giuffrida et al., 2004). This may demonstrate an increased tonic cannabinoid state occurring in schizophrenia, which could contribute to other neurotransmitter alterations in the disease. Low-frequency cannabis-using patients with schizophrenia exhibit >10-fold increased CSF anandamide levels than either healthy low- or high-frequency cannabis users. By comparison, high-frequency cannabis-using patients with schizophrenia had CSF anandamide downregulated to levels more similar to healthy controls, suggesting a means through which self-medication may be acting. In this study, no differences were observed in serum anandamide levels between any groups (Leweke et al., 2007). In contrast, an earlier study demonstrated significantly increased blood anandamide levels in patients with schizophrenia, which reduced when the patients were in clinical remission. Lowered mRNA transcripts for CB2 and FAAH were also seen in these blood samples, perhaps indicative of an immune component to the disease (De Marchi et al., 2003). The relationship between CSF and blood levels of anandamide, and that synthesized on demand in the synapse, remains to be established. Nevertheless, these results provide verification of endocannabinoid system alterations in schizophrenia and its potential to be manipulated.

Postmortem Schizophrenia Analysis

Postmortem studies have provided a range of conflicting data, most likely because of the complex cause of the disease, as well as a difficulty in controlling for both prescribed medicines and illicit drug use. Increased CB1 binding was found in the dorsolateral prefrontal cortex (Dean, Sundram, Bradbury, Scarr, & Copolov, 2001), the anterior cingulate cortex (Zavitsanou, Garrick, & Huang, 2004), and the superficial layers of the posterior cingulate cortex (Newell, Deng, & Huang, 2006). No changes were observed in the superior temporal gyrus (Deng, Han, & Huang, 2007). In contrast, an immunohistochemical analysis reported no change in the density of CB1-immunopositive cells in the anterior cingulate cortex from postmortem schizophrenic tissue (Koethe et al., 2007). Studies in the dorsolateral prefrontal cortex have found significant decreases in both CB1 mRNA and protein expression (Eggan, Hashimoto, & Lewis, 2008; Eggan, Stoyak, Verrico, & Lewis, 2010), whereas another study described CB1 downregulation in this region only following antipsychotic treatment (Uriguen et al., 2009).

Dean et al. later investigated presynaptic markers on dopaminergic neurons in postmortem caudate tissue from people with schizophrenia. Levels of autoradiographic binding to the dopamine transporter were decreased in patients who had no detectable blood THC upon autopsy, but there was no significant difference in patients who did have detectable THC, suggesting cannabis consumption to reverse a key neurochemical alteration in schizophrenia. There were no recorded differences in the catalyzing enzyme tyrosine hydroxylase (Dean, Bradbury, & Copolov, 2003).

Insight From Animal Studies

Neurological dysfunctions in psychosis are thought to be the result of impairments in the equilibrium of interactions between a number of different neurotransmitters with which the cannabinoid system interacts (Fakhoury, 2016). Cannabinoid signaling, which is commonly described as neuromodulatory, has been extensively studied in relationship to dopamine. Colocalization of CB1 with D1 dopamine receptors in the substantia nigra and globus pallidus internus, and D2 dopamine receptors in the globus pallidus externus, has been described (Le Moine & Bloch, 1995). Administration of the D2 agonist quinpirole has been found to stimulate the release of anandamide in the rat dorsal striatum (Giuffrida et al., 1999), and dopamine neurons release 2-AG, influencing afferent activity and their own firing pattern (Melis et al., 2004). Thus a direct consequence of the increased dopaminergic activity proposed in the dopamine hypothesis of schizophrenia might be a self-regulatory increase in endocannabinoid release. Functional interactions have been characterized whereby D2 receptors share the same inhibitory second messenger signaling pathway as CB1 (McAllister & Glass, 2002), and concurrent activation of striatal CB1 and D2 receptors, which are both normally coupled to Gi/Go, leads to a paradoxical Gs-linked stimulation of cAMP, implying a complex crosstalk between these two receptors (Glass & Felder, 1997), possibly as a consequence of functional dimerization (Kearn, Blake-Palmer, Daniel, Mackie, & Glass, 2005).

In vivo behavioral studies further outline potential cannabinoid involvement in dopamine neurotransmission; rat studies analyzing turning behavior having shown cannabinoid agonists to oppose the action of dopaminergic drugs (Sanudo-Pena, Patrick, Patrick, & Walker, 1996). THC increases dopamine release and utilization in the prefrontal cortex in a way similar to NMDA receptor antagonist phencyclidine (PCP), a compound used to produce a social withdrawal model of schizophrenia in rodents. This dopamine system activation is blocked by alpha-noradrenergic receptor agonists (Jentsch, Wise, Katz, & Roth, 1998). In contrast, long-term exposure to THC reduces dopamine transmission in the rat medial prefrontal cortex, with no change observed in striatal regions, suggesting that cognitive deficits are linked to cannabinoid signaling (Jentsch, Verrico, Le, & Roth, 1998). While inducing a schizophrenia-like syndrome in rats, PCP treatment also causes changes in CB1 function and increases levels of 2-AG in the prefrontal cortex (Vigano et al., 2008). Cannabinoid antagonist or FAAH inhibitor treatment is able to significantly counteract the PCP model, improving deficits in PPI independently and improving recognition memory, in a manner comparable to atypical antipsychotics; however, some adverse effects were apparent in control animals not treated with PCP (Ballmaier et al., 2007; Guidali et al., 2011; Seillier, Advani, Cassano, Hensler, & Giuffrida, 2010). Studies in CB1 receptor knockout mice show that disruption of the CNR1 gene significantly alters behavioral effects of PCP, further reducing locomotion, and increasing ataxia and stereotypy, implicating CB1 in both negative and positive symptoms of this model (Haller, Szirmai, Varga, Ledent, & Freund, 2005). An interesting study investigated voluntary cannabinoid consumption in rats treated with PCP, where cannabinoid agonist WIN55,212-2 led to attenuation of a range of PCP-induced behavioral deficits (Spano, Fadda, Frau, Fattore, & Fratta, 2010).

One rat model of schizophrenia uses maternal deprivation to induce specific symptoms of the disease mimicking the putative neurodevelopmental origin of schizophrenia. Maternal deprivation associates with dysregulation of the cannabinoid system, including CB1 downregulation and CB2 upregulation, in regionally selective ways, within the hippocampal structure (Suarez et al., 2009, 2010). When treated with inhibitors of endocannabinoid inactivation, the endocrine and cellular effects of maternal deprivation were reversed or attenuated in hippocampal and cerebellar regions (Llorente et al., 2008; Lopez-Gallardo et al., 2008).

Rearing of rodents in isolation also produces deficits similar to those seen in schizophrenia, including changes in components of the endocannabinoid system (Robinson, Loiacono, Christopoulos, Sexton, & Malone, 2010). THC appears to worsen sensorimotor gating in these already impaired animals in a receptor-dependent manner (Malone & Taylor, 2006). Chronic (but not acute) treatment with CB1 inverse agonist AM251 rescued cognitive impairments and CB1 functionality resulting from isolation rearing (Zamberletti, Vigano, Guidali, Rubino, & Parolaro, 2010). Social isolation also leads to significant decreases in CB1 receptor and increases in FAAH in the caudate putamen, but no changes in D2 receptors, suggesting the endocannabinoid system to be involved in this model of psychosis (Malone, Kearn, Chongue, Mackie, & Taylor, 2008).

Two gene knockout mice have been used to further explore the relationship between cannabinoids and schizophrenia. Dopamine transporter knockout mice, which model hyperdopaminergia similar to that seen in schizophrenia, exhibit dysfunction of the cannabinoid system with reduced anandamide levels. Inhibitors of endocannabinoid transport or degradation enhance endocannabinoid levels, reducing the spontaneous hyperlocomotion in these mice. This effect is prevented by antagonism of TRPV1, implicating this pathway (Tzavara et al., 2006). The nonpsychoactive cannabinoid CBD also reportedly exhibits antipsychotic activity, reversing disruptions in PPI, induced by an NMDA receptor antagonist, in a similar way to the antipsychotic clozapine, with evidence suggesting that this is also via a TRPV1 receptor mechanism (Long, Malone, & Taylor, 2006). Neuregulin-1 is a susceptibility gene for schizophrenia. Heterozygous neuregulin-1 knockout mice are more sensitive to the acute effects of THC treatment in various behavioral paradigms modeling symptoms of schizophrenia (Boucher, Arnold, et al., 2007); they express increased c-Fos in response to THC in brain regions important to schizophrenia (Boucher, Hunt, et al., 2007) and develop tolerance to CB1 agonist CP55,940 differentially (dependent on parameter measured) (Boucher et al., 2010). Studies in CB2 knockout mice have described schizophrenia-like behavioral and molecular features, identifying this receptor as a further potential cannabinoid system target in the disorder (Ortega-Alvaro, Aracil-Fernandez, Garcia-Gutierrez, Navarrete, & Manzanares, 2011).

Animal models have helped understand schizophrenia and the risk associated with cannabis use in adolescence, with a variety of rodent paradigms designed to model adolescent cannabis exposure. In these models animals have persistent deficits in learning and memory as well as ongoing mood changes. Taken together, this work suggests that the maturation of the endocannabinoid system in the adolescent brain is influenced by cannabinoid exposure, and that this may dysregulate the development of pivotal neurotransmitter systems in the cortex, in a way that resembles some aspects of schizophrenia (Rubino & Parolaro, 2016).

Also relevant are rodent paradigms that model psychotomimetic-induced hyperactivity and latent inhibition deficit models. In these models, CB1 antagonist AVE1625 improved memory and some cognitive deficits. When coadministered with antipsychotic drugs to positive symptom models, AVE1625 decreased antipsychotic-associated catalepsy and weight gain, suggesting a potential cotherapeutic strategy (Black et al., 2010).

Although largely the involvement of the endocannabinoid system remains contentious, overactivity is suggested in the pathology of schizophrenia. With cannabis use commonly linked to both disease triggering in susceptible individuals as well as for self-medication, and evidence of elevations in endocannabinoid levels and receptors in specific brain regions, this is clearly an important neuromodulatory system worthy of further investigation in the condition, particularly with respect to cannabinoid exposure in adolescence. The rise of potent synthetic cannabinoids as drugs of abuse makes this particularly pertinent.

CONCLUSIONS

This review demonstrates that a wealth of important information on normal and pathophysiological function of the endocannabinoid system has been gained from studying a range of human diseases and disease models. While many questions remain unanswered, it is clear that the endocannabinoid system is engaged in a complex interplay with other neurotransmitters, and that many cannabinoid compounds act at additional noncannabinoid targets. Animal models, although useful tools, often model only certain aspects of a disease and so it can be difficult to pull together findings from a range of models into a cohesive picture of cannabinoid function. Similarly, direct studies on human brain tissue often provide only a snapshot of end-stage disease and therefore present only limited opportunity to determine mechanism and progression. While acknowledging these limitations, current evidence clearly shows that the endocannabinoid system is critical to normal brain function and holds immense potential for the treatment of a range of disorders.

REFERENCES

Abood, M. E., Rizvi, G., Sallapudi, N., & McAllister, S. D. (2001). Activation of the CB1 cannabinoid receptor protects cultured mouse spinal neurons against excitotoxicity. *Neuroscience Letters, 309*(3), 197–201.

Aguado, T., Monory, K., Palazuelos, J., Stella, N., Cravatt, B., Lutz, B., ... Galve-Roperh, I. (2005). The endocannabinoid system drives neural progenitor proliferation. *FASEB Journal: Official Publication of the Federation of American Societies for Experimental Biology, 19*(12), 1704–1706.

Ahmad, R., Goffin, K., Van den Stock, J., De Winter, F. L., Cleeren, E., Bormans, G., ... Vandenbulcke, M. (2014). In vivo type 1 cannabinoid receptor availability in Alzheimer's disease. *European Neuropsychopharmacology: The Journal of the European College of Neuropsychopharmacology, 24*(2), 242–250.

Allen, K. L., Waldvogel, H. J., Glass, M., & Faull, R. L. (2009). Cannabinoid (CB₁), GABA_A and GABA_B receptor subunit changes in the globus pallidus in Huntington's disease. *Journal of Chemical Neuroanatomy, 37*(4), 266–281.

Alzheimer's Association. (2015). 2015 Alzheimer's disease facts and figures. *Alzheimer's & Dementia: The Journal of the Alzheimer's Association, 11*(3), 332–384.

Andreasson, S., Allebeck, P., Engstrom, A., & Rydberg, U. (1987). Cannabis and schizophrenia: A longitudinal study of Swedish conscripts. *The Lancet, 2*, 1483–1486.

Arevalo-Martin, A., Vela, J. M., Molina-Holgado, E., Borrell, J., & Guaza, C. (2003). Therapeutic action of cannabinoids in a murine model of multiple sclerosis. *The Journal of Neuroscience: The Official Journal of the Society for Neuroscience, 23*(7), 2511–2516.

Arroyo, R., Vila, C., & Dechant, K. L. (2014). Impact of Sativex® on quality of life and activities of daily living in patients with multiple sclerosis spasticity. *Journal of Comparative Effectiveness Research, 3*(4), 435–444.

Ashton, J. C., Friberg, D., Darlington, C. L., & Smith, P. F. (2006). Expression of the cannabinoid CB2 receptor in the rat cerebellum: An immunohistochemical study. *Neuroscience Letters, 396*(2), 113–116.

Aso, E., Andres-Benito, P., Carmona, M., Maldonado, R., & Ferrer, I. (2016). Cannabinoid receptor 2 participates in amyloid-beta processing in a mouse model of Alzheimer's disease but plays a minor role in the therapeutic properties of a cannabis-based medicine. *Journal of Alzheimer's Disease.*

Aso, E., Sanchez-Pla, A., Vegas-Lozano, E., Maldonado, R., & Ferrer, I. (2015). Cannabis-based medicine reduces multiple pathological processes in AβPP/PS1 mice. *Journal of Alzheimer's Disease, 43*(3), 977–991.

Atwood, B. K., & Mackie, K. (2010). CB₂: A cannabinoid receptor with an identity crisis. *British Journal of Pharmacology, 160*(3), 467–479.

Avraham, H. K., Jiang, S., Fu, Y., Rockenstein, E., Makriyannis, A., Zvonok, A., … Avraham, S. (2014). The cannabinoid CB₂ receptor agonist AM1241 enhances neurogenesis in GFAP/Gp120 transgenic mice displaying deficits in neurogenesis. *British Journal of Pharmacology, 171*(2), 468–479.

Baker, D., Gerritsen, W., Rundle, J., & Amor, S. (2011). Critical appraisal of animal models of multiple sclerosis. *Multiple Sclerosis: Clinical and Laboratory Research.*

Baker, D., Pryce, G., Croxford, J. L., Brown, P., Pertwee, R. G., Huffman, J. W., & Layward, L. (2000). Cannabinoids control spasticity and tremor in a multiple sclerosis model. *Nature, 404*(6773), 84–87.

Baker, D., Pryce, G., Croxford, J. L., Brown, P., Pertwee, R. G., Makriyannis, A., … Di Marzo, V. (2001). Endocannabinoids control spasticity in a multiple sclerosis model. *FASEB Journal: Official Publication of the Federation of American Societies for Experimental Biology, 15*(2), 300–302.

Ball, S., Vickery, J., Hobart, J., Wright, D., Green, C., Shearer, J., … Zajicek, J. (2015). The Cannabinoid Use in Progressive Inflammatory brain Disease (CUPID) trial: A randomised double-blind placebo-controlled parallel-group multicentre trial and economic evaluation of cannabinoids to slow progression in multiple sclerosis. *Health Technology Assessment, 19*(12), vii-viii, xxv-xxxi, 1–187.

Ballmaier, M., Bortolato, M., Rizzetti, C., Zoli, M., Gessa, G., Heinz, A., & Spano, P. (2007). Cannabinoid receptor antagonists counteract sensorimotor gating deficits in the phencyclidine model of psychosis. *Neuropsychopharmacology: Official Publication of the American College of Neuropsychopharmacology, 32*(10), 2098–2107.

Battista, N., Bari, M., Tarditi, A., Mariotti, C., Bachoud-Levi, A. C., Zuccato, C., … Maccarrone, M. (2007). Severe deficiency of the fatty acid amide hydrolase (FAAH) activity segregates with the Huntington's disease mutation in peripheral lymphocytes. *Neurobiology of Disease, 27*(1), 108–116.

Bekris, L. M., Yu, C. E., Bird, T. D., & Tsuang, D. W. (2010). Genetics of Alzheimer disease. *Journal of Geriatric Psychiatry and Neurology, 23*(4), 213–227.

Benito, C., Nunez, E., Tolon, R. M., Carrier, E. J., Rabano, A., Hillard, C. J., & Romero, J. (2003). Cannabinoid CB₂ receptors and fatty acid amide hydrolase are selectively overexpressed in neuritic plaque-associated glia in Alzheimer's disease brains. *The Journal of Neuroscience: The Official Journal of the Society for Neuroscience, 23*(35), 11136–11141.

Benito, C., Romero, J. P., Tolon, R. M., Clemente, D., Docagne, F., Hillard, C. J., ... Romero, J. (2007). Cannabinoid CB$_1$ and CB$_2$ receptors and fatty acid amide hydrolase are specific markers of plaque cell subtypes in human multiple sclerosis. *The Journal of Neuroscience: The Official Journal of the Society for Neuroscience, 27*(9), 2396–2402.

Berrendero, F., Sanchez, A., Cabranes, A., Puerta, C., Ramos, J. A., Garcia-Merino, A., & Fernandez-Ruiz, J. (2001). Changes in cannabinoid CB$_1$ receptors in striatal and cortical regions of rats with experimental allergic encephalomyelitis, an animal model of multiple sclerosis. *Synapse, 41*(3), 195–202.

Bisogno, T., & Di Marzo, V. (2010). Cannabinoid receptors and endocannabinoids: Role in neuroinflammatory and neurodegenerative disorders. *CNS & Neurological Disorders Drug Targets, 9*(5), 564–573.

Black, M. D., Stevens, R. J., Rogacki, N., Featherstone, R. E., Senyah, Y., Giardino, O., ... Varty, G. B. (2010). AVE1625, a cannabinoid CB$_1$ receptor antagonist, as a co-treatment with antipsychotics for schizophrenia: Improvement in cognitive function and reduction of antipsychotic-side effects in rodents. *Psychopharmacology.*

Blazquez, C., Chiarlone, A., Sagredo, O., Aguado, T., Pazos, M. R., Resel, E., ... Guzman, M. (2011). Loss of striatal type 1 cannabinoid receptors is a key pathogenic factor in Huntington's disease. *Brain: A Journal of Neurology, 134*(Pt 1), 119–136.

Borner, C., Smida, M., Hollt, V., Schraven, B., & Kraus, J. (2009). Cannabinoid receptor type 1- and 2-mediated increase in cyclic AMP inhibits T cell receptor-triggered signaling. *The Journal of Biological Chemistry, 284*(51), 35450–35460.

Bossong, M. G., & Niesink, R. J. (2010). Adolescent brain maturation, the endogenous cannabinoid system and the neurobiology of cannabis-induced schizophrenia. *Progress in Neurobiology, 92*(3), 370–385.

Bouchard, J., Truong, J., Bouchard, K., Dunkelberger, D., Desrayaud, S., Moussaoui, S., ... Muchowski, P. J. (2012). Cannabinoid receptor 2 signaling in peripheral immune cells modulates disease onset and severity in mouse models of Huntington's disease. *The Journal of Neuroscience: The Official Journal of the Society for Neuroscience, 32*(50), 18259–18268.

Boucher, A. A., Arnold, J. C., Duffy, L., Schofield, P. R., Micheau, J., & Karl, T. (2007). Heterozygous neuregulin 1 mice are more sensitive to the behavioural effects of Δ^9-tetrahydrocannabinol. *Psychopharmacology, 192*(3), 325–336.

Boucher, A. A., Hunt, G. E., Karl, T., Micheau, J., McGregor, I. S., & Arnold, J. C. (2007). Heterozygous neuregulin 1 mice display greater baseline and Δ^9-tetrahydrocannabinol-induced c-Fos expression. *Neuroscience, 149*(4), 861–870.

Boucher, A. A., Hunt, G. E., Micheau, J., Huang, X., McGregor, I. S., Karl, T., & Arnold, J. C. (2010). The schizophrenia susceptibility gene neuregulin 1 modulates tolerance to the effects of cannabinoids. *International Journal of Neuropsychopharmacology*, 1–13.

Brady, C. M., DasGupta, R., Dalton, C., Wiseman, O. J., Berkley, K. J., & Fowler, C. J. (2004). An open-label pilot study of cannabis-based extracts for bladder dysfunction in advanced multiple sclerosis. *Multiple Sclerosis: Clinical and Laboratory Research, 10*(4), 425–433.

Brenowitz, S. D., Best, A. R., & Regehr, W. G. (2006). Sustained elevation of dendritic calcium evokes widespread endocannabinoid release and suppression of synapses onto cerebellar Purkinje cells. *The Journal of Neuroscience: The Official Journal of the Society for Neuroscience, 26*(25), 6841–6850.

Brooks, J. W., Pryce, G., Bisogno, T., Jaggar, S. I., Hankey, D. J., Brown, P., ... Baker, D. (2002). Arvanil-induced inhibition of spasticity and persistent pain: Evidence for therapeutic sites of action different from the vanilloid VR1 receptor and cannabinoid CB$_1$/CB$_2$ receptors. *European Journal of Pharmacology, 439*(1–3), 83–92.

Broyd, S. J., van Hell, H. H., Beale, C., Yucel, M., & Solowij, N. (2016). Acute and chronic effects of cannabinoids on human cognition-a systematic review. *Biological Psychiatry, 79*(7), 557–567.

Buhler, B., Hambrecht, M., Loffler, W., An der Heiden, W., & Hafner, H. (2002). Precipitation and determination of the onset and course of schizophrenia by substance abuse - a retrospective and prospective study of 232 population-based first illness episodes. *Schizophrenia Research, 54*(3), 243–251.

Cabranes, A.,Venderova, K., de Lago, E., Fezza, F., Sanchez, A., Mestre, L., ... Fernandez-Ruiz, J. (2005). Decreased endocannabinoid levels in the brain and beneficial effects of agents activating cannabinoid and/or vanilloid receptors in a rat model of multiple sclerosis. *Neurobiology of Disease, 20*(2), 207–217.

Campos, A. C., Ortega, Z., Palazuelos, J., Fogaca, M. V., Aguiar, D. C., Diaz-Alonso, J., ... Guimaraes, F. S. (2013).The anxiolytic effect of cannabidiol on chronically stressed mice depends on hippocampal neurogenesis: Involvement of the endocannabinoid system. *International Journal of Neuropsychopharmacology, 16*(6), 1407–1419.

Cao, Q., Martinez, M., Zhang, J., Sanders, A.R. , Badner, J.A. , Cravchik, A., ... Gejman, P.V. (1997). Suggestive evidence for a schizophrenia susceptibility locus on chromosome 6q and a confirmation in an independent series of pedigrees. *Genomics, 43*(1), 1–8.

Cao, X., Liang, L., Hadcock, J. R., Iredale, P. A., Griffith, D. A., Menniti, F. S., ... Papa, S. M. (2007). Blockade of cannabinoid type 1 receptors augments the antiparkinsonian action of levodopa without affecting dyskinesias in 1-methyl-4-phenyl-1,2,3,6-tetrahydropyridine-treated rhesus monkeys. *The Journal of Pharmacology and Experimental Therapeutics, 323*(1), 318–326.

Carletti, F., Gambino, G., Rizzo,V., Ferraro, G., & Sardo, P. (2016). Involvement of TRPV1 channels in the activity of the cannabinoid WIN 55,212-2 in an acute rat model of temporal lobe epilepsy. *Epilepsy Research, 122,* 56–65.

Carroll, C. B., Bain, P. G., Teare, L., Liu, X., Joint, C., Wroath, C., ... Zajicek, J. P. (2004). Cannabis for dyskinesia in Parkinson disease: A randomized double-blind crossover study. *Neurology, 63*(7), 1245–1250.

Caspi, A., Moffitt, T. E., Cannon, M., McClay, J., Murray, R., Harrington, H., ... Craig, I. W. (2005). Moderation of the effect of adolescent-onset cannabis use on adult psychosis by a functional polymorphism in the catechol-O-methyltransferase gene: Longitudinal evidence of a gene X environment interaction. *Biological Psychiatry, 57*(10), 1117–1127.

Casteels, C., Lauwers, E., Baitar, A., Bormans, G., Baekelandt, V., & Van Laere, K. (2010). In vivo type 1 cannabinoid receptor mapping in the 6-hydroxydopamine lesion rat model of Parkinson's disease. *Brain Research, 1316,* 153–162.

Centonze, D., Bari, M., Rossi, S., Prosperetti, C., Furlan, R., Fezza, F., ... Maccarrone, M. (2007).The endocannabinoid system is dysregulated in multiple sclerosis and in experimental autoimmune encephalomyelitis. *Brain: A Journal of Neurology, 130*(Pt 10), 2543–2553.

Centonze, D., Finazzi-Agro, A., Bernardi, G., & Maccarrone, M. (2007).The endocannabinoid system in targeting inflammatory neurodegenerative diseases. *Trends in Pharmacological Sciences, 28*(4), 180–187.

Centonze, D., Rossi, S., Prosperetti, C., Tscherter, A., Bernardi, G., Maccarrone, M., & Calabresi, P. (2005). Abnormal sensitivity to cannabinoid receptor stimulation might contribute to altered gamma-aminobutyric acid transmission in the striatum of R6/2 Huntington's disease mice. *Biological Psychiatry, 57*(12), 1583–1589.

Cepeda, C., Hurst, R. S., Calvert, C. R., Hernandez-Echeagaray, E., Nguyen, O. K., Jocoy, E., ... Levine, M. S. (2003). Transient and progressive electrophysiological alterations in the corticostriatal pathway in a mouse model of Huntington's disease. *The Journal of Neuroscience: The Official Journal of the Society for Neuroscience, 23*(3), 961–969.

Cepeda, C., Starling, A. J., Wu, N., Nguyen, O. K., Uzgil, B., Soda, T., … Levine, M. S. (2004). Increased GABAergic function in mouse models of Huntington's disease: Reversal by BDNF. *Journal of Neuroscience Research, 78*(6), 855–867.

Chagas, M. H., Eckeli, A. L., Zuardi, A. W., Pena-Pereira, M. A., Sobreira-Neto, M. A., Sobreira, E. T., … Crippa, J. A. (2014). Cannabidiol can improve complex sleep-related behaviours associated with rapid eye movement sleep behaviour disorder in Parkinson's disease patients: A case series. *Journal of Clinical Pharmacy and Therapeutics, 39*(5), 564–566.

Chagas, M. H., Zuardi, A. W., Tumas, V., Pena-Pereira, M. A., Sobreira, E. T., Bergamaschi, M. M., … Crippa, J. A. (2014). Effects of cannabidiol in the treatment of patients with Parkinson's disease: An exploratory double-blind trial. *Journal of Psychopharmacology, 28*(11), 1088–1098.

Chavarria-Siles, I., Contreras-Rojas, J., Hare, E., Walss-Bass, C., Quezada, P., Dassori, A., … Escamilla, M. A. (2008). Cannabinoid receptor 1 gene (CNR1) and susceptibility to a quantitative phenotype for hebephrenic schizophrenia. *American Journal of Medical Genetics. Part B, Neuropsychiatric Genetics: The Official Publication of the International Society of Psychiatric Genetics, 147*(3), 279–284.

Chen, B., Bromley-Brits, K., He, G., Cai, F., Zhang, X., & Song, W. (2010). Effect of synthetic cannabinoid HU210 on memory deficits and neuropathology in Alzheimer's disease mouse model. *Current Alzheimer Research, 7*(3), 255–261.

Chiarlone, A., Bellocchio, L., Blazquez, C., Resel, E., Soria-Gomez, E., Cannich, A., … Guzman, M. (2014). A restricted population of CB1 cannabinoid receptors with neuroprotective activity. *Proceedings of the National Academy of Sciences of the United States of America, 111*(22), 8257–8262.

Choi, I. Y., Ju, C., Anthony Jalin, A. M., Lee da, I., Prather, P. L., & Kim, W. K. (2013). Activation of cannabinoid CB2 receptor-mediated AMPK/CREB pathway reduces cerebral ischemic injury. *The American Journal of Pathology, 182*(3), 928–939.

Chong, M. S., Wolff, K., Wise, K., Tanton, C., Winstock, A., & Silber, E. (2006). Cannabis use in patients with multiple sclerosis. *Multiple Sclerosis: Clinical and Laboratory Research, 12*(5), 646–651.

Clifford, D. B. (1983). Tetrahydrocannabinol for tremor in multiple sclerosis. *Annals of Neurology, 13*(6), 669–671.

Collin, C., Davies, P., Mutiboko, I. K., & Ratcliffe, S. (2007). Randomized controlled trial of cannabis-based medicine in spasticity caused by multiple sclerosis. *European Journal of Neurology: The Official Journal of the European Federation of Neurological Societies, 14*(3), 290–296.

Collin, C., Ehler, E., Waberzinek, G., Alsindi, Z., Davies, P., Powell, K., … Ambler, Z. (2010). A double-blind, randomized, placebo-controlled, parallel-group study of Sativex, in subjects with symptoms of spasticity due to multiple sclerosis. *Neurological Research, 32*(5), 451–459.

Compagnucci, C., Di Siena, S., Bustamante, M. B., Di Giacomo, D., Di Tommaso, M., Maccarrone, M., … Sette, C. (2013). Type-1 (CB$_1$) cannabinoid receptor promotes neuronal differentiation and maturation of neural stem cells. *PLoS One, 8*(1), e54271.

Compston, A., & Coles, A. (2002). Multiple sclerosis. *Lancet, 359*(9313), 1221–1231.

Consroe, P., Laguna, J., Allender, J., Snider, S., Stern, L., Sandyk, R., … Schram, K. (1991). Controlled clinical trial of cannabidiol in Huntington's disease. *Pharmacology, Biochemistry & Behavior, 40*(3), 701–708.

Consroe, P., Musty, R., Rein, J., Tillery, W., & Pertwee, R. (1997). The perceived effects of smoked cannabis on patients with multiple sclerosis. *European Neurology, 38*(1), 44–48.

Conte, A., Bettolo, C. M., Onesti, E., Frasca, V., Iacovelli, E., Gilio, F., … Inghilleri, M. (2008). Cannabinoid-induced effects on the nociceptive system: A neurophysiological study in patients with secondary progressive multiple sclerosis. *European Journal of Pain*.

Croxford, J. L., & Miller, S. D. (2003). Immunoregulation of a viral model of multiple sclerosis using the synthetic cannabinoid R+WIN55,212. *Journal of Clinical Investigation, 111*(8), 1231–1240.

Curtis, A., Mitchell, I., Patel, S., Ives, N., & Rickards, H. (2009). A pilot study using nabilone for symptomatic treatment in Huntington's disease. *Movement Disorders: Official Journal of the Movement Disorder Society, 24*(15), 2254–2259.

Curtis, A., & Rickards, H. (2006). Nabilone could treat chorea and irritability in Huntington's disease. *The Journal of Neuropsychiatry and Clinical Neurosciences, 18*(4), 553–554.

Curtis, M.A., Faull, R.L., & Glass, M. (2006). A novel population of progenitor cells expressing cannabinoid receptors in the subependymal layer of the adult normal and Huntington's disease human brain. *Journal of Chemical Neuroanatomy, 31*(3), 210–215.

D'Souza, D. C., Abi-Saab, W. M., Madonick, S., Forselius-Bielen, K., Doersch, A., Braley, G., … Krystal, J. H. (2005). Delta-9-tetrahydrocannabinol effects in schizophrenia: Implications for cognition, psychosis, and addiction. *Biological Psychiatry, 57*(6), 594–608.

D'Souza, D. C., Perry, E., MacDougall, L., Ammerman, Y., Cooper, T., Wu, Y. T., … Krystal, J. H. (2004). The psychotomimetic effects of intravenous delta-9-tetrahydrocannabinol in healthy individuals: Implications for psychosis. *Neuropsychopharmacology: Official Publication of the American College of Neuropsychopharmacology, 29*(8), 1558–1572.

De Chiara, V., Angelucci, F., Rossi, S., Musella, A., Cavasinni, F., Cantarella, C., … Centonze, D. (2010). Brain-derived neurotrophic factor controls cannabinoid CB_1 receptor function in the striatum. *The Journal of Neuroscience: The Official Journal of the Society for Neuroscience, 30*(24), 8127–8137.

De March, Z., Zuccato, C., Giampa, C., Patassini, S., Bari, M., Gasperi, V., … Fusco, F.R. (2008). Cortical expression of brain derived neurotrophic factor and type-1 cannabinoid receptor after striatal excitotoxic lesions. *Neuroscience, 152*(3), 734–740.

De Marchi, N., De Petrocellis, L., Orlando, P., Daniele, F., Fezza, F., & Di Marzo, V. (2003). Endocannabinoid signalling in the blood of patients with schizophrenia. *Lipids in Health and Disease, 2*, 5.

Dean, B., Bradbury, R., & Copolov, D. L. (2003). Cannabis-sensitive dopaminergic markers in postmortem central nervous system: Changes in schizophrenia. *Biological Psychiatry, 53*(7), 585–592.

Dean, B., Sundram, S., Bradbury, R., Scarr, E., & Copolov, D. (2001). Studies on [^{3}H] CP-55940 binding in the human central nervous system: Regional specific changes in density of cannabinoid-1 receptors associated with schizophrenia and cannabis use. *Neuroscience, 103*(1), 9–15.

Degenhardt, L., Hall, W., & Lynskey, M. (2003). Testing hypotheses about the relationship between cannabis use and psychosis. *Drug and Alcohol Dependence, 71*(1), 37–48.

van Dellen, A., Blakemore, C., Deacon, R., York, D., & Hannan, A. J. (2000). Delaying the onset of Huntington's in mice. *Nature, 404*(6779), 721–722.

Demuth, D. G., & Molleman, A. (2006). Cannabinoid signalling. *Life Sciences, 78*(6), 549–563.

Dendrou, C. A., Fugger, L., & Friese, M. A. (2015). Immunopathology of multiple sclerosis. *Nature Reviews. Immunology, 15*(9), 545–558.

Deng, C., Han, M., & Huang, X. F. (2007). No changes in densities of cannabinoid receptors in the superior temporal gyrus in schizophrenia. *Neuroscience Bulletin, 23*(6), 341–347.

Denovan-Wright, E. M., & Robertson, H. A. (2000). Cannabinoid receptor messenger RNA levels decrease in a subset of neurons of the lateral striatum, cortex and hippocampus of transgenic Huntington's disease mice. *Neuroscience, 98*(4), 705–713.

Di Forti, M., Morrison, P. D., Butt, A., & Murray, R. M. (2007). Cannabis use and psychiatric and cognitive disorders: The chicken or the egg? *Current Opinion in Psychiatry, 20*(3), 228–234.

Di Iorio, G., Lupi, M., Sarchione, F., Matarazzo, I., Santacroce, R., Petruccelli, F., … Di Giannantonio, M. (2013). The endocannabinoid system: A putative role in neurodegenerative diseases. *International Journal of High Risk Behaviors and Addiction, 2*(3), 100–106.

Di Marzo, V., Fontana, A., Cadas, H., Schinelli, S., Cimino, G., Schwartz, J. C., & Piomelli, D. (1994). Formation and inactivation of endogenous cannabinoid anandamide in central neurons. *Nature, 372*(6507), 686–691.

Di Marzo, V., Hill, M. P., Bisogno, T., Crossman, A. R., & Brotchie, J. M. (2000). Enhanced levels of endogenous cannabinoids in the globus pallidus are associated with a reduction in movement in an animal model of Parkinson's disease. *FASEB Journal: Official Publication of the Federation of American Societies for Experimental Biology, 14*(10), 1432–1438.

Diez-Dacal, B., & Perez-Sala, D. (2010). Anti-inflammatory prostanoids: Focus on the interactions between electrophile signaling and resolution of inflammation. *The Scientific World Journal, 10*, 655–675.

van Dijk, G., van Heijningen, S., Reijne, A. C., Nyakas, C., van der Zee, E. A., & Eisel, U. L. (2015). Integrative neurobiology of metabolic diseases, neuroinflammation, and neurodegeneration. *Frontiers in Neuroscience, 9*, 173.

Docagne, F., Muneton, V., Clemente, D., Ali, C., Loria, F., Correa, F., … Guaza, C. (2007). Excitotoxicity in a chronic model of multiple sclerosis: Neuroprotective effects of cannabinoids through CB1 and CB2 receptor activation. *Molecular and Cellular Neurosciences, 34*(4), 551–561.

Dowie, M. J., Bradshaw, H. B., Howard, M. L., Nicholson, L. F., Faull, R. L., Hannan, A. J., & Glass, M. (2009). Altered CB1 receptor and endocannabinoid levels precede motor symptom onset in a transgenic mouse model of Huntington's disease. *Neuroscience, 163*(1), 456–465.

Dowie, M. J., Grimsey, N. L., Hoffman, T., Faull, R. L., & Glass, M. (2014). Cannabinoid receptor CB2 is expressed on vascular cells, but not astroglial cells in the post-mortem human Huntington's disease brain. *Journal of Chemical Neuroanatomy, 59–60*, 62–71.

Dowie, M. J., Howard, M. L., Nicholson, L. F., Faull, R. L., Hannan, A. J., & Glass, M. (2010). Behavioural and molecular consequences of chronic cannabinoid treatment in Huntington's disease transgenic mice. *Neuroscience, 170*(1), 324–336.

Downer, E. J., Clifford, E., Gran, B., Nel, H. J., Fallon, P. G., & Moynagh, P. N. (2011). Identification of the synthetic cannabinoid R(+)WIN55,212–2 as a novel regulator of IFN regulatory factor 3 (IRF3) activation and IFN-β expression: Relevance to therapeutic effects in models of multiple sclerosis. *The Journal of Biological Chemistry*.

Dubreucq, S., Koehl, M., Abrous, D. N., Marsicano, G., & Chaouloff, F. (2010). CB1 receptor deficiency decreases wheel-running activity: Consequences on emotional behaviours and hippocampal neurogenesis. *Experimental Neurology, 224*(1), 106–113.

Dwork, A. J. (1997). Postmortem studies of the hippocampal formation in schizophrenia. *Schizophrenia Bulletin, 23*(3), 385–402.

Eggan, S. M., Hashimoto, T., & Lewis, D. A. (2008). Reduced cortical cannabinoid 1 receptor messenger RNA and protein expression in schizophrenia. *Archives of General Psychiatry, 65*(7), 772–784.

Eggan, S. M., Stoyak, S. R., Verrico, C. D., & Lewis, D. A. (2010). Cannabinoid CB1 receptor immunoreactivity in the prefrontal cortex: Comparison of schizophrenia and major depressive disorder. *Neuropsychopharmacology: Official Publication of the American College of Neuropsychopharmacology, 35*(10), 2060–2071.

El-Banoua, F., Caraballo, I., Flores, J. A., Galan-Rodriguez, B., & Fernandez-Espejo, E. (2004). Effects on turning of microinjections into basal ganglia of D_1 and D_2 dopamine receptors agonists and the cannabinoid CB_1 antagonist SR141716A in a rat Parkinson's model. *Neurobiology of Disease, 16*(2), 377–385.

Eljaschewitsch, E., Witting, A., Mawrin, C., Lee, T., Schmidt, P. M., Wolf, S., … Ullrich, O. (2006). The endocannabinoid anandamide protects neurons during CNS inflammation by induction of MKP-1 in microglial cells. *Neuron, 49*(1), 67–79.

Elphick, M. R., & Egerova, M. (2001). The neurobiology and evolution of cannabinoid signalling. *Philosophical Transactions of the Royal Society of London B Biological Sciences, 356*(1407), 381–408.

Emrich, H. M., Leweke, F. M., & Schneider, U. (1997). Towards a cannabinoid hypothesis of schizophrenia: Cognitive impairments due to dysregulation of the endogenous cannabinoid system. *Pharmacology, Biochemistry, and Behavior, 56*(4), 803–807.

Esposito, G., De Filippis, D., Carnuccio, R., Izzo, A. A., & Iuvone, T. (2006). The marijuana component cannabidiol inhibits β-amyloid-induced tau protein hyperphosphorylation through Wnt/β-catenin pathway rescue in PC12 cells. *Journal of Molecular Medicine, 84*(3), 253–258.

Esposito, G., De Filippis, D., Maiuri, M. C., De Stefano, D., Carnuccio, R., & Iuvone, T. (2006). Cannabidiol inhibits inducible nitric oxide synthase protein expression and nitric oxide production in β-amyloid stimulated PC12 neurons through p38 MAP kinase and NF-κB involvement. *Neuroscience Letters, 399*(1–2), 91–95.

Esposito, G., Iuvone, T., Savani, C., Scuderi, C., De Filippis, D., Papa, M., … Steardo, L. (2007). Opposing control of cannabinoid receptor stimulation on amyloid-β-induced reactive gliosis: In vitro and in vivo evidence. *The Journal of Pharmacology and Experimental Therapeutics, 322*(3), 1144–1152.

Eubanks, L. M., Rogers, C. J., Beuscher, A. E., Koob, G. F., Olson, A. J., Dickerson, T. J., & Janda, K. D. (2006). A molecular link between the active component of marijuana and Alzheimer's disease pathology. *Molecular Pharmaceutics, 3*(6), 773–777.

Evans, J. R., & Barker, R. A. (2008). Neurotrophic factors as a therapeutic target for Parkinson's disease. *Expert Opinion on Therapeutic Targets, 12*(4), 437–447.

Fakhoury, M. (2016). Role of the endocannabinoid system in the pathophysiology of schizophrenia. *Molecular Neurobiology.*

Fattore, L. (2016). Synthetic cannabinoids–further evidence supporting the relationship between cannabinoids and psychosis. *Biological Psychiatry, 79*(7), 539–548.

Fattore, L., & Fratta, W. (2011). Beyond THC: The new generation of cannabinoid designer drugs. *Frontiers in Behavioral Neuroscience, 5*, 60.

Feinstein, A., Banwell, E., & Pavisian, B. (2015). What to make of cannabis and cognition in MS: In search of clarity amidst the haze. *Multiple Sclerosis: Clinical and Laboratory Research, 21*(14), 1755–1760.

Fernandez-Espejo, E., Caraballo, I., de Fonseca, F. R., El Banoua, F., Ferrer, B., Flores, J. A., & Galan-Rodriguez, B. (2005). Cannabinoid CB_1 antagonists possess antiparkinsonian efficacy only in rats with very severe nigral lesion in experimental parkinsonism. *Neurobiology of Disease, 18*(3), 591–601.

Fernandez-Espejo, E., Caraballo, I., Rodriguez de Fonseca, F., Ferrer, B., El Banoua, F., Flores, J. A., & Galan-Rodriguez, B. (2004). Experimental parkinsonism alters anandamide precursor synthesis, and functional deficits are improved by AM404: A modulator of endocannabinoid function. *Neuropsychopharmacology: Official Publication of the American College of Neuropsychopharmacology, 29*(6), 1134–1142.

Fernandez-Ruiz, J., Moro, M. A., & Martinez-Orgado, J. (2015). Cannabinoids in neurodegenerative disorders and stroke/brain trauma: From preclinical models to clinical applications. *Neurotherapeutics: The Journal of the American Society for Experimental Neurotherapeutics, 12*(4), 793–806.

Fernandez-Ruiz, J., Romero, J., & Ramos, J. A. (2015). Endocannabinoids and neurodegenerative disorders: Parkinson's disease, Huntington's chorea, Alzheimer's disease, and others. *Handbook of Experimental Pharmacology, 231*, 233–259.

Ferreira, I. L., Resende, R., Ferreiro, E., Rego, A. C., & Pereira, C. F. (2010). Multiple defects in energy metabolism in Alzheimer's disease. *Current Drug Targets, 11*(10), 1193–1206.

Fox, S. H., Henry, B., Hill, M., Crossman, A., & Brotchie, J. (2002). Stimulation of cannabi-noid receptors reduces levodopa-induced dyskinesia in the MPTP-lesioned nonhuman primate model of Parkinson's disease. *Movement Disorders: Official Journal of the Movement Disorder Society, 17*(6), 1180–1187.

Franco, R., & Fernandez-Suarez, D. (2015). Alternatively activated microglia and macro-phages in the central nervous system. *Progress in Neurobiology, 131,* 65–86.

Freeman, R. M., Adekanmi, O., Waterfield, M. R., Waterfield, A. E., Wright, D., & Zajicek, J. (2006). The effect of cannabis on urge incontinence in patients with multiple sclerosis: A multicentre, randomised placebo-controlled trial (CAMS-LUTS). *International Urogynecology Journal and Pelvic Floor Dysfunction, 17*(6), 636–641.

Fritzsche, M. (2001). Are cannabinoid receptor knockout mice animal models for schizo-phrenia? *Medical Hypotheses, 56*(6), 638–643.

Gage, S. H., Hickman, M., & Zammit, S. (2016). Association between cannabis and psychosis: Epidemiologic evidence. *Biological Psychiatry, 79*(7), 549–556.

Galve-Roperh, I., Aguado, T., Palazuelos, J., & Guzman, M. (2007). The endocannabinoid system and neurogenesis in health and disease. *Neuroscientist, 13*(2), 109–114.

Garcia-Arencibia, M., Ferraro, L., Tanganelli, S., & Fernandez-Ruiz, J. (2008). Enhanced striatal glutamate release after the administration of rimonabant to 6-hydroxydopamine-lesioned rats. *Neuroscience Letters, 438*(1), 10–13.

Garcia-Arencibia, M., Garcia, C., Kurz, A., Rodriguez-Navarro, J. A., Gispert-Sachez, S., Mena, M. A., ... Fernandez-Ruiz, J. (2009). Cannabinoid CB1 receptors are early downregulated followed by a further upregulation in the basal ganglia of mice with deletion of specific park genes. *Journal of Neural Transmission* (Suppl. 73), 269–275.

Garcia-Arencibia, M., Gonzalez, S., de Lago, E., Ramos, J. A., Mechoulam, R., & Fernandez-Ruiz, J. (2007). Evaluation of the neuroprotective effect of cannabinoids in a rat model of Parkinson's disease: Importance of antioxidant and cannabinoid receptor-independent properties. *Brain Research, 1134*(1), 162–170.

Garcia-Ovejero, D., Arevalo-Martin, A., Navarro-Galve, B., Pinteaux, E., Molina-Holgado, E., & Molina-Holgado, F. (2013). Neuroimmune interactions of cannabinoids in neurogenesis: Focus on interleukin-1β (IL-1β) signalling. *Biochemical Society Transactions, 41*(6), 1577–1582.

Garcia, C., Palomo, C., Garcia-Arencibia, M., Ramos, J. A., Pertwee, R. G., & Fernandez-Ruiz, J. (2011). Symptom-relieving and neuroprotective effects of the phytocannabinoid Δ⁹-THCV in animal models of Parkinson's disease. *British Journal of Pharmacology.*

Gerdeman, G., & Lovinger, D. M. (2001). CB1 cannabinoid receptor inhibits synaptic release of glutamate in rat dorsolateral striatum. *Journal of Neurophysiology, 85*(1), 468–471.

Gilgun-Sherki, Y., Melamed, E., Mechoulam, R., & Offen, D. (2003). The CB1 cannabinoid receptor agonist, HU-210, reduces levodopa-induced rotations in 6-hydroxydopamine-lesioned rats. *Pharmacology & Toxicology, 93*(2), 66–70.

Giuffrida, A., Leweke, F. M., Gerth, C. W., Schreiber, D., Koethe, D., Faulhaber, J., ... Piomelli, D. (2004). Cerebrospinal anandamide levels are elevated in acute schizophrenia and are inversely correlated with psychotic symptoms. *Neuropsychopharmacology: Official Publication of the American College of Neuropsychopharmacology, 29*(11), 2108–2114.

Giuffrida, A., Parsons, L. H., Kerr, T. M., Rodriguez de Fonseca, F., Navarro, M., & Piomelli, D. (1999). Dopamine activation of endogenous cannabinoid signaling in dorsal striatum. *Nature Neuroscience, 2*(4), 358–363.

Glass, M. (2001). The role of cannabinoids in neurodegenerative diseases. *Progress in Neuro-Psychopharmacology & Biological Psychiatry, 25*(4), 743–765.

Glass, M., Dragunow, M., & Faull, R. L. (2000). The pattern of neurodegeneration in Huntington's disease: A comparative study of cannabinoid, dopamine, adenosine and $GABA_A$ receptor alterations in the human basal ganglia in Huntington's disease. *Neuroscience, 97*(3), 505–519.

Glass, M., Faull, R. L., & Dragunow, M. (1993). Loss of cannabinoid receptors in the substantia nigra in Huntington's disease. *Neuroscience, 56*(3), 523–527.

Glass, M., & Felder, C. C. (1997). Concurrent stimulation of cannabinoid CB1 and dopamine D2 receptors augments cAMP accumulation in striatal neurons: Evidence for a Gs linkage to the CB1 receptor. *Journal of Neuroscience, 17*(14), 5327–5333.

Glass, M., van Dellen, A., Blakemore, C., Hannan, A. J., & Faull, R. L. (2004). Delayed onset of Huntington's disease in mice in an enriched environment correlates with delayed loss of cannabinoid CB1 receptors. *Neuroscience, 123*(1), 207–212.

Gonzalez, S., Mena, M. A., Lastres-Becker, I., Serrano, A., de Yebenes, J. G., Ramos, J. A., & Fernandez-Ruiz, J. (2005). Cannabinoid CB$_1$ receptors in the basal ganglia and motor response to activation or blockade of these receptors in parkin-null mice. *Brain Research, 1046*(1–2), 195–206.

Gonzalez, S., Scorticati, C., Garcia-Arencibia, M., de Miguel, R., Ramos, J. A., & Fernandez-Ruiz, J. (2006). Effects of rimonabant, a selective cannabinoid CB1 receptor antagonist, in a rat model of Parkinson's disease. *Brain Research, 1073–1074,* 209–219.

Grundy, R. I. (2002). The therapeutic potential of the cannabinoids in neuroprotection. *Expert Opinion on Investigational Drugs, 11*(10), 1365–1374.

Gubellini, P., Picconi, B., Bari, M., Battista, N., Calabresi, P., Centonze, D., … Maccarrone, M. (2002). Experimental parkinsonism alters endocannabinoid degradation: Implications for striatal glutamatergic transmission. *The Journal of Neuroscience: The Official Journal of the Society for Neuroscience, 22*(16), 6900–6907.

Guidali, C., Vigano, D., Petrosino, S., Zamberletti, E., Realini, N., Binelli, G., … Parolaro, D. (2011). Cannabinoid CB1 receptor antagonism prevents neurochemical and behavioural deficits induced by chronic phencyclidine. *International Journal of Neuropsychopharmacology, 14*(1), 17–28.

Gururajan, A., Manning, E. E., Klug, M., & van den Buuse, M. (2012). Drugs of abuse and increased risk of psychosis development. *The Australian and New Zealand Journal of Psychiatry, 46*(12), 1120–1135.

Gutierrez-Valdez, A. L., Garcia-Ruiz, R., Anaya-Martinez, V., Torres-Esquivel, C., Espinosa-Villanueva, J., Reynoso-Erazo, L., … Avila-Costa, M. R. (2013). The combination of oral L-DOPA/rimonabant for effective dyskinesia treatment and cytological preservation in a rat model of Parkinson's disease and L-DOPA-induced dyskinesia. *Behavioural Pharmacology, 24*(8), 640–652.

Haller, J., Szirmai, M., Varga, B., Ledent, C., & Freund, T. F. (2005). Cannabinoid CB1 receptor dependent effects of the NMDA antagonist phencyclidine in the social withdrawal model of schizophrenia. *Behavioural Pharmacology, 16*(5–6), 415–422.

Hamdani, N., Tabeze, J. P., Ramoz, N., Ades, J., Hamon, M., Sarfati, Y., … Gorwood, P. (2008). The CNR1 gene as a pharmacogenetic factor for antipsychotics rather than a susceptibility gene for schizophrenia. *European Neuropsychopharmacology: The Journal of the European College of Neuropsychopharmacology, 18*(1), 34–40.

Hampson, A. J., Grimaldi, M., Axelrod, J., & Wink, D. (1998). Cannabidiol and $(-)\Delta^9$-tetrahydrocannabinol are neuroprotective antioxidants. *Proceedings of the National Academy of Sciences of the United States of America, 95*(14), 8268–8273.

Harrison, P. J. (1999). The neuropathology of schizophrenia. A critical review of the data and their interpretation. *Brain: A Journal of Neurology, 122*(Pt 4), 593–624.

Hebert, L. E., Scherr, P. A., Bienias, J. L., Bennett, D. A., & Evans, D. A. (2003). Alzheimer disease in the US population: Prevalence estimates using the 2000 census. *Archives of Neurology, 60*(8), 1119–1122.

Herkenham, M., Lynn, A. B., Johnson, M. R., Melvin, L. S., de Costa, B. R., & Rice, K. C. (1991). Characterization and localization of cannabinoid receptors in rat brain: A quantitative in vitro autoradiographic study. *Journal of Neuroscience, 11*(2), 563–583.

Hill, M. N., Titterness, A. K., Morrish, A. C., Carrier, E. J., Lee, T. T., Gil-Mohapel, J., … Christie, B. R. (2010). Endogenous cannabinoid signaling is required for voluntary exercise-induced enhancement of progenitor cell proliferation in the hippocampus. *Hippocampus, 20*(4), 513–523.

Ho, B.C. , Wassink, T.H. , Ziebell, S., & Andreasen, N.C. (2011). Cannabinoid receptor 1 gene polymorphisms and marijuana misuse interactions on white matter and cognitive deficits in schizophrenia. *Schizophrenia Research.*

Ho, L.W., Carmichael, J., Swartz, J., Wyttenbach, A., Rankin, J., & Rubinsztein, D. C. (2001). The molecular biology of Huntington's disease. *Psychological Medicine, 31*(1), 3–14.

Hockly, E., Cordery, P. M., Woodman, B., Mahal, A., van Dellen, A., Blakemore, C., … Bates, G. P. (2002). Environmental enrichment slows disease progression in R6/2 Huntington's disease mice. *Annals of Neurology, 51*(2), 235–242.

Hohmann, A. G., & Herkenham, M. (2000). Localization of cannabinoid CB_1 receptor mRNA in neuronal subpopulations of rat striatum: A double-label in situ hybridization study. *Synapse, 37*(1), 71–80.

Huntington's Disease Collaborative Research Group. (1993). A novel gene containing a trinucleotide repeat that is expanded and unstable on Huntington's disease chromosomes. *Cell, 72*(6), 971–983.

Hurley, M. J., Mash, D. C., & Jenner, P. (2003). Expression of cannabinoid CB1 receptor mRNA in basal ganglia of normal and parkinsonian human brain. *Journal of Neural Transmission, 110*(11), 1279–1288.

Iuvone, T., Esposito, G., Esposito, R., Santamaria, R., Di Rosa, M., & Izzo, A. A. (2004). Neuroprotective effect of cannabidiol, a non-psychoactive component from *Cannabis sativa*, on β-amyloid-induced toxicity in PC12 cells. *Journal of Neurochemistry, 89*(1), 134–141.

Jackson, S. J., Pryce, G., Diemel, L. T., Cuzner, M. L., & Baker, D. (2005). Cannabinoid-receptor 1 null mice are susceptible to neurofilament damage and caspase 3 activation. *Neuroscience, 134*(1), 261–268.

Jacobsen, J. C., Bawden, C. S., Rudiger, S. R., McLaughlan, C. J., Reid, S. J., Waldvogel, H. J., … Snell, R. G. (2010). An ovine transgenic Huntington's disease model. *Human Molecular Genetics, 19*(10), 1873–1882.

Jentsch, J. D., Verrico, C. D., Le, D., & Roth, R. H. (1998). Repeated exposure to Δ^9-tetrahydrocannabinol reduces prefrontal cortical dopamine metabolism in the rat. *Neuroscience Letters, 246*(3), 169–172.

Jentsch, J. D., Wise, A., Katz, Z., & Roth, R. H. (1998). Alpha-noradrenergic receptor modulation of the phencyclidine- and Δ^9-tetrahydrocannabinol-induced increases in dopamine utilization in rat prefrontal cortex. *Synapse, 28*(1), 21–26.

Jiang, W., Zhang, Y., Xiao, L., Van Cleemput, J., Ji, S. P., Bai, G., & Zhang, X. (2005). Cannabinoids promote embryonic and adult hippocampus neurogenesis and produce anxiolytic- and antidepressant-like effects. *Journal of Clinical Investigation, 115*(11), 3104–3116.

Johnston, T. H., Huot, P., Fox, S. H., Wakefield, J. D., Sykes, K. A., Bartolini, W. P., … Brotchie, J. M. (2011). Fatty acid amide hydrolase (FAAH) inhibition reduces L-3,4-dihydroxyphenylalanine-induced hyperactivity in the 1-methyl-4-phenyl-1,2,3,6-tetrahydropyridine-lesioned non-human primate model of Parkinson's disease. *The Journal of Pharmacology and Experimental Therapeutics, 336*(2), 423–430.

Kaiya, H. (1992). Second messenger imbalance hypothesis of schizophrenia. *Prostaglandins, Leukotrienes, and Essential Fatty Acids, 46*(1), 33–38.

Karst, M., Wippermann, S., & Ahrens, J. (2010). Role of cannabinoids in the treatment of pain and (painful) spasticity. *Drugs, 70*(18), 2409–2438.

Katona, S., Kaminski, E., Sanders, H., & Zajicek, J. (2005). Cannabinoid influence on cytokine profile in multiple sclerosis. *Clinical and Experimental Immunology, 140*(3), 580–585.

Kavia, R. B., De Ridder, D., Constantinescu, C. S., Stott, C. G., & Fowler, C. J. (2010). Randomized controlled trial of Sativex to treat detrusor overactivity in multiple sclerosis. *Multiple Sclerosis: Clinical and Laboratory Research, 16*(11), 1349–1359.

Kearn, C. S., Blake-Palmer, K., Daniel, E., Mackie, K., & Glass, M. (2005). Concurrent stimulation of cannabinoid CB1 and dopamine D2 receptors enhances heterodimer formation: A mechanism for receptor cross-talk? *Molecular Pharmacology, 67*(5), 1697–1704.

Kedzior, K. K., & Martin-Iverson, M. T. (2006). Chronic cannabis use is associated with attention-modulated reduction in prepulse inhibition of the startle reflex in healthy humans. *Journal of Psychopharmacology, 20*(4), 471–484.

Kelsey, J. E., Harris, O., & Cassin, J. (2009). The CB_1 antagonist rimonabant is adjunctively therapeutic as well as monotherapeutic in an animal model of Parkinson's disease. *Behavioural Brain Research, 203*(2), 304–307.

Khaspekov, L. G., Brenz Verca, M. S., Frumkina, L. E., Hermann, H., Marsicano, G., & Lutz, B. (2004). Involvement of brain-derived neurotrophic factor in cannabinoid receptor-dependent protection against excitotoxicity. *The European Journal of Neuroscience, 19*(7), 1691–1698.

Killestein, J., Hoogervorst, E. L., Reif, M., Blauw, B., Smits, M., Uitdehaag, B. M., ... Polman, C. H. (2003). Immunomodulatory effects of orally administered cannabinoids in multiple sclerosis. *Journal of Neuroimmunology, 137*(1–2), 140–143.

Killestein, J., Hoogervorst, E. L., Reif, M., Kalkers, N. F., Van Loenen, A. C., Staats, P. G., ... Polman, C. H. (2002). Safety, tolerability, and efficacy of orally administered cannabinoids in MS. *Neurology, 58*(9), 1404–1407.

Kim, D. J., & Thayer, S. A. (2000). Activation of CB1 cannabinoid receptors inhibits neurotransmitter release from identified synaptic sites in rat hippocampal cultures. *Brain Research, 852*(2), 398–405.

Klumpers, L. E., Beumer, T. L., van Hasselt, J. G., Lipplaa, A., Karger, L. B., Kleinloog, H. D., ... van Gerven, J. M. (2012). Novel Δ^9-tetrahydrocannabinol formulation Namisol® has beneficial pharmacokinetics and promising pharmacodynamic effects. *British Journal of Clinical Pharmacology, 74*(1), 42–53.

Koethe, D., Giuffrida, A., Schreiber, D., Hellmich, M., Schultze-Lutter, F., Ruhrmann, S., ... Leweke, F. M. (2009). Anandamide elevation in cerebrospinal fluid in initial prodromal states of psychosis. *The British Journal of Psychiatry: The Journal of Mental Science, 194*(4), 371–372.

Koethe, D., Llenos, I. C., Dulay, J. R., Hoyer, C., Torrey, E. F., Leweke, F. M., & Weis, S. (2007). Expression of CB1 cannabinoid receptor in the anterior cingulate cortex in schizophrenia, bipolar disorder, and major depression. *Journal of Neural Transmission, 114*(8), 1055–1063.

Kong, W., Li, H., Tuma, R. F., & Ganea, D. (2014). Selective CB2 receptor activation ameliorates EAE by reducing Th17 differentiation and immune cell accumulation in the CNS. *Cellular Immunology, 287*(1), 1–17.

Koppel, J., Bradshaw, H., Goldberg, T. E., Khalili, H., Marambaud, P., Walker, M. J., ... Davies, P. (2009). Endocannabinoids in Alzheimer's disease and their impact on normative cognitive performance: A case-control and cohort study. *Lipids in Health and Disease, 8*, 2.

Koskderelioglu, A., Binbay, I. T., Arac, N., & Veznedaroglu, B. (2008). Cycling mood disturbances and progressive neurological symptoms in a patient with Huntington's disease. *Progress in Neuro-psychopharmacology & Biological Psychiatry, 32*(4), 1079–1081.

Kovasznay, B., Fleischer, J., Tanenberg-Karant, M., Jandorf, L., Miller, A. D., & Bromet, E. (1997). Substance use disorder and the early course of illness in schizophrenia and affetive psychosis. *Schizophrenia Bulletin, 23*(2), 195–201.

Kreitzer, A. C., & Malenka, R. C. (2007). Endocannabinoid-mediated rescue of striatal LTD and motor deficits in Parkinson's disease models. *Nature, 445*(7128), 643–647.

Krishnan, S., Cairns, R., & Howard, R. (2009). Cannabinoids for the treatment of dementia. *The Cochrane Database of Systematic Reviews, 2,* CD007204.

Kubajewska, I., & Constantinescu, C. S. (2010). Cannabinoids and experimental models of multiple sclerosis. *Immunobiology, 215*(8), 647–657.

Kuperberg, G., & Heckers, S. (2000). Schizophrenia and cognitive function. *Current Opinion in Neurobiology, 10*(2), 205–210.

Lagalwar, S., Bordayo, E. Z., Hoffmann, K. L., Fawcett, J. R., & Frey, W. H., 2nd (1999). Anandamides inhibit binding to the muscarinic acetylcholine receptor. *Journal of Molecular Neuroscience, 13*(1–2), 55–61.

de Lago, E., Fernandez-Ruiz, J., Ortega-Gutierrez, S., Cabranes, A., Pryce, G., Baker, D., ... Ramos, J. A. (2006). UCM707, an inhibitor of the anandamide uptake, behaves as a symptom control agent in models of Huntington's disease and multiple sclerosis, but fails to delay/arrest the progression of different motor-related disorders. *European Neuropsychopharmacology: The Journal of the European College of Neuropsychopharmacology, 16*(1), 7–18.

de Lago, E., Urbani, P., Ramos, J. A., Di Marzo, V., & Fernandez-Ruiz, J. (2005). Arvanil, a hybrid endocannabinoid and vanilloid compound, behaves as an antihyperkinetic agent in a rat model of Huntington's disease. *Brain Research, 1050*(1–2), 210–216.

Lang, H., Li, Q., Yu, H., Li, P., Lu, Z., Xiong, S., ... Zhu, Z. (2015). Activation of TRPV1 attenuates high salt-induced cardiac hypertrophy through improvement of mitochondrial function. *British Journal of Pharmacology, 172*(23), 5548–5558.

Lastres-Becker, I., Berrendero, F., Lucas, J. J., Martin-Aparicio, E., Yamamoto, A., Ramos, J. A., & Fernandez-Ruiz, J. J. (2002). Loss of mRNA levels, binding and activation of GTP-binding proteins for cannabinoid CB1 receptors in the basal ganglia of a transgenic model of Huntington's disease. *Brain Research, 929*(2), 236–242.

Lastres-Becker, I., Bizat, N., Boyer, F., Hantraye, P., Brouillet, E., & Fernandez-Ruiz, J. (2003). Effects of cannabinoids in the rat model of Huntington's disease generated by an intrastriatal injection of malonate. *Neuroreport, 14*(6), 813–816.

Lastres-Becker, I., Bizat, N., Boyer, F., Hantraye, P., Fernandez-Ruiz, J., & Brouillet, E. (2004). Potential involvement of cannabinoid receptors in 3-nitropropionic acid toxicity in vivo. *Neuroreport, 15*(15), 2375–2379.

Lastres-Becker, I., Cebeira, M., de Ceballos, M. L., Zeng, B. Y., Jenner, P., Ramos, J. A., & Fernandez-Ruiz, J. J. (2001). Increased cannabinoid CB1 receptor binding and activation of GTP-binding proteins in the basal ganglia of patients with Parkinson's syndrome and of MPTP-treated marmosets. *The European Journal of Neuroscience, 14*(11), 1827–1832.

Lastres-Becker, I., de Miguel, R., De Petrocellis, L., Makriyannis, A., Di Marzo, V., & Fernandez-Ruiz, J. (2003). Compounds acting at the endocannabinoid and/or endovanilloid systems reduce hyperkinesia in a rat model of Huntington's disease. *Journal of Neurochemistry, 84*(5), 1097–1109.

Lastres-Becker, I., Fezza, F., Cebeira, M., Bisogno, T., Ramos, J. A., Milone, A., ... Di Marzo, V. (2001). Changes in endocannabinoid transmission in the basal ganglia in a rat model of Huntington's disease. *Neuroreport, 12*(10), 2125–2129.

Lastres-Becker, I., Gomez, M., De Miguel, R., Ramos, J. A., & Fernandez-Ruiz, J. (2002). Loss of cannabinoid CB_1 receptors in the basal ganglia in the late akinetic phase of rats with experimental Huntington's disease. *Neurotoxicity Research, 4*(7–8), 601–608.

Lastres-Becker, I., Hansen, H. H., Berrendero, F., De Miguel, R., Perez-Rosado, A., Manzanares, J., ... Fernandez-Ruiz, J. (2002). Alleviation of motor hyperactivity and neurochemical deficits by endocannabinoid uptake inhibition in a rat model of Huntington's disease. *Synapse, 44*(1), 23–35.

Lastres-Becker, I., Molina-Holgado, F., Ramos, J. A., Mechoulam, R., & Fernandez-Ruiz, J. (2005). Cannabinoids provide neuroprotection against 6-hydroxydopamine toxicity in vivo and in vitro: Relevance to Parkinson's disease. *Neurobiology of Disease, 19*(1–2), 96–107.

Le Moine, C., & Bloch, B. (1995). D1 and D2 dopamine receptor gene expression in the rat striatum: Sensitive cRNA probes demonstrate prominent segregation of D1 and D2 mRNAs in distinct neuronal populations of the dorsal and ventral striatum. *Journal of Comparative Neurology, 355*(3), 418–426.

Lee, J. H., Agacinski, G., Williams, J. H., Wilcock, G. K., Esiri, M. M., Francis, P. T., … Lai, M. K. (2010). Intact cannabinoid CB1 receptors in the Alzheimer's disease cortex. *Neurochemistry International, 57*(8), 985–989.

Leroy, S., Griffon, N., Bourdel, M. C., Olie, J. P., Poirier, M. F., & Krebs, M. O. (2001). Schizophrenia and the cannabinoid receptor type 1 (CB1): Association study using a single-base polymorphism in coding exon 1. *American Journal of Medical Genetics, 105*(8), 749–752.

Leweke, F. M., Giuffrida, A., Koethe, D., Schreiber, D., Nolden, B. M., Kranaster, L., … Piomelli, D. (2007). Anandamide levels in cerebrospinal fluid of first-episode schizophrenic patients: Impact of cannabis use. *Schizophrenia Research, 94*(1–3), 29–36.

Leweke, F. M., Giuffrida, A., Wurster, U., Emrich, H. M., & Piomelli, D. (1999). Elevated endogenous cannabinoids in schizophrenia. *Neuroreport, 10*(8), 1665–1669.

Leweke, F. M., & Koethe, D. (2008). Cannabis and psychiatric disorders: It is not only addiction. *Addiction Biology, 13*(2), 264–275.

Leweke, F. M., Mueller, J. K., Lange, B., & Rohleder, C. (2016). Therapeutic potential of cannabinoids in psychosis. *Biological Psychiatry, 79*(7), 604–612.

Lewerenz, J., & Maher, P. (2015). Chronic glutamate toxicity in neurodegenerative diseases-what is the evidence? *Frontiers in Neuroscience, 9*, 469.

Lewis, D. A., & Lieberman, J. A. (2000). Catching up on schizophrenia: Natural history and neurobiology. *Neuron, 28*, 325–334.

Linszen, D., Dingemans, P., & Lenior, M. (1994). Cannabis abuse and the course of recent-onset schizophrenic disorders. *Archives of General Psychiatry, 51*, 273–279.

Llorente, R., Llorente-Berzal, A., Petrosino, S., Marco, E. M., Guaza, C., Prada, C., … Viveros, M. P. (2008). Gender-dependent cellular and biochemical effects of maternal deprivation on the hippocampus of neonatal rats: A possible role for the endocannabinoid system. *Developmental Neurobiology, 68*(11), 1334–1347.

Long, L. E., Malone, D. T., & Taylor, D. A. (2006). Cannabidiol reverses MK-801-induced disruption of prepulse inhibition in mice. *Neuropsychopharmacology: Official Publication of the American College of Neuropsychopharmacology, 31*(4), 795–803.

Lopez-Gallardo, M., Llorente, R., Llorente-Berzal, A., Marco, E. M., Prada, C., Di Marzo, V., & Viveros, M. P. (2008). Neuronal and glial alterations in the cerebellar cortex of maternally deprived rats: Gender differences and modulatory effects of two inhibitors of endocannabinoid inactivation. *Developmental Neurobiology, 68*(12), 1429–1440.

Lopez-Rodriguez, A. B., Siopi, E., Finn, D. P., Marchand-Leroux, C., Garcia-Segura, L. M., Jafarian-Tehrani, M., & Viveros, M. P. (2015). CB1 and CB2 cannabinoid receptor antagonists prevent minocycline-induced neuroprotection following traumatic brain injury in mice. *Cerebral Cortex, 25*(1), 35–45.

Lorenzetti, V., Solowij, N., & Yucel, M. (2016). The role of cannabinoids in neuroanatomic alterations in cannabis users. *Biological Psychiatry, 79*(7), e17–e31.

Loria, F., Petrosino, S., Mestre, L., Spagnolo, A., Correa, F., Hernangomez, M., … Docagne, F. (2008). Study of the regulation of the endocannabinoid system in a virus model of multiple sclerosis reveals a therapeutic effect of palmitoylethanolamide. *The European Journal of Neuroscience, 28*(4), 633–641.

Lotan, I., Treves, T. A., Roditi, Y., & Djaldetti, R. (2014). Cannabis (medical marijuana) treatment for motor and non-motor symptoms of Parkinson disease: An open-label observational study. *Clinical Neuropharmacology, 37*(2), 41–44.

Lourbopoulos, A., Grigoriadis, N., Lagoudaki, R., Touloumi, O., Polyzoidou, E., Mavromatis, I., … Simeonidou, C. (2011). Administration of 2-arachidonoylglycerol ameliorates both acute and chronic experimental autoimmune encephalomyelitis. *Brain Research.*

Lutz, B., Marsicano, G., Maldonado, R., & Hillard, C. J. (2015). The endocannabinoid system in guarding against fear, anxiety and stress. *Nature Reviews. Neuroscience, 16*(12), 705–718.

Lyman, W. D., Sonett, J. R., Brosnan, C. F., Elkin, R., & Bornstein, M. B. (1989). Δ^9-tetrahydrocannabinol: A novel treatment for experimental autoimmune encephalomyelitis. *Journal of Neuroimmunology, 23*(1), 73–81.

Ma, L., Jia, J., Niu, W., Jiang, T., Zhai, Q., Yang, L., … Xiong, L. (2015). Mitochondrial CB1 receptor is involved in ACEA-induced protective effects on neurons and mitochondrial functions. *Scientific Reports, 5,* 12440.

Mailleux, P., & Vanderhaeghen, J. J. (1993). Dopaminergic regulation of cannabinoid receptor mRNA levels in the rat caudate-putamen: An in situ hybridization study. *Journal of Neurochemistry, 61*(5), 1705–1712.

Malone, D. T., Kearn, C. S., Chongue, L., Mackie, K., & Taylor, D. A. (2008). Effect of social isolation on CB1 and D2 receptor and fatty acid amide hydrolase expression in rats. *Neuroscience, 152*(1), 265–272.

Malone, D. T., & Taylor, D. A. (2006). The effect of Δ^9-tetrahydrocannabinol on sensorimotor gating in socially isolated rats. *Behavioural Brain Research, 166*(1), 101–109.

Maneuf, Y. P., Crossman, A. R., & Brotchie, J. M. (1997). The cannabinoid receptor agonist WIN 55,212-2 reduces D2, but not D1, dopamine receptor-mediated alleviation of akinesia in the reserpine-treated rat model of Parkinson's disease. *Experimental Neurology, 148*(1), 265–270.

Marinelli, S., Di Marzo, V., Berretta, N., Matias, I., Maccarone, M., Bernardi, G., & Mercuri, N. B. (2003). Presynaptic facilitation of glutamatergic synapses to dopaminergic neurons of the rat substantia nigra by endogenous stimulation of vanilloid receptors. *The Journal of Neuroscience: The Official Journal of the Society for Neuroscience, 23*(8), 3136–3144.

Marsicano, G., Moosmann, B., Hermann, H., Lutz, B., & Behl, C. (2002). Neuroprotective properties of cannabinoids against oxidative stress: Role of the cannabinoid receptor CB1. *Journal of Neurochemistry, 80*(3), 448–456.

Martinez-Gras, I., Hoenicka, J., Ponce, G., Rodriguez-Jimenez, R., Jimenez-Arriero, M. A., Perez-Hernandez, E., … Rubio, G. (2006). (AAT)n repeat in the cannabinoid receptor gene, CNR1: Association with schizophrenia in a Spanish population. *European Archives of Psychiatry and Clinical Neuroscience, 256*(7), 437–441.

Martinez, A., Macheda, T., Morgese, M. G., Trabace, L., & Giuffrida, A. (2012). The cannabinoid agonist WIN55212-2 decreases L-DOPA-induced PKA activation and dyskinetic behavior in 6-OHDA-treated rats. *Neuroscience Research, 72*(3), 236–242.

Martorana, A., Esposito, Z., & Koch, G. (2010). Beyond the cholinergic hypothesis: Do current drugs work in Alzheimer's disease? *CNS Neuroscience & Therapeutics, 16*(4), 235–245.

Martyn, C. N., Illis, L. S., & Thom, J. (1995). Nabilone in the treatment of multiple sclerosis. *Lancet, 345*(8949), 579.

Matsuda, L. A., Lolait, S. J., Brownstein, M. J., Young, A. C., & Bonner, T. I. (1990). Structure of a cannabinoid receptor and functional expression of the cloned cDNA. *Nature, 346*(6284), 561–564.

Mauler, F., Mittendorf, J., Horvath, E., & De Vry, J. (2002). Characterization of the diarylether sulfonylester (-)-(R)-3-(2-hydroxymethylindanyl-4-oxy)phenyl-4,4,4-trifluoro-1-sulfonate (BAY 38-7271) as a potent cannabinoid receptor agonist with neuroprotective properties. *The Journal of Pharmacology and Experimental Therapeutics, 302*(1), 359–368.

McAllister, S. D., & Glass, M. (2002). CB1 and CB2 receptor-mediated signalling: A focus on endocannabinoids. *Prostaglandins Leukotrienes & Essential Fatty Acids, 66*(2–3), 161–171.

McCaw, E. A., Hu, H., Gomez, G. T., Hebb, A. L., Kelly, M. E., & Denovan-Wright, E. M. (2004). Structure, expression and regulation of the cannabinoid receptor gene (CB1) in Huntington's disease transgenic mice. *European Journal of Biochemistry, 271*(23–24), 4909–4920.

McGeer, P. L., Itagaki, S., Tago, H., & McGeer, E. G. (1987). Reactive microglia in patients with senile dementia of the Alzheimer type are positive for the histocompatibility glycoprotein HLA-DR. *Neuroscience Letters, 79*(1–2), 195–200.

Mechoulam, R., & Golan, D. (1998). Comment on 'Health aspects of cannabis: Revisited' (Hollister). *International Journal of Neuropsychopharmacology, 1*(1), 83–85.

Mechoulam, R., & Hanu, L. (2001). The cannabinoids: An overview. Therapeutic implications in vomiting and nausea after cancer chemotherapy, in appetite promotion, in multiple sclerosis and in neuroprotection. *Pain Research and Management, 6*(2), 67–73.

Meinck, H. M., Schonle, P. W., & Conrad, B. (1989). Effect of cannabinoids on spasticity and ataxia in multiple sclerosis. *Journal of Neurology, 236*(2), 120–122.

Melis, M., Perra, S., Muntoni, A. L., Pillolla, G., Lutz, B., Marsicano, G., ... Pistis, M. (2004). Prefrontal cortex stimulation induces 2-arachidonoyl-glycerol-mediated suppression of excitation in dopamine neurons. *The Journal of Neuroscience: The Official Journal of the Society for Neuroscience, 24*(47), 10707–10715.

Meltzer, H. Y., Arvanitis, L., Bauer, D., & Rein, W. (2004). Placebo-controlled evaluation of four novel compounds for the treatment of schizophrenia and schizoaffective disorder. *The American Journal of Psychiatry, 161*(6), 975–984.

Meschler, J. P., Howlett, A. C., & Madras, B. K. (2001). Cannabinoid receptor agonist and antagonist effects on motor function in normal and 1-methyl-4-phenyl-1,2,5,6-tetrahydropyridine (MPTP)-treated non-human primates. *Psychopharmacology, 156*(1), 79–85.

Mesnage, V., Houeto, J. L., Bonnet, A. M., Clavier, I., Arnulf, I., Cattelin, F., ... Agid, Y. (2004). Neurokinin B, neurotensin, and cannabinoid receptor antagonists and Parkinson disease. *Clinical Neuropharmacology, 27*(3), 108–110.

Mestre, L., Correa, F., Arevalo-Martin, A., Molina-Holgado, E., Valenti, M., Ortar, G., ... Guaza, C. (2005). Pharmacological modulation of the endocannabinoid system in a viral model of multiple sclerosis. *Journal of Neurochemistry, 92*(6), 1327–1339.

Mievis, S., Blum, D., & Ledent, C. (2011). Worsening of Huntington disease phenotype in CB1 receptor knockout mice. *Neurobiology of Disease.*

Milton, N. G. (2002). Anandamide and noladin ether prevent neurotoxicity of the human amyloid-beta peptide. *Neuroscience Letters, 332*(2), 127–130.

Molina-Holgado, F., Rubio-Araiz, A., Garcia-Ovejero, D., Williams, R. J., Moore, J. D., Arevalo-Martin, A., ... Molina-Holgado, E. (2007). CB2 cannabinoid receptors promote mouse neural stem cell proliferation. *The European Journal of Neuroscience, 25*(3), 629–634.

Monory, K., Massa, F., Egertova, M., Eder, M., Blaudzun, H., Westenbroek, R., ... Lutz, B. (2006). The endocannabinoid system controls key epileptogenic circuits in the hippocampus. *Neuron, 51*(4), 455–466.

Moore, T. H., Zammit, S., Lingford-Hughes, A., Barnes, T. R., Jones, P. B., Burke, M., & Lewis, G. (2007). Cannabis use and risk of psychotic or affective mental health outcomes: A systematic review. *Lancet, 370*(9584), 319–328.

Morgese, M. G., Cassano, T., Cuomo, V., & Giuffrida, A. (2007). Anti-dyskinetic effects of cannabinoids in a rat model of Parkinson's disease: Role of CB_1 and TRPV1 receptors. *Experimental Neurology, 208*(1), 110–119.

Morgese, M. G., Cassano, T., Gaetani, S., Macheda, T., Laconca, L., Dipasquale, P., ... Giuffrida, A. (2009). Neurochemical changes in the striatum of dyskinetic rats after administration of the cannabinoid agonist WIN55,212-2. *Neurochemistry International, 54*(1), 56–64.

Morrison, P. D., Zois, V., McKeown, D. A., Lee, T. D., Holt, D. W., Powell, J. F., … Murray, R. M. (2009). The acute effects of synthetic intravenous Δ^9-tetrahydrocannabinol on psychosis, mood and cognitive functioning. *Psychological Medicine, 39*(10), 1607–1616.

Moss, D. E., McMaster, S. B., & Rogers, J. (1981). Tetrahydrocannabinol potentiates reserpine-induced hypokinesia. *Pharmacology, Biochemistry, and Behavior, 15*(5), 779–783.

Mulder, J., Zilberter, M., Pasquare, S. J., Alpar, A., Schulte, G., Ferreira, S. G., … Harkany, T. (2011). Molecular reorganization of endocannabinoid signalling in Alzheimer's disease. *Brain: A Journal of Neurology, 134*(Pt 4), 1041–1060.

Muller-Vahl, K. R., & Emrich, H. M. (2008). Cannabis and schizophrenia: Towards a cannabinoid hypothesis of schizophrenia. *Expert Review of Neurotherapeutics, 8*(7), 1037–1048.

Muller-Vahl, K. R., Schneider, U., & Emrich, H. M. (1999). Nabilone increases choreatic movements in Huntington's disease. *Movement Disorders, 14*(6), 1038–1040.

Muller, H., Sperling, W., Kohrmann, M., Huttner, H. B., Kornhuber, J., & Maler, J. M. (2010). The synthetic cannabinoid Spice as a trigger for an acute exacerbation of cannabis induced recurrent psychotic episodes. *Schizophrenia Research, 118*(1–3), 309–310.

Muslimovic, D., Post, B., Speelman, J. D., Schmand, B., & de Haan, R. J. (2008). Determinants of disability and quality of life in mild to moderate Parkinson disease. *Neurology, 70*(23), 2241–2247.

Nagayama, T., Sinor, A. D., Simon, R. P., Chen, J., Graham, S. H., Jin, K., & Greenberg, D. A. (1999). Cannabinoids and neuroprotection in global and focal cerebral ischemia and in neuronal cultures. *The Journal of Neuroscience: The Official Journal of the Society for Neuroscience, 19*(8), 2987–2995.

Navarrete, M., & Araque, A. (2008). Endocannabinoids mediate neuron-astrocyte communication. *Neuron, 57*(6), 883–893.

Navarrete, M., & Araque, A. (2010). Endocannabinoids potentiate synaptic transmission through stimulation of astrocytes. *Neuron, 68*(1), 113–126.

Naver, B., Stub, C., Moller, M., Fenger, K., Hansen, A. K., Hasholt, L., & Sorensen, S. A. (2003). Molecular and behavioral analysis of the R6/1 Huntington's disease transgenic mouse. *Neuroscience, 122*(4), 1049–1057.

Newell, K. A., Deng, C., & Huang, X. F. (2006). Increased cannabinoid receptor density in the posterior cingulate cortex in schizophrenia. *Experimental Brain Research, 172*(4), 556–560.

Ni, X., Geller, E.B., Eppihimer, M.J., Eisenstein, T.K., Adler, M.W., & Tuma, R.F. (2004). Win 55212-2, a cannabinoid receptor agonist, attenuates leukocyte/endothelial interactions in an experimental autoimmune encephalomyelitis model. *Multiple Sclerosis: Clinical and Laboratory Research, 10*(2), 158–164.

Nicniocaill, B., Haraldsson, B., Hansson, O., O'Connor, W. T., & Brundin, P. (2001). Altered striatal amino acid neurotransmitter release monitored using microdialysis in R6/1 Huntington transgenic mice. *The European Journal of Neuroscience, 13*(1), 206–210.

Notcutt, W. G. (2015). Clinical use of cannabinoids for symptom control in multiple sclerosis. *Neurotherapeutics: The Journal of the American Society for Experimental Neurotherapeutics, 12*(4), 769–777.

Notcutt, W., Price, M., Miller, R., Newport, S., Phillips, C., Simmons, S., & Sansom, C. (2004). Initial experiences with medicinal extracts of cannabis for chronic pain: Results from 34 'N of 1' studies. *Anaesthesia, 59*(5), 440–452.

Novotna, A., Mares, J., Ratcliffe, S., Novakova, I., Vachova, M., Zapletalova, O., … Davies, P. (2011). A randomized, double-blind, placebo-controlled, parallel-group, enriched-design study of nabiximols★ (Sativex®), as add-on therapy, in subjects with refractory spasticity caused by multiple sclerosis. *European Journal of Neurology: The Official Journal of the European Federation of Neurological Societies, 18*(9), 1122–1131.

Nunez, E., Benito, C., Pazos, M. R., Barbachano, A., Fajardo, O., Gonzalez, S., ... Romero, J. (2004). Cannabinoid CB2 receptors are expressed by perivascular microglial cells in the human brain: An immunohistochemical study. *Synapse, 53*(4), 208–213.

Oliveira da Cruz, J. F., Robin, L. M., Drago, F., Marsicano, G., & Metna-Laurent, M. (2015). Astroglial type-1 cannabinoid receptor (CB): A new player in the tripartite synapse. *Neuroscience.*

Ortega-Alvaro, A., Aracil-Fernandez, A., Garcia-Gutierrez, M. S., Navarrete, F., & Manzanares, J. (2011). Deletion of CB_2 cannabinoid receptor induces schizophrenia-related behaviors in mice. *Neuropsychopharmacology: Official Publication of the American College of Neuropsychopharmacology.*

Ortega-Gutierrez, S., Molina-Holgado, E., Arevalo-Martin, A., Correa, F., Viso, A., Lopez-Rodriguez, M. L., ... Guaza, C. (2005). Activation of the endocannabinoid system as therapeutic approach in a murine model of multiple sclerosis. *FASEB Journal: Official Publication of the Federation of American Societies for Experimental Biology, 19*(10), 1338–1340.

Osuna-Zazuetal, M. A., Ponce-Gomez, J. A., & Perez-Neri, I. (2015). Neuroprotective mechanisms of cannabinoids in brain ischemia and neurodegenerative disorders. *Investigacion Clinica, 56*(2), 188–200.

Page, K. J., Besret, L., Jain, M., Monaghan, E. M., Dunnett, S. B., & Everitt, B. J. (2000). Effects of systemic 3-nitropropionic acid-induced lesions of the dorsal striatum on cannabinoid and mu-opioid receptor binding in the basal ganglia. *Experimental Brain Research, 130*(2), 142–150.

Palazuelos, J., Aguado, T., Egia, A., Mechoulam, R., Guzman, M., & Galve-Roperh, I. (2006). Non-psychoactive CB2 cannabinoid agonists stimulate neural progenitor proliferation. *FASEB Journal: Official Publication of the Federation of American Societies for Experimental Biology, 20*(13), 2405–2407.

Palazuelos, J., Aguado, T., Pazos, M. R., Julien, B., Carrasco, C., & Resel, E. (2009). Microglial CB2 cannabinoid receptors are neuroprotective in Huntington's disease excitotoxicity. *Brain: A Journal of Neurology, 132*(Pt 11), 3152–3164.

Palazuelos, J., Davoust, N., Julien, B., Hatterer, E., Aguado, T., Mechoulam, R., ... Galve-Roperh, I. (2008). The CB_2 cannabinoid receptor controls myeloid progenitor trafficking: Involvement in the pathogenesis of an animal model of multiple sclerosis. *The Journal of Biological Chemistry, 283*(19), 13320–13329.

Palazuelos, J., Ortega, Z., Diaz-Alonso, J., Guzman, M., & Galve-Roperh, I. (2012). CB2 cannabinoid receptors promote neural progenitor cell proliferation via mTORC1 signaling. *The Journal of Biological Chemistry, 287*(2), 1198–1209.

Panikashvili, D., Mechoulam, R., Beni, S. M., Alexandrovich, A., & Shohami, E. (2005). CB_1 cannabinoid receptors are involved in neuroprotection via NF-κB inhibition. *Journal of Cerebral Blood Flow & Metabolism, 25*(4), 477–484.

Panikashvili, D., Simeonidou, C., Ben-Shabat, S., Hanus, L., Breuer, A., Mechoulam, R., & Shohami, E. (2001). An endogenous cannabinoid (2-AG) is neuroprotective after brain injury. *Nature, 413*(6855), 527–531.

Passmore, M. J. (2008). The cannabinoid receptor agonist nabilone for the treatment of dementia-related agitation. *International Journal of Geriatric Psychiatry, 23*(1), 116–117.

Pegorini, S., Zani, A., Braida, D., Guerini-Rocco, C., & Sala, M. (2006). Vanilloid VR1 receptor is involved in rimonabant-induced neuroprotection. *British Journal of Pharmacology, 147*(5), 552–559.

Perras, C. (2005). Sativex for the management of multiple sclerosis symptoms. *Issues in Emerging Health Technologies, 72*, 1–4.

Petro, D. J., & Ellenberger, C., Jr. (1981). Treatment of human spasticity with Δ^9-tetrahydrocannabinol. *The Journal of Clinical Pharmacology, 21*(8–9 Suppl.), 413S–416S.

Petschner, P., Tamasi, V., Adori, C., Kirilly, E., Ando, R. D., Tothfalusi, L., & Bagdy, G. (2013). Gene expression analysis indicates CB1 receptor upregulation in the hippocampus and neurotoxic effects in the frontal cortex 3 weeks after single-dose MDMA administration in Dark Agouti rats. *BMC Genomics, 14,* 930.

Pillai, A. (2008). Brain-derived neurotropic factor/TrkB signaling in the pathogenesis and novel pharmacotherapy of schizophrenia. *Neurosignals, 16*(2–3), 183–193.

Pintor, A., Tebano, M. T., Martire, A., Grieco, R., Galluzzo, M., Scattoni, M. L., ... Popoli, P. (2006). The cannabinoid receptor agonist WIN 55,212-2 attenuates the effects induced by quinolinic acid in the rat striatum. *Neuropharmacology.*

Pisani, A., Fezza, F., Galati, S., Battista, N., Napolitano, S., Finazzi-Agro, A., ... Maccarrone, M. (2005). High endogenous cannabinoid levels in the cerebrospinal fluid of untreated Parkinson's disease patients. *Annals of Neurology, 57*(5), 777–779.

Porter, B. E., & Jacobson, C. (2013). Report of a parent survey of cannabidiol-enriched cannabis use in pediatric treatment-resistant epilepsy. *Epilepsy & Behavior, 29*(3), 574–577.

Potvin, S., Kouassi, E., Lipp, O., Bouchard, R. H., Roy, M. A., Demers, M. F., ... Stip, E. (2008). Endogenous cannabinoids in patients with schizophrenia and substance use disorder during quetiapine therapy. *Journal of Psychopharmacology, 22*(3), 262–269.

Prenderville, J. A., Kelly, A. M., & Downer, E. J. (2015). The role of cannabinoids in adult neurogenesis. *British Journal of Pharmacology, 172*(16), 3950–3963.

Price, D. A., Martinez, A. A., Seillier, A., Koek, W., Acosta, Y., Fernandez, E., ... Giuffrida, A. (2009). WIN55,212-2, a cannabinoid receptor agonist, protects against nigrostriatal cell loss in the 1-methyl-4-phenyl-1,2,3,6-tetrahydropyridine mouse model of Parkinson's disease. *The European Journal of Neuroscience, 29*(11), 2177–2186.

Pryce, G., Ahmed, Z., Hankey, D. J., Jackson, S. J., Croxford, J. L., Pocock, J. M., ... Baker, D. (2003). Cannabinoids inhibit neurodegeneration in models of multiple sclerosis. *Brain: A Journal of Neurology, 126*(Pt 10), 2191–2202.

Pryce, G., & Baker, D. (2007). Control of spasticity in a multiple sclerosis model is mediated by CB1, not CB2, cannabinoid receptors. *British Journal of Pharmacology, 150*(4), 519–525.

Pryce, G., & Baker, D. (2015). Endocannabinoids in multiple sclerosis and amyotrophic lateral sclerosis. *Handbook of Experimental Pharmacology, 231,* 213–231.

Pryce, G., Cabranes, A., Fernandez-Ruiz, J., Bisogno, T., Di Marzo, V., Long, J. Z., ... Baker, D. (2013). Control of experimental spasticity by targeting the degradation of endocannabinoids using selective fatty acid amide hydrolase inhibitors. *Multiple Sclerosis: Clinical and Laboratory Research, 19*(14), 1896–1904.

Pryor, S. R. (2000). Is platelet release of 2-arachidonoyl-glycerol a mediator of cognitive deficits? An endocannabinoid theory of schizophrenia and arousal. *Medical Hypotheses, 55*(6), 494–501.

Rahimi, A., Faizi, M., Talebi, F., Noorbakhsh, F., Kahrizi, F., & Naderi, N. (2015). Interaction between the protective effects of cannabidiol and palmitoylethanolamide in experimental model of multiple sclerosis in C57BL/6 mice. *Neuroscience, 290,* 279–287.

Ramil, E., Sanchez, A. J., Gonzalez-Perez, P., Rodriguez-Antiguedad, A., Gomez-Lozano, N., Ortiz, P., ... Garcia-Merino, A. (2010). The cannabinoid receptor 1 gene (CNR1) and multiple sclerosis: An association study in two case-control groups from Spain. *Multiple Sclerosis: Clinical and Laboratory Research, 16*(2), 139–146.

Ramirez, B. G., Blazquez, C., Gomez del Pulgar, T., Guzman, M., & de Ceballos, M. L. (2005). Prevention of Alzheimer's disease pathology by cannabinoids: Neuroprotection mediated by blockade of microglial activation. *The Journal of Neuroscience: The Official Journal of the Society for Neuroscience, 25*(8), 1904–1913.

Reddy, D. S., & Golub, V. M. (2016). The pharmacological basis of cannabis therapy for epilepsy. *The Journal of Pharmacology and Experimental Therapeutics, 357*(1), 45–55.

Richfield, E. K., & Herkenham, M. (1994). Selective vulnerability in Huntington's disease: Preferential loss of cannabinoid receptors in lateral globus pallidus. *Annals of Neurology, 36*(4), 577–584.

Rivers, J. R., & Ashton, J. C. (2010). The development of cannabinoid CBII receptor agonists for the treatment of central neuropathies. *Central Nervous System Agents in Medicinal Chemistry, 10*(1), 47–64.

Rivers-Auty, J. R., Smith, P. F., & Ashton, J. C. (2014). The cannabinoid CB2 receptor agonist GW405833 does not ameliorate brain damage induced by hypoxia-ischemia in rats. *Neuroscience Letters, 569*, 104–109.

Rizzo, M. A., Hadjimichael, O. C., Preiningerova, J., & Vollmer, T. L. (2004). Prevalence and treatment of spasticity reported by multiple sclerosis patients. *Multiple Sclerosis: Clinical and Laboratory Research, 10*(5), 589–595.

Robinson, P. A. (2008). Protein stability and aggregation in Parkinson's disease. *The Biochemical Journal, 413*(1), 1–13.

Robinson, S. A., Loiacono, R. E., Christopoulos, A., Sexton, P. M., & Malone, D. T. (2010). The effect of social isolation on rat brain expression of genes associated with endocannabinoid signaling. *Brain Research, 1343*, 153–167.

Rodgman, C., Kinzie, E., & Leimbach, E. (2011). Bad Mojo: Use of the new marijuana substitute leads to more and more ED visits for acute psychosis. *The American Journal of Emergency Medicine, 29*(2), 232.

Rog, D. J., Nurmikko, T. J., Friede, T., & Young, C. A. (2005). Randomized, controlled trial of cannabis-based medicine in central pain in multiple sclerosis. *Neurology, 65*(6), 812–819.

Rog, D. J., Nurmikko, T. J., & Young, C. A. (2007). Oromucosal Δ^9-tetrahydrocannabinol/cannabidiol for neuropathic pain associated with multiple sclerosis: An uncontrolled, open-label, 2-year extension trial. *Clinical Therapeutics, 29*(9), 2068–2079.

Rom, S., & Persidsky, Y. (2013). Cannabinoid receptor 2: Potential role in immunomodulation and neuroinflammation. *Journal of Neuroimmune Pharmacology: The Official Journal of the Society on Neuroimmune Pharmacology, 8*(3), 608–620.

Rom, S., Zuluaga-Ramirez, V., Dykstra, H., Reichenbach, N. L., Pacher, P., & Persidsky, Y. (2013). Selective activation of cannabinoid receptor 2 in leukocytes suppresses their engagement of the brain endothelium and protects the blood–brain barrier. *The American Journal of Pathology, 183*(5), 1548–1558.

Romero, J., Berrendero, F., Perez-Rosado, A., Manzanares, J., Rojo, A., Fernandez-Ruiz, J.J., … Ramos, J.A. (2000). Unilateral 6-hydroxydopamine lesions of nigrostriatal dopaminergic neurons increased CB1 receptor mRNA levels in the caudate-putamen. *Life Sciences, 66*(6), 485–494.

Roser, P., Vollenweider, F. X., & Kawohl, W. (2008). Potential antipsychotic properties of central cannabinoid (CB$_1$) receptor antagonists. *The World Journal of Biological Psychiatry*, 1–12.

Rossi, S., Buttari, F., Studer, V., Motta, C., Gravina, P., Castelli, M., …Centonze, D. (2011). The (AAT)n repeat of the cannabinoid CB1 receptor gene influences disease progression in relapsing multiple sclerosis. *Multiple Sclerosis: Clinical and Laboratory Research, 17*(3), 281–288.

Rubino, T., & Parolaro, D. (2016). The impact of exposure to cannabinoids in adolescence: Insights from animal models. *Biological Psychiatry, 79*(7), 578–585.

Rubino, T., Zamberletti, E., & Parolaro, D. (2012). Adolescent exposure to cannabis as a risk factor for psychiatric disorders. *Journal of Psychopharmacology, 26*(1), 177–188.

Russo, E., & Guy, G. W. (2006). A tale of two cannabinoids: The therapeutic rationale for combining tetrahydrocannabinol and cannabidiol. *Medical Hypotheses, 66*(2), 234–246.

Russo, E. B., Guy, G. W., & Robson, P. J. (2007). Cannabis, pain, and sleep: Lessons from therapeutic clinical trials of Sativex, a cannabis-based medicine. *Chemistry & Biodiversity, 4*(8), 1729–1743.

Russo, M., Naro, A., Leo, A., Sessa, E., D'Aleo, G., Bramanti, P., & Calabro, R. S. (2016). Evaluating Sativex® in neuropathic pain management: A clinical and neurophysiological assessment in multiple sclerosis. *Pain Medicine: The Official Journal of the American Academy of Pain Medicine*.

Saez, T. M., Aronne, M. P., Caltana, L., & Brusco, A. H. (2014). Prenatal exposure to the CB1 and CB2 cannabinoid receptor agonist WIN 55,212-2 alters migration of early-born glutamatergic neurons and GABAergic interneurons in the rat cerebral cortex. *Journal of Neurochemistry*, *129*(4), 637–648.

Sagredo, O., Gonzalez, S., Aroyo, I., Pazos, M. R., Benito, C., Lastres-Becker, I., … Fernandez-Ruiz, J. (2009). Cannabinoid CB2 receptor agonists protect the striatum against malonate toxicity: Relevance for Huntington's disease. *Glia*, *57*(11), 1154–1167.

Sagredo, O., Ramos, J. A., Decio, A., Mechoulam, R., & Fernandez-Ruiz, J. (2007). Cannabidiol reduced the striatal atrophy caused 3-nitropropionic acid in vivo by mechanisms independent of the activation of cannabinoid, vanilloid TRPV1 and adenosine A_{2A} receptors. *The European Journal of Neuroscience*, *26*(4), 843–851.

Sanchez, A. J., Gonzalez-Perez, P., Galve-Roperh, I., & Garcia-Merino, A. (2006). R-(+)-[2,3-Dihydro-5-methyl-3-(4-morpholinylmethyl)-pyrrolo-[1,2,3-de]-1,4 -benzoxazin-6-yl]-1-naphtalenylmethanone (WIN-2) ameliorates experimental autoimmune encephalomyelitis and induces encephalitogenic T cell apoptosis: Partial involvement of the CB_2 receptor. *Biochemical Pharmacology*, *72*(12), 1697–1706.

Sandyk, R., Consroe, P., Stern, L., Snider, S. R., & Bliklen, D. (1988). Preliminary trial of cannabidiol in Huntington's disease. Marijuana: An international research report. In G. Chesher, P. Consroe, & R. Musty (Eds.), *National campaign against drug abuse monograph series no. 7* (pp. 157–162). Canberra: Australian Government Publishing Service.

Sanudo-Pena, M. C., Patrick, S. L., Patrick, R. L., & Walker, J. M. (1996). Effects of intranigral cannabinoids on rotational behavior in rats: Interactions with the dopaminergic system. *Neuroscience Letters*, *206*(1), 21–24.

Sanudo-Pena, M. C., Tsou, K., & Walker, J. M. (1999). Motor actions of cannabinoids in the basal ganglia output nuclei. *Life Sciences*, *65*(6–7), 703–713.

Schwarcz, G., Karajgi, B., & McCarthy, R. (2009). Synthetic delta-9-tetrahydrocannabinol (dronabinol) can improve the symptoms of schizophrenia. *Journal of Clinical Psychopharmacology*, *29*(3), 255–258.

Schwartz, R. H., Gruenewald, P. J., Klitzner, M., & Fedio, P. (1989). Short-term memory impairment in cannabis-dependent adolescents. *The American Journal of Diseases of Children*, *143*(10), 1214–1219.

Scotter, E. L., Goodfellow, C. E., Graham, E. S., Dragunow, M., & Glass, M. (2010). Neuroprotective potential of CB1 receptor agonists in an in vitro model of Huntington's disease. *British Journal of Pharmacology*, *160*(3), 747–761.

Scuderi, C., Steardo, L., & Esposito, G. (2014). Cannabidiol promotes amyloid precursor protein ubiquitination and reduction of beta amyloid expression in SHSY5YAPP+ cells through PPARgamma involvement. *Phytotherapy Research*, *28*(7), 1007–1013.

Segovia, G., Mora, F., Crossman, A. R., & Brotchie, J. M. (2003). Effects of CB1 cannabinoid receptor modulating compounds on the hyperkinesia induced by high-dose levodopa in the reserpine-treated rat model of Parkinson's disease. *Movement Disorders: Official Journal of the Movement Disorder Society*, *18*(2), 138–149.

Seifert, J., Ossege, S., Emrich, H. M., Schneider, U., & Stuhrmann, M. (2007). No association of CNR1 gene variations with susceptibility to schizophrenia. *Neuroscience Letters*, *426*(1), 29–33.

Seillier, A., Advani, T., Cassano, T., Hensler, J. G., & Giuffrida, A. (2010). Inhibition of fatty-acid amide hydrolase and CB1 receptor antagonism differentially affect behavioural responses in normal and PCP-treated rats. *International Journal of Neuropsychopharmacology*, *13*(3), 373–386.

Sekar, A., Bialas, A. R., de Rivera, H., Davis, A., Hammond, T. R., Kamitaki, N., ... McCarroll, S. A. (2016). Schizophrenia risk from complex variation of complement component 4. *Nature, 530*(7589), 177–183.

Serpa, A., Correia, S., Ribeiro, J. A., Sebastiao, A. M., & Cascalheira, J. F. (2015). The combined inhibitory effect of the adenosine A1 and cannabinoid CB1 receptors on cAMP accumulation in the hippocampus is additive and independent of A1 receptor desensitization. *BioMed Research International,* 2015, 872684.

Serpa, A., Pinto, I., Bernardino, L., & Cascalheira, J. F. (2015). Combined neuroprotective action of adenosine A1 and cannabinoid CB1 receptors against NMDA-induced excitotoxicity in the hippocampus. *Neurochemistry International, 87,* 106–109.

Shakespeare, D. T., Boggild, M., & Young, C. (2003). Anti-spasticity agents for multiple sclerosis. *The Cochrane Database of Systematic Reviews, 4,* CD001332.

Shen, M., & Thayer, S. A. (1998). Cannabinoid receptor agonists protect cultured rat hippocampal neurons from excitotoxicity. *Molecular Pharmacology, 54*(3), 459–462.

Shenton, M. E., Kikinis, R., Jolesz, F. A., Pollak, S. D., LeMay, M., Wible, C. G., ... Coleman, M. (1992). Abnormalities of the left temporal lobe and thought disorder in schizophrenia. A quantitative magnetic resonance imaging study. *New England Journal of Medicine, 327*(9), 604–612.

Sherif, M., Radhakrishnan, R., D'Souza, D. C., & Ranganathan, M. (2016). Human laboratory studies on cannabinoids and psychosis. *Biological Psychiatry, 79*(7), 526–538.

Shinjyo, N., & Di Marzo, V. (2013). The effect of cannabichromene on adult neural stem/progenitor cells. *Neurochemistry International, 63*(5), 432–437.

Sieradzan, K. A., Fox, S. H., Hill, M., Dick, J. P., Crossman, A. R., & Brotchie, J. M. (2001). Cannabinoids reduce levodopa-induced dyskinesia in Parkinson's disease: A pilot study. *Neurology, 57*(11), 2108–2111.

Silverdale, M. A., McGuire, S., McInnes, A., Crossman, A. R., & Brotchie, J. M. (2001). Striatal cannabinoid CB1 receptor mRNA expression is decreased in the reserpine-treated rat model of Parkinson's disease. *Experimental Neurology, 169*(2), 400–406.

Simmons, J., Cookman, L., Kang, C., & Skinner, C. (2011). Three cases of "spice" exposure. *Clinical Toxicology (Philadelphia), 49*(5), 431–433.

Simmons, J. R., Skinner, C. G., Williams, J., Kang, C. S., Schwartz, M. D., & Wills, B. K. (2011). Intoxication from smoking "spice". *Annals of Emergency Medicine, 57*(2), 187–188.

Skaper, S. D., Buriani, A., Dal Toso, R., Petrelli, L., Romanello, S., Facci, L., & Leon, A. (1996). The ALIAmide palmitoylethanolamide and cannabinoids, but not anandamide, are protective in a delayed postglutamate paradigm of excitotoxic death in cerebellar granule neurons. *Proceedings of the National Academy of Sciences of the United States of America, 93*(9), 3984–3989.

Skosnik, P. D., Krishnan, G. P., Aydt, E. E., Kuhlenshmidt, H. A., & O'Donnell, B. F. (2006). Psychophysiological evidence of altered neural synchronization in cannabis use: Relationship to schizotypy. *The American Journal of Psychiatry, 163*(10), 1798–1805.

Skosnik, P. D., Park, S., Dobbs, L., & Gardner, W. L. (2008). Affect processing and positive syndrome schizotypy in cannabis users. *Psychiatry Research, 157*(1–3), 279–282.

Skosnik, P. D., Spatz-Glenn, L., & Park, S. (2001). Cannabis use is associated with schizotypy and attentional disinhibition. *Schizophrenia Research, 48*(1), 83–92.

Smith, P. F. (2007). Symptomatic treatment of multiple sclerosis using cannabinoids: Recent advances. *Expert Review of Neurotherapeutics, 7*(9), 1157–1163.

Solas, M., Francis, P. T., Franco, R., & Ramirez, M. J. (2013). CB2 receptor and amyloid pathology in frontal cortex of Alzheimer's disease patients. *Neurobiology of Aging, 34*(3), 805–808.

Soltys, J., Yushak, M., & Mao-Draayer, Y. (2010). Regulation of neural progenitor cell fate by anandamide. *Biochemical and Biophysical Research Communications, 400*(1), 21–26.

Spano, M. S., Fadda, P., Frau, R., Fattore, L., & Fratta, W. (2010). Cannabinoid self-administration attenuates PCP-induced schizophrenia-like symptoms in adult rats. *European Neuropsychopharmacology: The Journal of the European College of Neuropsychopharmacology*, *20*(1), 25–36.

Spires, T. L., Grote, H. E., Varshney, N. K., Cordery, P. M., van Dellen, A., Blakemore, C., & Hannan, A. J. (2004). Environmental enrichment rescues protein deficits in a mouse model of Huntington's disease, indicating a possible disease mechanism. *The Journal of Neuroscience: The Official Journal of the Society for Neuroscience*, *24*(9), 2270–2276.

Stadelmann, C., Wegner, C., & Bruck, W. (2011). Inflammation, demyelination, and degeneration - recent insights from MS pathology. *Biochimica et Biophysica Acta*, *1812*(2), 275–282.

Steffens, M., Szabo, B., Klar, M., Rominger, A., Zentner, J., & Feuerstein, T. J. (2003). Modulation of electrically evoked acetylcholine release through cannabinoid CB1 receptors: Evidence for an endocannabinoid tone in the human neocortex. *Neuroscience*, *120*(2), 455–465.

van der Stelt, M., Fox, S. H., Hill, M., Crossman, A. R., Petrosino, S., Di Marzo, V., & Brotchie, J. M. (2005). A role for endocannabinoids in the generation of parkinsonism and levodopa-induced dyskinesia in MPTP-lesioned non-human primate models of Parkinson's disease. *FASEB Journal: Official Publication of the Federation of American Societies for Experimental Biology*, *19*(9), 1140–1142.

van der Stelt, M., Mazzola, C., Esposito, G., Matias, I., Petrosino, S., De Filippis, D., ... Di Marzo, V. (2006). Endocannabinoids and beta-amyloid-induced neurotoxicity in vivo: Effect of pharmacological elevation of endocannabinoid levels. *Cellular and Molecular Life Sciences*, *63*(12), 1410–1424.

Struble, R. G., Ala, T., Patrylo, P. R., Brewer, G. J., & Yan, X. X. (2010). Is brain amyloid production a cause or a result of dementia of the Alzheimer's type? *Journal of Alzheimer's Disease*, *22*(2), 393–399.

Suarez, J., Llorente, R., Romero-Zerbo, S. Y., Mateos, B., Bermudez-Silva, F. J., de Fonseca, F. R., & Viveros, M. P. (2009). Early maternal deprivation induces gender-dependent changes on the expression of hippocampal CB_1 and CB_2 cannabinoid receptors of neonatal rats. *Hippocampus*, *19*(7), 623–632.

Suarez, J., Rivera, P., Llorente, R., Romero-Zerbo, S. Y., Bermudez-Silva, F. J., de Fonseca, F. R., & Viveros, M. P. (2010). Early maternal deprivation induces changes on the expression of 2-AG biosynthesis and degradation enzymes in neonatal rat hippocampus. *Brain Research*, *1349*, 162–173.

Sullivan, J. M. (2000). Cellular and molecular mechanisms underlying learning and memory impairments produced by cannabinoids. *Learning & Memory*, *7*(3), 132–139.

Sutterlin, P., Williams, E. J., Chambers, D., Saraf, K., von Schack, D., Reisenberg, M., ... Williams, G. (2013). The molecular basis of the cooperation between EGF, FGF and eCB receptors in the regulation of neural stem cell function. *Molecular and Cellular Neurosciences*, *52*, 20–30.

Svendsen, K. B., Jensen, T. S., & Bach, F. W. (2004). Does the cannabinoid dronabinol reduce central pain in multiple sclerosis? Randomised double blind placebo controlled crossover trial. *British Medical Journal*, *329*(7460), 253.

Tadaiesky, M. T., Dombrowski, P. A., Da Cunha, C., & Takahashi, R. N. (2010). Effects of SR141716A on cognitive and depression-related behavior in an animal model of premotor Parkinson's disease. *Parkinson's Disease*, *2010*, 238491.

Tapia-Arancibia, L., Aliaga, E., Silhol, M., & Arancibia, S. (2008). New insights into brain BDNF function in normal aging and Alzheimer disease. *Brain Research Reviews*.

Thaker, G. K., & Carpenter, W. T., Jr. (2001). Advances in schizophrenia. *Nature Medicine*, *7*(6), 667–671.

Tzavara, E. T., Li, D. L., Moutsimilli, L., Bisogno, T., Di Marzo, V., Phebus, L. A., … Giros, B. (2006). Endocannabinoids activate transient receptor potential vanilloid 1 receptors to reduce hyperdopaminergia-related hyperactivity: Therapeutic implications. *Biological Psychiatry, 59*(6), 508–515.

Ujike, H., Takaki, M., Nakata, K., Tanaka, Y., Takeda, T., Kodama, M., … Kuroda, S. (2002). CNR1, central cannabinoid receptor gene, associated with susceptibility to hebephrenic schizophrenia. *Molecular Psychiatry, 7*(5), 515–518.

Ungerleider, J. T., Andyrsiak, T., Fairbanks, L., Ellison, G. W., & Myers, L. W. (1987). Delta-9-THC in the treatment of spasticity associated with multiple sclerosis. *Advances in Alcohol & Substance Abuse, 7*(1), 39–50.

Uriguen, L., Garcia-Fuster, M. J., Callado, L. F., Morentin, B., La Harpe, R., Casado, V., … Meana, J. J. (2009). Immunodensity and mRNA expression of A_{2A} adenosine, D_2 dopamine, and CB_1 cannabinoid receptors in postmortem frontal cortex of subjects with schizophrenia: Effect of antipsychotic treatment. *Psychopharmacology, 206*(2), 313–324.

Van Dam, D., & De Deyn, P. P. (2011). Animal models in the drug discovery pipeline for Alzheimer's disease. *British Journal of Pharmacology*.

Van Laere, K., Casteels, C., Dhollander, I., Goffin, K., Grachev, I., Bormans, G., & Vandenberghe, W. (2010). Widespread decrease of type 1 cannabinoid receptor availability in Huntington disease in vivo. *Journal of Nuclear Medicine: Official Publication, Society of Nuclear Medicine, 51*(9), 1413–1417.

Van Sickle, M. D., Duncan, M., Kingsley, P. J., Mouihate, A., Urbani, P., Mackie, K., … Sharkey, K. A. (2005). Identification and functional characterization of brainstem cannabinoid CB_2 receptors. *Science, 310*(5746), 329–332.

Vaney, C., Heinzel-Gutenbrunner, M., Jobin, P., Tschopp, F., Gattlen, B., Hagen, U., … Reif, M. (2004). Efficacy, safety and tolerability of an orally administered cannabis extract in the treatment of spasticity in patients with multiple sclerosis: A randomized, double-blind, placebo-controlled, crossover study. *Multiple Sclerosis: Clinical and Laboratory Research, 10*(4), 417–424.

Varvel, S. A., Wise, L. E., & Lichtman, A. H. (2009). Are CB_1 receptor antagonists nootropic or cognitive impairing agents? *Drug Development Research, 70*(8), 555–565.

Vendel, E., & de Lange, E. C. (2014). Functions of the CB_1 and CB_2 receptors in neuroprotection at the level of the blood–brain barrier. *Neuromolecular Medicine, 16*(3), 620–642.

Venderova, K., Ruzicka, E., Vorisek, V., & Visnovsky, P. (2004). Survey on cannabis use in Parkinson's disease: Subjective improvement of motor symptoms. *Movement Disorders: Official Journal of the Movement Disorder Society, 19*(9), 1102–1106.

Vigano, D., Guidali, C., Petrosino, S., Realini, N., Rubino, T., Di Marzo, V., & Parolaro, D. (2008). Involvement of the endocannabinoid system in phencyclidine-induced cognitive deficits modelling schizophrenia. *International Journal of Neuropsychopharmacology*, 1–16.

van Vliet, S. A., Vanwersch, R. A., Jongsma, M. J., Olivier, B., & Philippens, I. H. (2008). Therapeutic effects of $Δ^9$-THC and modafinil in a marmoset Parkinson model. *European Neuropsychopharmacology: The Journal of the European College of Neuropsychopharmacology, 18*(5), 383–389.

Volicer, L., Stelly, M., Morris, J., McLaughlin, J., & Volicer, B. J. (1997). Effects of dronabinol on anorexia and disturbed behavior in patients with Alzheimer's disease. *International Journal of Geriatric Psychiatry, 12*(9), 913–919.

Volk, D. W., & Lewis, D. A. (2016). The role of endocannabinoid signaling in cortical inhibitory neuron dysfunction in schizophrenia. *Biological Psychiatry, 79*(7), 595–603.

Voruganti, L. N., Slomka, P., Zabel, P., Mattar, A., & Awad, A. G. (2001). Cannabis induced dopamine release: An in-vivo SPECT study. *Psychiatry Research, 107*(3), 173–177.

Wade, D. T., Collin, C., Stott, C., & Duncombe, P. (2010). Meta-analysis of the efficacy and safety of Sativex (nabiximols), on spasticity in people with multiple sclerosis. *Multiple Sclerosis: Clinical and Laboratory Research, 16*(6), 707–714.

Wade, D. T., Makela, P. M., House, H., Bateman, C., & Robson, P. (2006). Long-term use of a cannabis-based medicine in the treatment of spasticity and other symptoms in multiple sclerosis. *Multiple Sclerosis: Clinical and Laboratory Research, 12*(5), 639–645.

Wade, D. T., Makela, P., Robson, P., House, H., & Bateman, C. (2004). Do cannabis-based medicinal extracts have general or specific effects on symptoms in multiple sclerosis? A double-blind, randomized, placebo-controlled study on 160 patients. *Multiple Sclerosis: Clinical and Laboratory Research, 10*(4), 434–441.

Wade, D. T., Robson, P., House, H., Makela, P., & Aram, J. (2003). A preliminary controlled study to determine whether whole-plant cannabis extracts can improve intractable neurogenic symptoms. *Clinical Rehabilitation, 17*(1), 21–29.

Waksman, Y., Olson, J. M., Carlisle, S. J., & Cabral, G. A. (1999). The central cannabinoid receptor (CB1) mediates inhibition of nitric oxide production by rat microglial cells. *The Journal of Pharmacology and Experimental Therapeutics, 288*(3), 1357–1366.

Walsh, S., Gorman, A. M., Finn, D. P., & Dowd, E. (2010). The effects of cannabinoid drugs on abnormal involuntary movements in dyskinetic and non-dyskinetic 6-hydroxydopamine lesioned rats. *Brain Research, 1363*, 40–48.

Walter, L., & Stella, N. (2004). Cannabinoids and neuroinflammation. *British Journal of Pharmacology, 141*(5), 775–785.

Walther, S., Mahlberg, R., Eichmann, U., & Kunz, D. (2006). Delta-9-tetrahydrocannabinol for nighttime agitation in severe dementia. *Psychopharmacology, 185*(4), 524–528.

Ware, M. A., Adams, H., & Guy, G. W. (2005). The medicinal use of cannabis in the UK: Results of a nationwide survey. *International Journal of Clinical Practice, 59*(3), 291–295.

Webb, M., Luo, L., Ma, J. Y., & Tham, C. S. (2008). Genetic deletion of Fatty Acid Amide Hydrolase results in improved long-term outcome in chronic autoimmune encephalitis. *Neuroscience Letters, 439*(1), 106–110.

Westlake, T. M., Howlett, A. C., Bonner, T. I., Matsuda, L. A., & Herkenham, M. (1994). Cannabinoid receptor binding and messenger RNA expression in human brain: An in vitro receptor autoradiography and in situ hybridization histochemistry study of normal aged and Alzheimer's brains. *Neuroscience, 63*(3), 637–652.

Wexler, N. S., Lorimer, J., Porter, J., Gomez, F., Moskowitz, C., Shackell, E., & Landwehrmeyer, B. (2004). Venezuelan kindreds reveal that genetic and environmental factors modulate Huntington's disease age of onset. *Proceedings of the National Academy of Sciences of the United States of America, 101*(10), 3498–3503.

Wilkinson, J. D., Whalley, B. J., Baker, D., Pryce, G., Constanti, A., Gibbons, S., & Williamson, E. M. (2003). Medicinal cannabis: Is Δ9-tetrahydrocannabinol necessary for all its effects? *The Journal of Pharmacy and Pharmacology, 55*(12), 1687–1694.

Wilson, R. I., & Nicoll, R. A. (2002). Neuroscience - endocannabinoid signaling in the brain. *Science, 296*(5568), 678–682.

Wissel, J., Haydn, T., Muller, J., Brenneis, C., Berger, T., Poewe, W., & Schelosky, L. D. (2006). Low dose treatment with the synthetic cannabinoid Nabilone significantly reduces spasticity-related pain: A double-blind placebo-controlled cross-over trial. *Journal of Neurology, 253*(10), 1337–1341.

Wolf, S. A., Bick-Sander, A., Fabel, K., Leal-Galicia, P., Tauber, S., Ramirez-Rodriguez, G., … Kempermann, G. (2010). Cannabinoid receptor CB1 mediates baseline and activity-induced survival of new neurons in adult hippocampal neurogenesis. *Cell Communication and Signaling, 8*, 12.

Wong, D. F., Kuwabara, H., Horti, A. G., Raymont, V., Brasic, J., Guevara, M., … Cascella, N. (2010). Quantification of cerebral cannabinoid receptors subtype 1 (CB1) in healthy subjects and schizophrenia by the novel PET radioligand [^{11}C]OMAR. *Neuroimage, 52*(4), 1505–1513.

Xapelli, S., Agasse, F., Sarda-Arroyo, L., Bernardino, L., Santos, T., Ribeiro, F. F., … Malva, J. O. (2013). Activation of type 1 cannabinoid receptor (CB1R) promotes neurogenesis in murine subventricular zone cell cultures. *PLoS One, 8*(5), e63529.

Xu, C., Loh, H. H., & Law, P.Y. (2015). Effects of addictive drugs on adult neural stem/progenitor cells. *Cellular and Molecular Life Sciences.*

Yiangou, Y., Facer, P., Durrenberger, P., Chessell, I. P., Naylor, A., Bountra, C., ... Anand, P. (2006). COX-2, CB2 and P2X7-immunoreactivities are increased in activated microglial cells/macrophages of multiple sclerosis and amyotrophic lateral sclerosis spinal cord. *BMC Neurology, 6*, 12.

Yu, S. J., Reiner, D., Shen, H., Wu, K. J., Liu, Q. R., & Wang, Y. (2015). Time-dependent protection of CB2 receptor agonist in stroke. *PLoS One, 10*(7), e0132487.

Zajicek, J. (2002). Cannabinoids on trial for multiple sclerosis. *Lancet Neurology, 1*(3), 147.

Zajicek, J. P., & Apostu, V. I. (2011). Role of cannabinoids in multiple sclerosis. *CNS Drugs, 25*(3), 187–201.

Zajicek, J., Fox, P., Sanders, H., Wright, D., Vickery, J., Nunn, A., & Thompson, A. (2003). Cannabinoids for treatment of spasticity and other symptoms related to multiple sclerosis (CAMS study): Multicentre randomised placebo-controlled trial. *Lancet, 362*(9395), 1517–1526.

Zajicek, J. P., Sanders, H. P., Wright, D. E., Vickery, P. J., Ingram, W. M., Reilly, S. M., ... Thompson, A. J. (2005). Cannabinoids in multiple sclerosis (CAMS) study: Safety and efficacy data for 12 months follow up. *Journal of Neurology, Neurosurgery & Psychiatry, 76*(12), 1664–1669.

Zamberletti, E., Vigano, D., Guidali, C., Rubino, T., & Parolaro, D. (2010). Long-lasting recovery of psychotic-like symptoms in isolation-reared rats after chronic but not acute treatment with the cannabinoid antagonist AM251. *International Journal of Neuropsychopharmacology*, 1–14.

Zammit, S., Spurlock, G., Williams, H., Norton, N., Williams, N., O'Donovan, M. C., & Owen, M. J. (2007). Genotype effects of CHRNA7, CNR1 and COMT in schizophrenia: Interactions with tobacco and cannabis use. *The British Journal of Psychiatry: The Journal of Mental Science, 191*, 402–407.

Zavitsanou, K., Garrick, T., & Huang, X. F. (2004). Selective antagonist [^{3}H]SR141716A binding to cannabinoid CB1 receptors is increased in the anterior cingulate cortex in schizophrenia. *Progress in Neuro-psychopharmacology & Biological Psychiatry, 28*(2), 355–360.

Zeng, B. Y., Dass, B., Owen, A., Rose, S., Cannizzaro, C., Tel, B. C., & Jenner, P. (1999). Chronic L-DOPA treatment increases striatal cannabinoid CB1 receptor mRNA expression in 6-hydroxydopamine-lesioned rats. *Neuroscience Letters, 276*(2), 71–74.

Zhang, H., Hilton, D. A., Hanemann, C. O., & Zajicek, J. (2011). Cannabinoid receptor and N-acyl phosphatidylethanolamine phospholipase D-evidence for altered expression in multiple sclerosis. *Brain Pathology.*

Zhang, M., Martin, B. R., Adler, M. W., Razdan, R. K., Ganea, D., & Tuma, R. F. (2008). Modulation of the balance between cannabinoid CB_1 and CB_2 receptor activation during cerebral ischemic/reperfusion injury. *Neuroscience, 152*(3), 753–760.

Zuardi, A. W., Crippa, J. A., Hallak, J. E., Moreira, F. A., & Guimaraes, F. S. (2006). Cannabidiol, a *Cannabis sativa* constituent, as an antipsychotic drug. *Brazilian Journal of Medical and Biological Research, 39*(4), 421–429.

Zuardi, A. W., Crippa, J. A., Hallak, J. E., Pinto, J. P., Chagas, M. H., Rodrigues, G. G., ... Tumas, V. (2009). Cannabidiol for the treatment of psychosis in Parkinson's disease. *Journal of Psychopharmacology, 23*(8), 979–983.

Zuardi, A. W., Hallak, J. E., Dursun, S. M., Morais, S. L., Sanches, R. F., Musty, R. E., & Crippa, J. A. (2006). Cannabidiol monotherapy for treatment-resistant schizophrenia. *Journal of Psychopharmacology, 20*(5), 683–686.

CHAPTER 6

The Role of the Endocannabinoid System in Addiction

Jose M. Trigo[1], Bernard Le Foll[1,2]
[1]Centre for Addiction and Mental Health (CAMH), Toronto, ON, Canada; [2]University of Toronto, Toronto, ON, Canada

INTRODUCTION

Addiction has been defined as a chronic, relapsing disorder characterized by a compulsion to seek and take drugs, loss of control over drug intake, and the emergence of a negative emotional state (i.e., withdrawal syndrome) when the drug is not present (Koob & Le Moal, 2008). A major feature of drug addiction is the persistent vulnerability to relapse to drug use even after prolonged abstinence, which seems to be correlated with long-lasting functional and physical changes in brain structures following chronic drug intake (Wang et al., 2016; Zou et al., 2015). Therefore a major goal of addiction research is the search of pharmacological tools able to prevent relapse.

Multiple neurotransmitter systems have been found to be related to drug addiction. In fact, because of these multiple interactions it seems very unlikely that just one single neurotransmitter is responsible for all the neurobiological adaptations underlying substance use disorders. Even though a great number of studies have focused on how drugs of abuse increase extracellular dopamine levels in the mesolimbic system, a myriad of other studies have also shown the involvement of other neurotransmitters including serotonin, glutamate, γ-aminobutyric acid (GABA), acetylcholine (ACh), and several peptides. Among the different neurotransmitter systems studied, research in the endogenous cannabinoid system (ECS) has gained importance in the past decades. In this chapter, we will first provide an overview of the ECS and the neurobiology of reward followed by a review of preclinical evidence on the involvement of the different components of the ECS in diverse aspects of the addiction process, including acquisition, maintenance, and relapse to different drugs including alcohol, nicotine, opioids, psychostimulants, and cannabinoids.

The Endocannabinoid System
ISBN 978-0-12-809666-6
http://dx.doi.org/10.1016/B978-0-12-809666-6.00006-X

THE ENDOCANNABINOID SYSTEM: OVERVIEW

Cannabis sativa and its botanical varieties have been used for a long time; there are recorded uses of hemp dated between 10,000 and 3000 BC (Booth, 2003). However, the scientific study of the pharmacological properties of cannabis only started not long ago. The main psychoactive component of *C. sativa* Δ^9-tetrahydrocannabinol (THC) was isolated in 1964 (Mechoulam & Gaoni, 1964). On the other hand, the discovery of the structure and the expression of the cloned complementary DNA of the cannabinoid receptor 1 (CB$_1$) occurred in 1990 (Matsuda, Lolait, Brownstein, Young, & Bonner, 1990). The discovery of the CB$_1$ receptors was rapidly followed by the identification of other components of the ECS. Thus the cannabinoid receptor 2 (CB$_2$) was cloned 3 years later (Munro, Thomas, & Abu-Shaar, 1993) and the existence of the endogenous cannabinoids *N*-arachidonoylethanolamine (anandamide) and 2-arachidonoylglycerol (2-AG) were also revealed on the 1990s (Devane et al., 1992; Mechoulam et al., 1995; Sugiura et al., 1995). The ECS comprehends the cannabinoid receptors and endogenous cannabinoids together with the enzymes responsible for their synthesis, uptake, and metabolism. Since the initial identification of the cannabinoid receptors and the first endogenous cannabinoids, the list of components of the ECS and the knowledge of the molecular mechanisms of the ECS have grown significantly (see Ronan, Wongngamnit, & Beresford, 2016 for review). In contrast, the study of the pharmacological properties of the newer endogenous cannabinoids (e.g., virodhamine and 2-arachidonoylether) has not been fully developed and will not be covered here. We will focus mostly on the role of CB$_1$ and CB$_2$ receptors and of the main endogenous cannabinoids (anandamide and 2-AG).

Both CB$_1$ and CB$_2$ receptors are G protein–coupled receptors coupled to similar transduction systems (i.e., G$_i$ or G$_o$ proteins). In fact, the CB$_1$ receptors are the most abundant G protein–coupled receptors in the human brain. Traditionally, CB$_2$ receptors have been mostly identified on the periphery (e.g., in the immune system). However, new reports indicate that CB$_2$ receptors are also present centrally in neurons and glia (e.g., in brain microglia during neuroinflammation) (Atwood & Mackie, 2010). The presence of CB$_2$ centrally might explain why manipulating CB$_2$ can affect responses to several addictive substances.

An important characteristic of the endogenous cannabinoids is that they are synthesized on demand from phospholipid precursors located in the membrane and are not stored in vesicles but immediately released. The precursor of anandamide is *N*-arachidonoylphosphatidylethanolamine (NAPE), and 2-AG is synthesized from arachidonic-acid (AA)-containing diacylglycerol. Once released anandamide is degraded by the fatty acid amide hydrolase (FAAH) enzyme, whereas 2-AG is metabolized by the monoacylglycerol lipase (MAGL), which seems to be differently located in the cell (i.e., pre/post synaptic for MAGL and FAAH, respectively).

As transmembrane uptake processes do exist in other neurotransmitter systems, it has been hypothesized that the ECS might also have a dedicated transport process contributing to inactivate endogenous cannabinoids once released. However, such a transporter or reuptake system has not been yet clearly identified (i.e., cloned) yet for the ECS. On the other hand, there are studies suggesting the existence of a bidirectional endocannabinoid transport across membranes (Chicca, Marazzi, Nicolussi, & Gertsch, 2012; Nicolussi & Gertsch, 2015). The development of new experimental approaches to study the endocannabinoid transport process (Rau, Nicolussi, Chicca, & Gertsch, 2016) might help to put light on the existence or not of endocannabinoid uptake in the near future. In this chapter, we will focus on the role of the ECS on addictive properties of drugs. For more information on the endocannabinoid synthesis and degradation processes, we would like to refer the reader to new reviews on this topic (Lu & Mackie, 2016) or to the Chapter 1 in this book.

A first glance at the distribution of the receptors of the ECS (e.g., CB_1 receptors) in the brain (i.e., presence in areas involved in emotional and cognitive processing) provides a hint for its possible involvement in such emotional and cognitive processes, which are highly related to the neurobiological basis of reward. Indeed, the presence of the CB_1 receptor is especially dense in brain areas related to the neurobiology of drug reward (Tsou, Brown, Sanudo-Pena, Mackie, & Walker, 1998). Therefore CB_1 receptors are highly expressed in important regions for the processing of rewarding stimuli including ventral tegmental area (VTA), ventral pallidum, caudate putamen, and nucleus accumbens (NAc). There is also a dense expression of CB_1 receptors in areas related to cognitive function, emotion formation, and motivation such as the prefrontal cortex, amygdala, cingulate cortex, and hypothalamus. As it has been extensively reported, the mesocorticolimbic dopamine pathways play an essential role in the addiction processes (Di Chiara & Imperato, 1988). It should be noted, however, that CB_1 receptors are not expressed on the dopamine neurons in the midbrain; instead the endocannabinoids are able to modulate VTA synaptic signaling through the glutamatergic and GABAergic afferences to the VTA dopaminergic neurons (see Parsons & Hurd, 2015 for review). Thus CB_1

receptors are located in both presynaptic glutamatergic neurons projecting to VTA from cortical areas and in GABAergic terminals projecting from globus pallidus or medium spiny neurons projecting from the NAc to VTA. On the other hand, CB_1 receptors located in glutamatergic terminals projecting to the NAc correspondingly modulate NAc medium spiny neurons and fast-spiking interneurons within the NAc.

This review will focus on preclinical findings obtained with the conditioned place preference (CPP) paradigm and with drug self-administration paradigm. The CPP paradigm is a standard preclinical behavioral model generally used to study the rewarding (but also aversive) effects of drugs. The basic characteristics of this paradigm involve the association (conditioning) of a particular environment (place) with drug treatment, followed by the association of a different environment with the absence of the drug (i.e., the drug's vehicle). In contrast, the self-administration paradigm consists of a procedure that allows the experimental animal to activate an operandum (e.g., by a lever press) to obtain a reward (e.g., an intravenous infusion of a drug). Under the self-administration paradigm the drug is voluntarily taken, whereas under CPP paradigm the drug administration is controlled by the experimenter. The drug self-administration paradigm also allows to measure drug taking under various schedules of reinforcement (e.g., fixed ratio schedules or progressive ratio schedules), and drug-seeking behaviors after a phase of extinction (reinstatement of previously extinguished drug seeking behavior, a model of relapse) (Panlilio & Goldberg, 2007).

In the following sections, we will describe the role of the ECS in the addictive properties of various drugs of abuse.

ALCOHOL

Alcohol is currently the most widely used drug, contributing to a variety of medical and socioeconomic problems (Global-Drug-Survey, 2016). Alcohol produces its depressant effects on the central nervous system (CNS) by acting on different neurotransmitters including GABA and glutamate (Grobin, Matthews, Devaud, & Morrow, 1998; Morrow, Suzdak, Karanian, & Paul, 1988; Most, Ferguson, & Harris, 2014). The repeated administration of alcohol results in neuroadaptive changes leading to the development of tolerance and withdrawal symptoms when alcohol is not present. Accordingly, studies in rodents have suggested the existence of adaptations on the ECS in response to the repeated exposure to alcohol that might be distinct of those occurring during acute exposure to alcohol (Serrano et al., 2012).

The existence of interactions between the ECS and alcohol was evidenced in the 1970s when cross-tolerance effects between THC and alcohol in rodents was reported (Jones & Stone, 1970; Newman, Lutz, Gould, & Domino, 1972; Siemens & Doyle, 1979; Sprague & Craigmill, 1976). Subsequent observations both in humans and rodents showed that the prolonged administration of alcohol increased the formation of fatty acid ethyl esters and the anandamide precursor NAPE in different organs including the brain (Basavarajappa & Hungund, 1999a; Hungund, Goldstein, Villegas, & Cooper, 1988; Laposata & Lange, 1986), further supporting the idea that alcohol exposure might have an impact on the ECS at the same time. Therefore it has been shown that prolonged administration of alcohol increased brain levels of both anandamide and 2-AG in rodents (Basavarajappa, Saito, Cooper, & Hungund, 2000, 2003). On the other hand, chronic administration of ethanol also downregulated CB_1 receptors, agonist-stimulated [35S]GTP gamma S binding, and its CB_1 gene expression in rodents (Basavarajappa, Cooper, & Hungund, 1998; Basavarajappa & Hungund, 1999b; Ortiz, Oliva, Perez-Rial, Palomo, & Manzanares, 2004a). Indeed, repeated administration of alcohol is able to modify the acute response to cannabinoids in experimental animals. Therefore behavioral studies have shown that repeated administration of alcohol was able to alter three of the four "tetrad" behaviors (hypothermic, hypolocomotive, and antinociceptive but not cataleptic) typically induced by the acute administration of the CB_1 agonist WIN 55,212-2 in rats (Pava et al., 2012).

In view of the effects that the repeated exposure to alcohol has in the ECS and the cross-tolerance effects between THC and alcohol, some investigators hypothesized that the ECS might actually play an active role in the pharmacological effects of alcohol and in the neuroadaptive changes occurring during the development of tolerance to alcohol. A growing body of data over the past 20 years has evidenced the involvement of the ECS in regulating both the acute effects of alcohol and the development of tolerance and susceptibility to relapse (see Henderson-Redmond, Guindon, & Morgan, 2016; Pava & Woodward, 2012 for detailed reviews on this topic). Importantly, the ability of alcohol to induce changes in the ECS might be distinct in the different components of the ECS (e.g., in anandamide vs. 2-AG or CB_1 vs. CB_2 receptors) (Alvarez-Jaimes, Stouffer, & Parsons, 2009; Mitrirattanakul et al., 2007; Zheng, Wu, Dong, Ding, & Song, 2015), and might also differ between brain regions (Ceccarini, Casteels, Koole, Bormans, & Van Laere, 2013; Ortiz, Oliva, Perez-Rial, Palomo, & Manzanares, 2004b), and consequently, this could have reflect on how the response of the ECS is affected and region-specifically recovers over time during alcohol abstinence (Ceccarini et al., 2013; Moranta, Esteban, & Garcia-Sevilla, 2006; Vinod, Yalamanchili, Xie, Cooper, & Hungund, 2006). In the following subsections, we will describe the involvement of the different components of the ECS in the acute and chronic effects of alcohol.

CB_1 Modulation

The absence of CB_1 receptors significantly reduced voluntary alcohol consumption (Hungund, Szakall, Adam, Basavarajappa, & Vadasz, 2003), alcohol self-administration, and alcohol-induced CPP in mice (Houchi et al., 2005; Thanos, Dimitrakakis, Rice, Gifford, & Volkow, 2005). Similarly, the administration of the CB_1 inverse agonist/antagonist rimonabant reduced voluntary alcohol intake in mice (Arnone et al., 1997; Marinho et al., 2015) and in rats (Colombo et al., 1998). In contrast, the $CB_{1/2}$ receptor agonists WIN 55,212-2 or CP 55,940 seem to facilitate bingelike alcohol intake and increase voluntary alcohol drinking in rodents (Colombo et al., 2002; Linsenbardt & Boehm, 2009). Moreover, these facilitating effects of CB_1 receptor agonists are blocked by the CB_1 receptor inverse agonist/antagonist rimonabant (Colombo et al., 2002). Blocking CB_1 receptor with either rimonabant or surinabant (SR 147778) also abolished the alcohol deprivation effect (i.e., the extra intake of alcohol occurring after a period of abstinence) in rats (Gessa, Serra, Vacca, Carai, & Colombo, 2005; Serra et al., 2002) suggesting that blocking the CB_1 receptor might have antirelapse properties (see Maccioni, Colombo, & Carai, 2010 for review). Administration of alcohol increases the extracellular endocannabinoid levels (anandamide and 2-AG) in the NAc (Caille, Alvarez-Jaimes, Polis, Stouffer, & Parsons, 2007). Indeed, acute alcohol administration interferes with endocannabinoid-mediated mechanisms on different brain structures such as the striatum (Clarke & Adermark, 2010), the basolateral amygdala (Perra, Pillolla, Luchicchi, & Pistis, 2008; Talani & Lovinger, 2015), and the hippocampus (Basavarajappa, Ninan, & Arancio, 2008). The ability of alcohol to elicit dopamine release in the NAc is abolished in CB_1 knockout mice (Hungund et al., 2003). Additionally, the pharmacological blockade of CB_1 receptor reversed alcohol stimulated firing in VTA dopaminergic neurons and NAc neuron spiking responses (Perra et al., 2005). The release of endogenous cannabinoids might enhance/potentiate previous activations of the dopaminergic neurons induced by alcohol and other drugs (Cheer, Wassum, Heien, Phillips, & Wightman, 2004). Therefore the involvement of the CB_1 receptors might be required to obtain a full response on the phasic release of dopamine in response to alcohol and other abused substances

(Cheer et al., 2007). Indeed, it has been proposed that the release of endocannabinoids might serve to modulate dopamine neuron excitability and busting, providing with a "fine-tune mechanism" for dopamine release (Riegel & Lupica, 2004).

The effects of alcohol on endocannabinoid levels might result in changes in the endocannabinoid signal. Different studies have suggested that the effects of chronic alcohol on the ECS are mediated by downregulation of CB_1 receptors or its gene expression (Basavarajappa et al., 1998; Basavarajappa & Hungund, 1999b; Mitrirattanakul et al., 2007; Ortiz et al., 2004a; Pava et al., 2012; Vinod et al., 2006). Indeed, it has been shown that the absence of CB_1 receptors impaired the neuroadaptations induced by chronic alcohol in key players on the development of alcohol dependence as the glutamatergic NMDA receptors and the GABA (A) receptors in mice (Warnault et al., 2007). Interestingly, in line with the idea of CB_1 receptors mediating neuroadaptations induced by chronic ethanol, behavioral studies have shown that treatment with the CB_1 antagonist/inverse agonists AM251 and SR147778 normalized some of the neurochemical alterations induced by chronic alcohol (e.g., AM251 blocked or reduced changes produced by alcohol in tyrosine hydroxylase and corticotrophin-releasing gene expression) and reduced ethanol consumption in rats (Femenia, Garcia-Gutierrez, & Manzanares, 2010; Lallemand & De Witte, 2006). On the other hand, it has been observed that CB_1 agonists (e.g., THC) are able to ameliorate the severity of alcohol withdrawal in mice (Blum, Hudson, Friedman, & Wallace, 1975; Sprague & Craigmill, 1978).

Remarkably, some of the effects of chronic alcohol on the ECS in experimental animals seem to be reversible and time-dependent. Several preclinical studies have suggested that the cross-tolerance effects between cannabinoids and chronic alcohol can be reversed after some time of alcohol abstinence. Therefore Pava et al. (2012) showed that the tolerance to the tetrad behaviors induced by the CB_1 agonist WIN 55,212-2 in alcohol-dependent mice was reversed after 10 days of alcohol abstinence. However, it has also been shown that other effects (e.g., the modulating effects of WIN 55,212-2 on monoamine synthesis) are not fully recovered following short abstinence periods (e.g., 24 h) in rats (Moranta et al., 2006), which might suggest the existence of a time dependent recovery on the ECS following chronic alcohol.

Fatty Acid Amide Hydrolase/Anandamide Modulation

As described in CB_1 Modulation section, exogenous CB_1 agonists seem to facilitate alcohol intake and, on the contrary, the administration of CB_1

antagonists reduce alcohol intake in preclinical models. Therefore it might be expected that increasing the endogenous cannabinoid tone (e.g., absence of FAAH) might also facilitate alcohol intake. Studies in FAAH knockout mice have revealed increased consumption and preference for alcohol in these animals (Basavarajappa, Yalamanchili, Cravatt, Cooper, & Hungund, 2006; Blednov, Cravatt, Boehm, Walker, & Harris, 2007). Accordingly, pharmacological studies have shown that the administration of the FAAH inhibitor URB597 resulted in increased alcohol intake in mice (Blednov et al., 2007; Vinod, Sanguino, Yalamanchili, Manzanares, & Hungund, 2008). In contrast, URB597 neither facilitated alcohol self-administration nor facilitated reinstatement of alcohol seeking in alcohol-preferring rats (Cippitelli et al., 2008).

The relevance of FAAH/anandamide on the effects of alcohol has also been revealed in studies looking at changes in genes related to endocannabinoid signaling following both acute-intermittent versus chronic-continuous alcohol consumption. Thus, a reduction in messenger RNA (mRNA) expression for FAAH following both continuous and intermittent alcohol exposure in rats has been observed (Serrano et al., 2012). Indeed, chronic alcohol seems to induce region-specific changes in anandamide concentrations in different brain regions (i.e., decrease in the midbrain vs. increase in the limbic forebrain) in rats (Gonzalez et al., 2002). The changes in anandamide in response to chronic alcohol suggest that anandamide might play a relevant role during alcohol withdrawal. However, the findings regarding a possible involvement of anandamide in alcohol withdrawal are scarce and seem contradictory. On one hand, it has been shown that the absence of FAAH, as obtained in FAAH constitutional knockout mice, did not affect acute alcohol withdrawal (Blednov et al., 2007). On the other hand, a reduction in the severity of convulsions during withdrawal following chronic ethanol exposure in FAAH knockout mice it has been observed (Vinod et al., 2008).

Monoacylglycerol Lipase/2-Arachidonoylglycerol Modulation

The possible involvement of 2-AG in the rewarding properties of alcohol has been suggested both in genetic and pharmacological studies. Therefore it has been observed that the genetic predisposition for enhanced alcohol preference (i.e., Sardinian alcohol-preferring rats) includes a different rate of 2-AG degradation, as compared to alcohol-nonpreferring rats (Melis et al., 2014). Interestingly, the study by Melis et al. (2014) revealed that the dopaminergic cells from alcohol-naive Sardinian alcohol-preferring rats displayed a decreased probability of GABA release and a larger 2-AG-dependent form of short-term plasticity (i.e., a larger depolarization-induced suppression of inhibition) due to a different rate of 2-AG degradation. Therefore the changes in endocannabinoid-mediated transmission on Sardinian alcohol-preferring rats might result in decreasing the aversive consequences of alcohol intake, resulting in a net increase of the rewarding or salient properties of alcohol in these animals. Exposure to alcohol might also affect the endogenous tone of cannabinoid 2-AG and its hydrolytic enzyme MAGL. Several studies have revealed that the effects of alcohol on 2-AG levels might be region specific. Thus increased 2-AG levels in the NAc of rats self-administering ethanol has been reported (Caille et al., 2007). In another study, chronic exposure to alcohol decreased tissue levels of 2-AG in the midbrain but not in the NAc (Gonzalez et al., 2002). Additionally, ethanol dependence and withdrawal have also been associated with dysregulated mRNA expression of MAGL in the amygdala (Serrano et al., 2012).

Inhibition of MAGL has not been explored much. It should be noted that N-arachidonoylmaleimide, an MAGL inhibitor, increased alcohol consumption and preference in mice (Gutierrez-Lopez et al., 2010).

CB$_2$ Modulation

Contrary to a more distinctive role played by CB$_1$ receptors in modulating the addictive properties of alcohol, the effects of CB$_2$ receptors on alcohol preference have been less investigated. Administration of the CB$_2$ agonist JWH015 enhanced alcohol preference only in stressed rodents, but not in control animals (Ishiguro et al., 2007; Onaivi, Carpio, et al., 2008). In contrast, other studies have shown that both the systemic or intracranial local administration of CB$_2$ receptor agonists inhibits alcohol self-administration, alcohol-induced CPP, and alcohol-induced locomotor sensitization in mice (Al Mansouri et al., 2014). In line with these findings, genetic studies have shown increased voluntary intake, self-administration, motivation, and CPP for alcohol in CB$_2$ knockout mice (Ortega-Alvaro et al., 2015). Indeed, some studies seem to indicate a link between the enhanced preference for alcohol and a reduced CB$_2$ gene expression in ventral midbrain in mice (Ishiguro et al., 2007).

In summary, preclinical studies suggest a strong involvement of the ECS in acute and chronic effects of alcohol. More specifically, cannabinoid CB$_1$ agonists seem to facilitate the reinforcing/rewarding properties of alcohol, whereas CB$_1$ antagonists seem to block or decrease alcohol taking and alcohol seeking. Altogether, these studies suggest the existence of complex interactions between ECS and the acute/chronic effects of alcohol. Therefore alcohol exposure might modify the homeostasis of the ECS and at the same time changes in the ECS might affect the ability of alcohol to produce reward.

NICOTINE

Nicotine is a frequently abused drug causing addiction and relapse during abstinence. Current approved medications for smoking cessation have not solved the problem regarding the high rates of relapse. Preclinical and human clinical studies have shown a pivotal role of the cannabinoid system in nicotine addiction. Indeed, increasing preclinical evidence supports the idea of a bidirectional crosstalk between the ECS and the nicotinic cholinergic system. The nicotinic ACh receptors, as CB_1 receptors, are abundantly expressed in the CNS and closely overlap different brain regions (Herkenham et al., 1991; Nashmi & Lester, 2006). Similarly, the exogenous ligands THC and nicotine exert similar pharmacological effects including hypothermia, impaired locomotion, antinociception, and rewarding effects (Goldberg, Spealman, & Goldberg, 1981; Justinova, Goldberg, Heishman, & Tanda, 2005; Little, Compton, Johnson, Melvin, & Martin, 1988; Valjent, Mitchell, Besson, Caboche, & Maldonado, 2002).

In the following subsections we will describe the involvement of the different components of the ECS in the acute and chronic effects of nicotine.

CB_1 Modulation

The selective CB_1 receptor inverse agonist/antagonist rimonabant decreases intravenous nicotine self-administration behavior in rats (Cohen, Perrault, Voltz, Steinberg, & Soubrie, 2002). We and others subsequently reported that rimonabant decreases the motivation to self-administer nicotine, as measured using progressive ratio schedules of reinforcement (Forget, Coen, & Le Foll, 2009) (see Fig. 6.1).

Rimonabant also blocks the development of nicotine-induced CPP (Forget, Hamon, & Thiebot, 2005; Hashemizadeh, Sardari, & Rezayof, 2014; Le Foll & Goldberg, 2004) and the reinstatement of previously extinguished nicotine-seeking behavior in rats (Cohen, Perrault, Griebel, & Soubrie, 2005; Diergaarde, de Vries, Raaso, Schoffelmeer, & De Vries, 2008; Forget et al., 2009) (see Fig. 6.2).

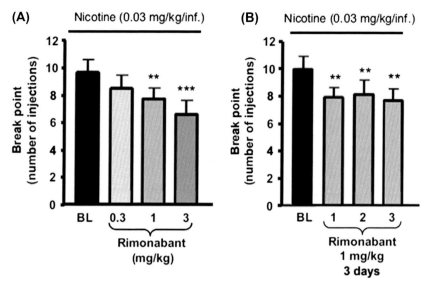

Figure 6.1 Effects of rimonabant on nicotine (0.03 mg/kg/injection) self-administration under a progressive ratio schedule. Data are expressed as means (± standard error of mean). (A) The number of injections (break point) during baseline conditions (BL) and during sessions with rimonabant [0.3–3 mg/kg, intraperitoneally (IP) H-60 min] pretreatment. (B) Effects of rimonabant (1 mg/kg, IP H-60 min) on self-administration under a progressive ratio schedule were tested during three consecutive sessions. *Adapted and reproduced with permission from Forget, B., Coen, K. M., & Le Foll, B. (2009). Inhibition of fatty acid amide hydrolase reduces reinstatement of nicotine seeking but not break point for nicotine self-administration–comparison with CB(1) receptor blockade.* Psychopharmacology (Berlin), 205(4), 613–624. http://dx.doi.org/10.1007/s00213-009-1569-5.

In agreement with these findings in rats, it has been shown that genetic deletion of cannabinoid CB_1 receptors reduces nicotine-induced CPP (Castane et al., 2002; Merritt, Martin, Walters, Lichtman, & Damaj, 2008). New studies have reported that rimonabant administration significantly decreased nicotine self-administration under a fixed ratio 10 schedule of reinforcement in squirrel monkeys (Schindler, Redhi, et al., 2016). Rimonabant was also very effective in inhibiting reinstatement of extinguished drug-seeking behavior induced by nicotine priming or nicotine cue presentation in squirrel monkeys (Schindler, Redhi, et al., 2016). Therefore there is a clear picture emerging across species (mice, rats, and squirrel monkeys) indicating the critical role of CB_1 in mediating nicotine taking and nicotine seeking. However, rimonabant was associated with an increased risk of anxiety and depression.

The CB1 neutral antagonist AM4113 has been lately compared to rimonabant for its effects on the abuse-related effects of nicotine and its

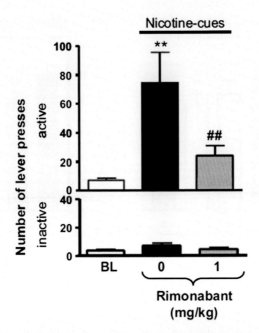

Figure 6.2 Effects of rimonabant (1 mg/kg, intraperitoneally H-60 min) on responses made at the active (top) and the inactive (below) levers during the cue-induced rein-statement of nicotine seeking after extinction. *Adapted and reproduced with permission from Forget, B., Coen, K. M., & Le Foll, B. (2009). Inhibition of fatty acid amide hydrolase reduces reinstatement of nicotine seeking but not break point for nicotine self-administra-tion–comparison with CB(1) receptor blockade.* Psychopharmacology (Berlin), 205(4), 613–624. http://dx.doi.org/10.1007/s00213-009-1569-5.

effects on anxiety and depression-like behavior in rats (Gueye et al., 2016). AM4113 significantly attenuated nicotine taking, motivation for nicotine (see Fig. 6.3), as well as cue-, priming- and stress-induced reinstatement of nicotine-seeking behavior (see Fig. 6.4).

These effects were accompanied by a decrease of the firing and burst rates in the VTA dopamine neurons in response to nicotine. On the other hand, AM4113 pretreatment did not have effects on operant responding for food. Importantly, AM4113 did not have effects on anxiety and showed antidepressant like effects (see Fig. 6.5). More specifically, AM4113 did not significantly modify the time spent in the open arm in the elevated plus maze test, which evaluates anxiety and, contrary to rimonabant, decreased immobility in the forced swimming test, which evaluates depression-like behaviors in animal models (Gueye et al., 2016).

Nicotine infusions

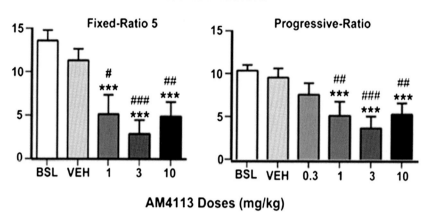

Figure 6.3 Effects of acute AM4113 (60 min pretreatment time) on nicotine-taking behavior under fixed ratio 5 (FR-5) and progressive ratio (PR) schedules of reinforcement. *Bars* represent average responding (± standard error of mean) during FR-5 (1-h sessions) and PR (≤ 4-h sessions) schedules following treatment with saline (baseline, BSL), vehicle (VEH), or AM4113 (1, 3, and 10 mg/kg). *Reproduced with permission from Gueye, A. B., Pryslawsky, Y., Trigo, J. M., Poulia, N., Delis, F., Antoniou, K., … Le Foll, B. (2016). The CB1 neutral antagonist AM4113 retains the therapeutic efficacy of the inverse agonist rimonabant for nicotine dependence and weight loss with better psychiatric tolerability.* International Journal of Neuropsychopharmacology. http://dx.doi.org/10.1093/ijnp/pyw068.

It should be noted that the neutral CB_1 antagonist AM4113 has also been shown to be effective to decrease nicotine self-administration under a fixed ratio 10 schedule of reinforcement in squirrel monkeys and to inhibit reinstatement of extinguished drug-seeking behavior induced by nicotine priming or nicotine cue presentation in squirrel monkeys (Schindler, Redhi, et al., 2016). Further studies are required in humans to determine if the neutral CB_1 antagonists will be devoid of the psychiatric side effects that limited the use of rimonabant.

The opposite modulation (i.e., stimulating CB_1 receptors) led, in contrast, to an increased motivation to self-administer nicotine as measured using a progressive ratio schedule of reinforcement (Gamaleddin, Wertheim, et al., 2012), enhanced cue-induced reinstatement of nicotine-seeking behavior, and the discriminative stimulus effects of low doses of nicotine in rats (Gamaleddin, Wertheim, et al., 2012). Several of these responses in rats were blocked by the CB_1 inverse agonist/antagonist rimonabant.

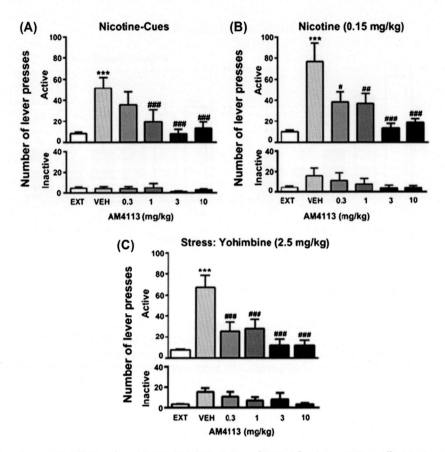

Figure 6.4 Effects of acute AM4113 (0.3, 1, 3, and 10 mg/kg, intraperitoneally 60 min pretreatment time) on reinstatement of nicotine-seeking behavior induced by presentation of nicotine-associated cues (A), by a priming dose of nicotine (B), and pharmacological stressor yohimbine (C). Data are expressed as the number of active (A–C, top) and inactive (A–C, bottom) lever presses (±standard error of mean). Effects of presentation of nicotine-associated cues alone, a priming dose of nicotine, or stress by yohimbine under vehicle treatment on reinstatement of nicotine-seeking behavior [***$P < .001$ compared with baseline behavior (EXT)]. Pretreatment with AM4113 blocked reinstatement of nicotine-seeking behavior (#$P < .05$, ##$P < .01$, and ###$P < .001$). No differences were observed on inactive lever presses. *Reproduced with permission from Gueye, A. B., Pryslawsky, Y., Trigo, J. M., Poulia, N., Delis, F., Antoniou, K., ... Le Foll, B. (2016). The CB1 neutral antagonist AM4113 retains the therapeutic efficacy of the inverse agonist rimonabant for nicotine dependence and weight loss with better psychiatric tolerability.* International Journal of Neuropsychopharmacology. http://dx.doi.org/10.1093/ijnp/pyw068.

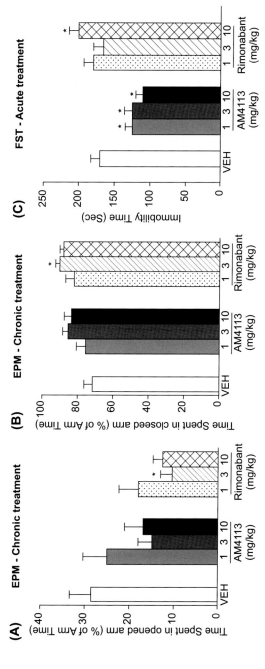

Figure 6.5 Effects of chronic AM4113 and rimonabant treatments on anxietylike behavior using the elevated plus maze (EPM) test. *Bars represent means (± standard error of mean). Rimonabant (1, 3, and 10 mg/kg) dose dependently decreased time spent in the open arm (A) and increased time spent in the closed arm (B). No effects were observed following AM4113 (1, 3, and 10 mg/kg). *Significantly different versus vehicle P < .05. Effect of acute (C–D) and chronic (E–F) AM4113 treatment on immobility and climbing time in the forced swimming test (FST). *P < .05, **P < .01, ***P < .001 significant difference versus vehicle, Dunnett's multiple comparison test. Reproduced with permission from Gueye, A. B., Pryslawsky, Y., Trigo, J. M., Poulia, N., Delis, F., Antoniou, K., … Le Foll, B. (2016). The CB1 neutral antagonist AM4113 retains the therapeutic efficacy of the inverse agonist rimonabant for nicotine dependence and weight loss with better psychiatric tolerability. International Journal of Neuropsychopharmacology. http://dx.doi.org/10.1093/ijnp/pyw068.*

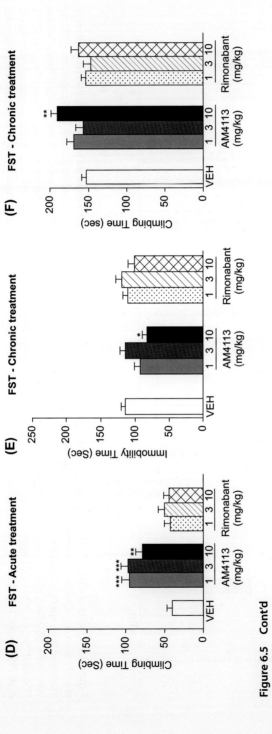

Figure 6.5 Cont'd

Taken together, these findings indicate that CB_1 receptors have a bidirectional role on both nicotine reward/reinforcement and on relapse to nicotine-seeking behavior in abstinent subjects.

Fatty Acid Amide Hydrolase/Anandamide Modulation

FAAH inhibition by URB597 has been shown to reverse some addiction-related behavioral and neurochemical effects of nicotine in rats (Forget et al., 2009; Scherma et al., 2008). The inhibition of FAAH by URB597 was able to prevent the development of nicotine-induced CPP, reduced acquisition of nicotine self-administration behavior, and inhibited reinstatement of nicotine-seeking behavior induced by nicotine priming and cue-induced reinstatement in abstinent rats, while demonstrating no rewarding effects per se (Scherma et al., 2008). However, there was no impact on nicotine-taking assessed using a fixed ratio schedule or a progressive ratio schedule of reinforcement (Forget et al., 2009). In support of the results obtained with the FAAH inhibitor URB597, the anandamide uptake inhibitor VDM11, (5Z, 8Z, 11Z, 14Z)-N-(4- hydroxy-2-methylphenyl)-5, 8, 11, 14-eicosatetraenamide, was able to reduce both cue- and nicotine-induced reinstatement of nicotine-seeking behavior. However, VDM11 did not affect responding for nicotine under fixed ratio or progressive ratio schedules of reinforcement (Gamaleddin, Guranda, Goldberg, & Le Foll, 2011). Findings with VDM11 were further confirmed using AM404, another anandamide uptake inhibitor, which also attenuated cue and nicotine priming reinstatement of nicotine-seeking behavior (Gamaleddin et al., 2013). Additionally, AM404 decreased nicotine CPP and its reinstatement. AM404 has been also shown to attenuate dopamine increase on the NAc shell following nicotine injection (Scherma et al., 2012). The fact that two different ligands, developed to elevate anandamide levels, produce similar effects, strongly suggest that those effects are mediated by anandamide elevation. These results have been expanded by some experiments performed in squirrel monkeys testing the impact of two FAAH inhibitors (URB 597 and URB 694) (Justinova et al., 2015). These ligands were able to shift nicotine self-administration dose–response functions in a manner consistent with reduced nicotine reward and blocked reinstatement of nicotine seeking induced by reexposure to either nicotine priming or nicotine-associated cues. In contrast, there was no effect on cocaine or food self-administration. It should be noted that some of the effects of FAAH inhibition have been reversed by the peroxisome proliferator-activated receptor (PPAR)-α antagonist, MK886 in nonhuman primates (Justinova et al., 2015) and by CB_1 blockade (Forget, Guranda, Gamaleddin, Goldberg, & Le Foll, 2016) in rats, indicating that not all effects mediated by these FAAH antagonists may be mediated by anandamide.

In contrast to those consistent findings in rats and nonhuman primates, findings obtained in mice suggest a different role. FAAH knockout mice display enhanced nicotine CPP (Merritt et al., 2008). Also, pretreatment with URB597 enhanced nicotine CPP in mice (Merritt et al., 2008). Administration of a subthreshold dose of nicotine that did not produce CPP in wild-type mice effectively produced CPP in FAAH knockout mice, and this effect was mediated by CB_1 receptors (Merritt et al., 2008). However, no differences on nicotine CPP were observed between FAAH knockout and wild-type mice when higher doses of nicotine were tested (Merritt et al., 2008). It is unclear why the different species used produced different results at this point.

Monoacylglycerol Lipase/2-Arachidonoylglycerol Modulation

Very limited work has been performed exploring the role of MAGL on nicotine's effects. In a new study, JZL184 (0, 8, and 16 mg/kg) did not affect food taking, nicotine taking, or motivation for nicotine. However, MAGL inhibition by JZL184 (16 mg/kg) increased reinstatement of previously extinguished nicotine seeking induced by presentation of nicotine-associated cues, but did not produce reinstatement on its own (Trigo & Le Foll, 2016). This study suggests that 2-AG could mediate reinstatement of nicotine seeking. The influence of MAGL on nicotine withdrawal has been explored in mice. JZL184 dose dependently reduced somatic and aversive withdrawal signs (Muldoon et al., 2015). Interestingly, this effect was blocked by rimonabant, indicating a CB_1-mediated mechanism. The impact of the genetic deletion of the enzyme (MAGL knockout mice) also indicates attenuation of nicotine withdrawal (Muldoon et al., 2015). It would be important that further studies explore the role of 2-AG, as this may be an important contributor of nicotine's effects.

CB_2 Modulation

In rats, the activation of CB_2 receptors by the selective CB_2 agonist AM1241 did not have effects on motivation to obtain nicotine or nicotine intake (Gamaleddin, Zvonok, Makriyannis, Goldberg, & Le Foll, 2012). Similarly, the selective CB_2 antagonist AM630 did not modify nicotine taking or motivation to obtain nicotine under the progressive ratio schedule of reinforcement. Moreover, the CB_2 agonist and antagonist were not able to affect cue- or nicotine-induced reinstatement of nicotine-seeking behavior (Gamaleddin, Zvonok, et al., 2012). CB_2 antagonist AM630 was not able to block $CB_{1/2}$ stimulation effects induced by WIN 55,212-2 mediating nicotine seeking, whereas CB_1 antagonist was effective in blocking these effects (Gamaleddin, Wertheim, et al., 2012; Gamaleddin, Zvonok, et al., 2012).

On the other hand, some studies have documented the relevance of CB_2 receptors on the rewarding/reinforcing properties of nicotine in mice (Ignatowska-Jankowska, Muldoon, Lichtman, & Damaj, 2013; Navarrete et al., 2013). Altogether, these results suggest that there may be important species differences that mediate these effects. Further studies performed in nonhuman primates or human subjects would allow better exploration of these discrepancies.

In summary, initial studies showing the effectiveness of CB_1 receptor inverse agonist/antagonists such as rimonabant have led to new advances in our understanding of how different components of the ECS modulate nicotine dependence, opening the possibilities for the pharmacological manipulation of the ECS and the development of promising alternative strategies for the treatment of nicotine dependence and relapse prevention.

OPIOIDS

Opioids and cannabinoids share some common characteristics on their ability to modulate pain, and produce sedation, hypotension, and motor depression (Corchero, Manzanares, & Fuentes, 2004; Maldonado & Valverde, 2003; Manzanares et al., 1999; Massi, Vaccani, Romorini, & Parolaro, 2001). In fact, cannabinoids have been shown to modulate the opioid function at many different levels including cross-tolerance, mutual potentiation, and receptor cross talk (Cichewicz & McCarthy, 2003; Corchero, Manzanares, et al., 2004; Maldonado & Valverde, 2003; Manzanares et al., 1999; Parolaro et al., 2010; Vigano, Rubino, & Parolaro, 2005). Moreover, specific receptors for both the ECS and the endogenous opioid system (i.e., CB_1 and μ-opioid receptors) have an overlapped distribution in the limbic system (Pickel, Chan, Kash, Rodriguez, & MacKie, 2004; Rodriguez, Mackie, & Pickel, 2001; Salio et al., 2001). The intracellular signal response (i.e., G-protein activation) of μ-opioid receptors can be modulated by the stimulation of CB_1 receptors (Corchero, Romero, et al., 1999; Rios, Gomes, & Devi, 2006). The administration of CB_1 agonists is able to regulate the endogenous release of opioids (Corchero, Avila, Fuentes, & Manzanares, 1997; Corchero,

Fuentes, & Manzanares, 1997, 1999; Corchero, Romero, et al., 1999; Manzanares et al., 1998; Valverde et al., 2001) and μ-opioid receptor density (Corchero, Oliva, et al., 2004).

Early studies showed that the opioid antagonist naloxone was able to block THC-induced dopamine efflux in the NAc of rats (Chen et al., 1990). On the other hand, intra-accumbens infusions of the cannabinoid antagonist rimonabant decreased intravenous self-administration of heroin in rats (Caille & Parsons, 2006), implicating NAc CB_1 receptors in the modulation of opiate reinforcement. In this section, we will describe the interactions between the different components of the ECS and the endogenous opioid system, and more specifically, we will focus on the ability of the ECS to modulate the addictive properties of opioids and in the development of tolerance and dependence in opioid addiction.

CB_1 Modulation

THC and heroin exert their effects on mesolimbic dopamine transmission through a common μ-opioid receptor mechanism (Tanda, Pontieri, & Di Chiara, 1997). However, the blockade of CB_1 did not affect heroin-induced elevation of dopamine (Tanda et al., 1997). In contrast, later studies using behavioral paradigms in rodents have shown that the CB_1 inverse agonist/ antagonist rimonabant administration decreased the rewarding properties of opioids (Caille & Parsons, 2006; Navarro et al., 2001). On the other hand, administration of the exogenous $CB_{1/2}$ agonists THC and WIN 55,212-2 facilitated the reinforcing properties of heroin (Solinas et al., 2005). Interestingly, it has been observed that rats repeatedly preexposed to THC self-administered significantly more heroin and showed significantly shorter postinjection pauses, suggesting the existence of tolerance to heroin effects (Solinas, Panlilio, & Goldberg, 2004), whereas an acute pretreatment with THC and WIN 55,212-2 was not able to modify heroin self-administration under a fixed ratio schedule and only affected the breaking point reached under a progressive ratio schedule of reinforcement (Solinas et al., 2005). Altogether, these results suggest that the exposure to CB_1 agonists facilitate the reinforcing effects of opioid drugs. Blockade of CB_1 receptors by rimonabant has been shown to have effects in both the acute effects induced by morphine (e.g., increased hyperlocomotion) and the effects produced after the repeated exposure to opiates [e.g., attenuated morphine-induced behavioral sensitization (Marinho et al., 2015)]. Indeed, the administration of CB_1 receptor antagonists decrease the rewarding effects of opioids in different behavioral paradigms including intravenous self-administration (Caille &

Parsons, 2003; Navarro et al., 2001; Solinas, Panlilio, Antoniou, Pappas, & Goldberg, 2003) and CPP in rodents (Chaperon, Soubrie, Puech, & Thiebot, 1998; Mas-Nieto et al., 2001; Navarro et al., 2001; Singh, Verty, McGregor, & Mallet, 2004). Interestingly, contrary to CB_1 receptor agonists, the CB_1 receptor antagonist rimonabant not only modified heroin self-administration under progressive ratio schedules but also under fixed ratio and continuous reinforcement schedules (Solinas et al., 2003). It has been suggested that the effects of CB_1 receptor antagonists in reducing the rewarding properties of heroin are region specific. In a study by Caille and Parsons (2006), it was shown that the local administration of rimonabant in the NAc but not in the intraventral pallidal region reduced the intravenous self-administration of heroin.

Genetic studies also seem to suggest the relevance of CB_1 receptors in the rewarding properties of opioids. CB_1 knockout mice display decreased or abolished rewarding properties of morphine (Cossu et al., 2001; De Vries, Homberg, Binnekade, Raaso, & Schoffelmeer, 2003; Ledent et al., 1999; Martin, Ledent, Parmentier, Maldonado, & Valverde, 2000) (but see Rice, Gordon, & Gifford, 2002 for a negative study).

Interestingly, interventions in the ECS might also modulate opiate-seeking behavior in dependent animals and the reinforcing and motivational properties of opioid-associated cues. Therefore the administration of the CB_1 receptor agonist HU-210 reinstated heroin-seeking behavior in dependent rats (De Vries et al., 2003). In the same study, by De Vries et al. (2003), rimonabant decreased heroin seeking elicited by a priming injection of heroin in rats. Similar studies have shown that WIN 55,212-2, but not THC, effectively restored heroin-seeking behavior and that both cannabinoid and heroin-induced reinstatement of heroin seeking was blocked by the coadministration of rimonabant (Fattore, Spano, Melis, Fadda, & Fratta, 2011; Fattore, Spano, Cossu, Deiana, & Fratta, 2003).

Fatty Acid Amide Hydrolase/Anandamide Modulation

Anandamide levels are increased following the acute administration of morphine and during heroin self-administration (Caille et al., 2007; Vigano et al., 2004). However, neither the inhibitor of anandamide transport N-(4-hydroxyphenyl) arachidonylethanolamide (AM404) nor the inhibitor of the enzyme degrading anandamide (FAAH) were able to alter the reinforcing efficacy of heroin in rats (Solinas et al., 2005).

Preclinical studies suggest that increasing anandamide levels might alleviate opiate withdrawal symptoms. A study in opiate-dependent mice showed that AM404 was able to attenuate the spontaneous, but not the naloxone-precipitated, withdrawal syndrome in morphine-dependent mice (Del Arco et al., 2002). On the other hand, the FAAH inhibitor PF-3845 reduced the intensity of naloxone-precipitated withdrawal in morphine-dependent mice (Ramesh et al., 2013, 2011). In a similar study, the administration of URB597 reduced most of the withdrawal symptoms to morphine precipitated by an injection of naloxone (Shahidi & Hasanein, 2011). Interestingly, the dual FAAH/MAGL inhibitor SA-57, which has greater potency in inhibiting FAAH over MAGL, reduced withdrawal signs at doses that only partially elevated 2-AG while fully elevating anandamide (Gamage, 2013). Other studies have revealed that increasing anandamide levels with FAAH inhibitors (e.g., URB597 or PF-3845) only decreased some morphine withdrawal symptoms but did not affect the aversive-affective state of naloxone-precipitated morphine withdrawal as investigated using the conditioned place aversion (CPA) paradigm (Gamage et al., 2015; Wills et al., 2014).

Monoacylglycerol Lipase/2-Arachidonoylglycerol Modulation

The selective increase in brain levels of 2-AG with the MAGL inhibitor MJN110 produced opioid-sparing effects in mice (Wilkerson et al., 2016), suggesting that increasing 2-AG endogenous tone might affect the response to opiates in rodents. Previous studies have shown that an acute injection of morphine decreased tissue levels of 2-AG in the striatum but increased the levels of anandamide in several other brain structures including NAc, prefrontal cortex, caudate putamen, and hippocampus (Vigano et al., 2004). Additionally, in these studies, administration of rimonabant was able to attenuate the expression of morphine sensitization (Vigano et al., 2004), further suggesting the involvement of CB_1 receptors in this process.

Some studies have suggested that 2-AG might also be involved in opiate withdrawal. Therefore similar to the effects of increasing the endogenous tone of anandamide (i.e., with FAAH inhibitors), the increase in 2-AG levels (e.g., with MAGL inhibitors or 2-AG administration) seems also effective in alleviating somatic opiate withdrawal symptoms. A study administering 2-AG (i.e., intracerobroventricular injection) found reduced signs of withdrawal (jumping and forepaw tremor) following naloxone challenge in morphine-dependent mice (Yamaguchi et al., 2001). Accordingly, the MAGL inhibitor JZL184 attenuated both spontaneous and naloxone-precipitated withdrawal in morphine-dependent mice (Ramesh et al., 2013, 2011). In contrast, in another study, JZL184 only reduced some withdrawal symptoms (e.g., jumping), but did not affect the acquisition of morphine withdrawal CPA in mice (Gamage et al., 2015).

CB$_2$ Modulation

The involvement of the CB$_2$ receptor in the interactions with the opioid system has not been widely explored yet (see Befort, 2015 for review). However, some studies have suggested that both the activation and the blockade of CB$_2$ receptors might modulate the signaling of specific opioid receptors (e.g., μ-opioid receptor). Therefore the administration of the CB$_2$ receptor antagonist SR144528 in mice decreased μ-opioid receptor signaling both in vivo and in vitro (Paldy et al., 2008). The fact that SR144528 decreased μ-opioid receptor signaling both in CB$_1$ knockout and wild-type mice suggests that this effect was mediated via CB$_2$ receptors (Paldy et al., 2008). On the other hand, the CB$_{1/2}$ agonist noladin ether (NE) decreased μ-opioid receptor mRNA expression (Paldyova et al., 2008). Since NE is also an agonist of CB$_1$ receptors, the authors of this study performed the experiments in both CB$_1$ knockout and wild-type mice showing that these effects were present in both genotypes and concluding that the effect was mediated via CB$_2$ receptors (Paldyova et al., 2008). Additionally, NE also reduced μ-opioid receptor activation in vitro and this effect was antagonized by the CB$_2$ receptor antagonist SR144528 (Paldyova et al., 2008). Conversely, other studies have suggested that chronic treatment with heroin might modify CB$_2$ gene expression in mice (Onaivi, Ishiguro, et al., 2008). However, the differences found in the study by Onaivi, Ishiguro, et al. (2008) were not statistically significant.

In summary, in spite of the overwhelming evidence of the interconnection between the ECS and the opioid system, the interactions between opiates and endogenous cannabinoids seem complex (e.g., morphine differentially modifies anandamide and 2-AG levels in diverse brain structures) and more studies are needed to further investigate the role played by the endogenous cannabinoids in the rewarding properties of opiates. Additionally, little is currently known about the interactions between CB$_2$ receptors and the endogenous opioid system and the possible modulatory effects of CB$_2$ receptors in the addictive properties of opiates.

PSYCHOSTIMULANTS

The ECS seems to be involved in some properties of the most commonly abused illegal psychostimulants including cocaine, amphetamines, and their derivatives methamphetamine and 3,4-methylenedioxymethamphetamine (MDMA).

CB$_1$ Modulation

On one hand, it has been reported that the administration of the CB$_1$ antagonist rimonabant blocked the acquisition of CPP to cocaine, methamphetamine, and MDMA (Braida, Iosue, Pegorini, & Sala, 2005; Chaperon et al., 1998; Yu et al., 2011), and on the other hand, mice lacking the CB$_1$ receptor readily acquired CPP to cocaine, amphetamine, and MDMA (Houchi et al., 2005; Martin et al., 2000; Miller, Ward, & Dykstra, 2008; Rodriguez-Arias et al., 2013). Similarly, conflicting findings have been obtained using the self-administration paradigm and transgenic mice: mice lacking CB$_1$ were able to acquire cocaine and amphetamine self-administration (Cossu et al., 2001). Importantly, in the study by Cossu et al., the animals were held by the tail while self-administering the drug. Later studies in freely moving animals have shown an impairment in the acquisition of cocaine self-administration in mice lacking CB$_1$ receptor (Soria et al., 2005). Similarly, MDMA self-administration was abolished in CB$_1$ knockout mice (Tourino, Ledent, Maldonado, & Valverde, 2008). Divergent findings have been reported on the impact of this CB$_1$ gene deletion on the dopamine response to psychostimulant drug administration (Cheer et al., 2007; Soria et al., 2005).

On the other hand, the CPP induced by cocaine was dose-dependently blocked by rimonabant (Chaperon et al., 1998). Therefore rimonabant blocked CPP to cocaine when it was injected during the pairing of the visual/spatial cue stimuli and rewarding effects of the drug, but not when it was administered during the expression of cocaine-induced CPP (Chaperon et al., 1998), suggesting that the CB$_1$ receptor played a relevant role on the acquisition of the association between the visual/spatial cue stimuli and the rewarding properties of the drug.

The intravenous self-administration paradigm has clearly shown that maintenance of cocaine intravenous self-administration was not affected by CB$_1$ antagonists (i.e., rimonabant or AM251) neither in rodents (Caille & Parsons, 2006; De Vries et al., 2001; Filip et al., 2006; Lesscher, Hoogveld, Burbach, van Ree, & Gerrits, 2005; Xi et al., 2008) or nonhuman primates (Tanda, Munzar, & Goldberg, 2000). Conversely, it has been shown that the CB$_1$ antagonist AM 251 significantly decreased methamphetamine self-administration (Vinklerova, Novakova, & Sulcova, 2002).

On the other hand, blocking CB$_1$ receptors might have effects in the reinstatement of psychostimulant seeking following extinction. Therefore the administration of rimonabant disrupted the reinstatement of extinguished methamphetamine- or cocaine-induced CPP in mice (Yu et al., 2011). Similarly, other studies have shown that both rimonabant and AM251 were able to attenuate cue- and priming-induced reinstatement of cocaine self-administration in rodents (Adamczyk et al., 2012; Filip et al., 2006; Ward, Rosenberg, Dykstra, & Walker, 2009; Xi et al., 2006). Additionally, rimonabant reduced the initial burst of responding during cocaine extinction and the mice treated with rimonabant required fewer daily sessions to reach criterion for extinction of cocaine-maintained responding than vehicle-treated animals (Ward et al., 2009). Interestingly, the effects of CB_1 antagonists in stress induced reinstatement of cocaine self-administration are more controversial, with studies showing both effects (Vaughn et al., 2012) and no effects of CB_1 antagonists on reinstatement of cocaine seeking by stress (DeVries et al., 2001; Kupferschmidt, Klas, & Erb, 2012). These apparent contradictory results might be due the type of stress used (i.e., footshock vs. forced swim) or species (rats vs. mice). In fact, some studies have suggested that the CB_1 antagonist AM251 might interfere with CRF-, but not footshock- or cocaine-induced reinstatement (Kupferschmidt, Klas, et al., 2012) and that low doses of the CB_1 antagonist AM251 might reverse the behavioral anxiety during cocaine withdrawal (Kupferschmidt, Newman, Boonstra, & Erb, 2012). Alternatively, other authors have proposed that the mechanism by which the blockade of CB_1 receptors (i.e., by AM251) is able to block cocaine-induced reinstatement is related to the ability of the CB_1 antagonist AM251 to inhibit the increase in NAc glutamate release induced by cocaine (Xi et al., 2006), suggesting the existence of tonic inhibition over NAc glutamate release by CB_1 receptors under cocaine-extinction conditions. As can be inferred from these studies, there is a certain agreement in the ability of CB_1 receptor antagonists to block cocaine-seeking behaviors. Therefore we might expect that the administration of CB_1 receptor agonists will facilitate the reinstatement of cocaine-seeking behavior. Accordingly, it has been observed that the synthetic CB_1 agonist HU-210 induced relapse to cocaine seeking even after prolonged withdrawal periods in rats (DeVries et al., 2001).

Fatty Acid Amide Hydrolase/Anandamide Modulation

Increasing the endogenous tone of anandamide (i.e., by FAAH inhibition) modulates cocaine effects. In a study by Luchicchi et al. (2010), the inhibition of FAAH with URB597 revealed region-specific effects in the neuronal responses to cocaine. More specifically, the administration of URB597 blocked the effects of cocaine in the medium spiny neurons in the NAc but not in the VTA dopaminergic neurons (i.e., as measured using single-unit electrophysiological recordings) in rats (Luchicchi et al., 2010). The authors of this study have proposed that the effects of URB597 on cocaine's mesolimbic dopaminergic transmission modulation were exerted through activation of both surface cannabinoid CB_1 and PPAR-α nuclear receptors (Luchicchi et al., 2010). In this regard, the fact that other studies have shown no differences between mice lacking CB_1 receptors and wild-type controls in cocaine-induced increase in extracellular dopamine in mesolimbic pathways (Soria et al., 2005) suggests that the PPAR-α nuclear receptors might have a preeminent role in the effects of URB597 on cocaine's mesolimbic dopaminergic response.

Behavioral studies have shown no effects of the FAAH inhibitor URB597 on cocaine self-administration in rats (Adamczyk et al., 2009) and nonhuman primates (Justinova, Mangieri, et al., 2008). However, URB597 attenuated both cocaine- and cue-induced drug-seeking behaviors in rats (Adamczyk et al., 2009), suggesting a potential therapeutic utility of URB597. Interestingly, the administration of URB597 did not promote reinstatement of extinguished cocaine-seeking behavior in nonhuman primates (Justinova, Mangieri, et al., 2008), suggesting that the increase in the endogenous cannabinoids produced by URB597 (and possibly the increased brain levels of PPAR-α, oleoylethanolamine, and palmitoylethanolamide, commented in Fatty Acid Amide Hydrolase/Anandamide Modulation section) tone was not sufficient to trigger cocaine-seeking behavior in primates.

On the other hand, the exposure to psychostimulants modulates the endogenous tone of cannabinoids. In a study by Palomino and collaborators, the behavioral sensitization to the locomotor effects after repeated exposure to cocaine was associated with an increased N-acyl phosphatidylethanolamine phospholipase D/FAAH ratio, suggesting an enhanced production of anandamide in the cerebellum (Palomino et al., 2014). Conversely, another study using the self-administration paradigm showed a significant decrease in anandamide levels in the cerebellum (but a significant increase in 2-AG in the cerebellum, frontal cortex, and hippocampus; see Monoacylglycerol Lipase/2-Arachidonoylglycerol Modulation section) following repeated exposure to cocaine both in rats self-administering the drug and in yoked animals (i.e., passively receiving cocaine infusions) (Bystrowska, Smaga, Frankowska, & Filip, 2014). Interestingly, in the study by Bystrowska et al. (2014), reduced levels of anandamide were found in the frontal cortex, the hippocampus, and the NAc during extinction in yoked cocaine rats as compared to the saline control group. Altogether, these results suggest the existence of a region-specific interplay in endogenous cannabinoids levels in response to the exposure and withdrawal of cocaine.

Monoacylglycerol Lipase/2-Arachidonoylglycerol Modulation
Similar to anandamide, some studies suggest the existence of a modulation between 2-AG levels and psychostimulant effects. Therefore, it has been

shown that the acute administration of cocaine increases 2-AG levels in the VTA, resulting in a net increase of the dopaminergic signal (i.e., by reducing the GABAergic input), which might lastly regulate the rewarding properties of cocaine (Wang, Treadway, Covey, Cheer, & Lupica, 2015). Interestingly, a new study has shown that exenatide (Ex-4), a glucagonlike peptide that reduces both the expression of 2-AG and the product of its presynaptic degradation AA, abolished cocaine-induced elevation of dopamine (Reddy et al., 2016), suggesting that 2-AG levels might alter presynaptic dopaminergic homeostasis and cocaine actions through the product of its presynaptic degradation (i.e., AA).

Further supporting the interactions between cocaine and 2-AG levels, it has been shown that acute exposure to cocaine decreased diacylglycerol lipase alpha/beta (DAGLα/β) expression, suggesting a downregulation of 2-AG production, as DAGLα/β are required to generate 2-AG (Bisogno et al., 2003; Palomino et al., 2014). Similarly, in the study by Palomino et al. (2014), repeated cocaine exposure (i.e., sensitization paradigm) was accompanied by a decreased DAGLβ/MAGL ratio, suggesting also reduced 2-AG generation after repeated exposure to cocaine in mice. On the other hand, repeated cocaine exposure (i.e., 14 days self-administration) resulted in increased levels of 2-AG in several structures including the frontal cortex, the hippocampus, and the cerebellum and decreased levels in the hippocampus and the dorsal striatum in rats (also in yoked animals) (Bystrowska et al., 2014). The use of different paradigms, species, and time frame for measurements might account for the differences between the above studies with regard to the effects of psychostimulants on 2-AG levels.

Behavioral studies suggest that stress-induced cocaine relapse might be mediated by endocannabinoids. More specifically, in a study by Tung et al. (2016), acute restraint stress increased the levels of 2-AG in the VTA, suggesting that 2-AG might play a relevant role in the reinstatement of extinguished cocaine CPP in mice. Interestingly, a reduction in 2-AG levels in the hippocampus, the dorsal striatum, and the cerebellum during extinction (10 days) in rats has been reported (Bystrowska et al., 2014). Therefore it would be interesting to investigate if increasing 2-AG levels (no associated to stress) during extinction would be sufficient to reinstate cocaine-seeking behaviors.

CB$_2$ Modulation

The capacity of CB$_2$ receptors to modulate cocaine-reinforcing properties has been revealed using behavioral genetic and pharmacological studies in mice. Thus mice overexpressing CB$_2$ receptors (i.e., CB2xP mice) showed decreased cocaine motor responses and cocaine self-administration (Aracil-Fernandez et al., 2012). In agreement, the CB_2 receptor agonist O-1966 blocked cocaine-induced CPP in wild type but did not have effects in CB_2 knockout mice (Ignatowska-Jankowska et al., 2013). In an elegant study by Xi et al. (2011), both systemic and local (i.e., intra-accumbens) administration of the selective CB_2 receptor agonists JWH133 and GW405833 dose dependently inhibited intravenous cocaine self-administration and cocaine-enhanced locomotion in $CB_{1,2}$ wild-type and CB_1 knockout but not in CB_2 knockout mice. Moreover, the effects of the CB_2 receptor agonists on cocaine-reinforcing properties were blocked by the selective CB_2 receptor antagonist AM630 (Xi et al., 2011). Additionally, the intra-accumbens administration of JWH133 dose dependently decreased extracellular dopamine, whereas intra-accumbens administration of AM630 elevated dopamine levels and blocked the reduction in cocaine self-administration and extracellular dopamine produced by JWH133 (Xi et al., 2011), suggesting that the effects of CB_2 receptors on cocaine-rewarding properties are associated with a dopamine-dependent mechanism. Further supporting these findings, microinjections of JWH133 into the VTA inhibited cocaine self-administration in wild-type or CB_1 knockout mice, but these effects were absent in CB_2 knockout mice and blocked by CB_2 antagonist AM630 (Zhang et al., 2014). The involvement of CB_2 receptors in cocaine-induced activity in midbrain dopaminergic neurons in mice is further supported by electrophysiological studies showing that CB_2 receptor agonists (e.g., JWH133) inhibited VTA dopaminergic neuronal firing in vivo and ex vivo.

On the other hand, the capacity of CB_2 receptors to modulate cocaine-reinforcing properties in rats is less evident. Interestingly, it has been reported important species differences in CB_2 mRNA splicing and expression, protein sequences, and receptor responses to CB_2 ligands between mice and rats (Zhang et al., 2015). Moreover, this study revealed a significantly higher CB_2 mRNA expression in dopaminergic neurons of the VTA in mice as compared to rats, signifying the involvement of CB_2 receptors in cocaine-induced activity in midbrain dopaminergic neurons in mice (Zhang et al., 2015). Remarkably, it has been shown that the systemic administration of the CB_2 agonist JWH133 inhibited intravenous cocaine self-administration in mice but not in rats (Zhang et al., 2015). However, new studies have revealed that local administration of JWH133 into the NAc inhibited both cocaine-enhanced extracellular dopamine and cocaine self-administration in rats (Zhang et al., 2016). Additionally, the activation of CB_2 receptors with JWH133 inhibited VTA dopaminergic firing, suggesting that CB_2 receptors are expressed in VTA dopaminergic neurons and functionally modulate dopaminergic neuronal activities and cocaine self-administration behavior in rats (Zhang et al., 2016).

The cannabinoid receptor CB_2 has been related to the adaptive response induced by repeated exposure to cocaine. Therefore the CB_2 receptor seems be involved in the sensitization effects of psychostimulants in mice (Aracil-Fernandez et al., 2012; Xi et al., 2011), but not in rats (Blanco-Calvo et al., 2014). Interestingly, the pharmacological blockade of CB_2 receptors with AM630 prevented the cocaine-induced conditioned locomotion response in rats (Blanco-Calvo et al., 2014). On the other hand, studies in dependent animals have shown that blocking CB_2 receptors might prevent reinstatement of cocaine-seeking behavior. Thus the CB_2 antagonist SR144528 selectively decreased cocaine-induced reinstatement of cocaine-seeking behavior but not cocaine self-administration in rats (Adamczyk et al., 2012).

In summary, studies on the role of the ECS on the effects of psychostimulants seem quite contradictory even when using the same experimental paradigm or even within same species (Rivera et al., 2013). Some reviews on the role of CB_1 in psychostimulant addictive properties have concluded that the involvement of CB_1 receptors in the rewarding properties of psychostimulants is marginal (Oliere, Joliette-Riopel, Potvin, & Jutras-Aswad, 2013; Wiskerke, Pattij, Schoffelmeer, & De Vries, 2008). On the other hand, new studies have provided interesting evidence on the involvement of CB_2 receptors on the reinforcing properties of certain psychostimulants such as cocaine, particularly in mice.

CANNABINOID

Cannabis preparations are readily self-administered in humans both in laboratory and in real-life conditions (see Justinova, Goldberg, et al., 2005 for review). Indeed, the prevalence of cannabis use has shown a sustained increase in the past years, being currently the first most illegal used substance (recreational use of cannabis is currently illegal in the majority of countries/US states) and the second most prevalent used substance among those reporting use of different drugs in the past 12 months (Global-Drug-Survey, 2016). On the other hand, reliable self-administration of THC by laboratory animals has been challenging and only demonstrated in certain species (e.g., nonhuman primates vs. rodents) or using particular experimental conditions (e.g., food restriction, synthetic vs. naturally occurring cannabinoids). Therefore self-administration behavior can be maintained in squirrel monkeys by doses of THC comparable to those in marijuana smoke inhaled by humans (Justinova, Goldberg, et al., 2005; Justinova, Tanda, Munzar, & Goldberg, 2004; Justinova, Tanda, Redhi, & Goldberg, 2003; Tanda et al., 2000). On the other hand, only 2-AG and the synthetic CB_1 cannabinoid agonists WIN 55,212-2, CP 55,940, and HU210 seem to maintain intravenous or intracerebroventricular self-administration in rats and mice (Braida, Pozzi, Parolaro, & Sala, 2001; De Luca et al., 2014; Fattore, Cossu, Martellotta, & Fratta, 2001; Martellotta, Cossu, Fattore, Gessa, & Fratta, 1998; Navarro et al., 2001). The synthetic cannabinoid CB_1 receptor agonist CP 55,940 has not yet been reported to be self-administered intravenously by experimental animals, but maintained self-administration behavior when injected intracerebroventricularly (Braida, Pozzi, Parolaro, et al., 2001), produced CPP in rats (Braida et al., 2001), and reduced intracranial self-stimulation thresholds in a dose-related fashion in mice (Grim, Wiebelhaus, Morales, Negus, & Lichtman, 2015).

Remarkably, intravenous self-administration in rodents only happened under certain experimental conditions. Therefore in the studies by Martellotta et al. (1998) and Navarro et al. (2001), mice were restrained on their movement as they received the drug by a tail injection. On the other hand, in the study by Fattore et al. (2001) rats were unrestrained in movement but required to be food restricted (body weight was maintained at 80% of normal) to acquire WIN 55,212-2 self-administration. A later study in mice showed intravenous self-administration of WIN 55,212-2 in mice food restricted to reduce their body weight by 10% of their initial weight (Mendizabal, Zimmer, & Maldonado, 2006). Interestingly, mice required as many as 15 sessions to attain this behavior, which is similar to the 16 sessions reported in the aforementioned study by Fattore et al. in rats. Remarkably, in the study by Mendizabal et al. (2006), it was observed that administering WIN 55,212-2 in the home cage 24 h before the first self-administration session, expedited the acquisition of cannabinoid self-administration to the fourth session. This initial preexposure priming with THC seems to be required as well to establish CPP to THC in mice (Castane, Robledo, Matifas, Kieffer, & Maldonado, 2003; Valjent & Maldonado, 2000), suggesting that the initial exposure to cannabinoids might have aversive effects in rodents and therefore might be related to the difficulty to establish operant self-administration and CPP for cannabinoids in rodents. Other studies have also suggested that both CPP and self-administration of the synthetic agonist WIN 55,212-2 might be restricted to certain rodent subpopulations (e.g., high novelty seekers) (Rodriguez-Arias et al., 2016) or species strains

(Deiana et al., 2007). Therefore different preclinical studies seem to suggest the existence of rewarding properties of cannabinoids in rodents. However, the rewarding properties of cannabinoids in rodents might only be revealed under certain experimental conditions or only in particular subpopulations (e.g., high-novelty-seeking animals). An important consideration in these studies is the observation that THC was not able to substitute more potent synthetic CB_1 cannabinoid agonists (i.e., WIN 55,212-2) (Lefever, Marusich, Antonazzo, & Wiley, 2014), suggesting that synthetic CB_1 cannabinoid agonist self-administration may be of limited use (e.g., as a screening tool for detection of the reinforcing effects of potential cannabinoid medications) but might not be used to test the reinforcing properties of THC in rodents.

CB_1 Modulation

The cannabinoid CB_1 receptor antagonist rimonabant blocked THC, anandamide, and methanandamide self-administration in squirrel monkeys (Justinova, Munzar, et al., 2008; Justinova, Solinas, Tanda, Redhi, & Goldberg, 2005; Tanda et al., 2000). Moreover, rimonabant blocked THC but not cocaine self-administration, suggesting the specific involvement of CB_1 receptor on the reinforcing effects of THC. New studies have further confirmed that blocking CB_1 receptor with either rimonabant or the neutral CB_1 antagonist AM4113 dose-dependently blocked THC but not cocaine or food self-administration in squirrel monkeys (Schindler, Redhi, et al., 2016). Additionally, pretreatment with both rimonabant and AM4113 blocked reinstatement of THC seeking induced by both THC-associated cues and priming in squirrel monkeys (Justinova, Munzar, et al., 2008; Schindler, Redhi, et al., 2016). The effects of rimonabant opposing the facilitating effects of CB_1 receptor agonists have been also observed using other paradigms.

Fatty Acid Amide Hydrolase/Anandamide Modulation

Similarly to THC, the endogenous cannabinoid anandamide and its synthetic analog R(+)-methanandamide are intravenously self-administered by squirrel monkeys (Justinova, Solinas, et al., 2005). New studies have shown that the anandamide transport inhibitors AM404 and VDM11 were able to substitute and maintain intravenous self-administration in monkeys with a history of anandamide or cocaine (only for AM404) self-administration (Schindler, Scherma, et al., 2016). Additionally, THC/anandamide-seeking behavior can be reinstated by a priming injection of AM404 (Schindler, Scherma, et al., 2016). The effects of AM404 were blocked by rimonabant, and, conversely, the FAAH inhibitor URB597 was shown to potentiate AM404 effects (Schindler, Scherma, et al., 2016), suggesting that these effects were mediated by anandamide and CB_1 receptors. On the other hand, URB597 was not able to maintain intravenous self-administration but potentiated anandamide self-administration in squirrel monkeys (Justinova, Mangieri, et al., 2008). In the study by Justinova et al., URB597 did not promote reinstatement of extinguished drug-seeking behavior previously maintained by THC or anandamide. Clinical studies are currently investigating the effectiveness of the FAAH inhibitor PF-04457845 for the treatment of cannabis withdrawal (clinicaltrials.gov: NCT01618656), although no results have been yet published.

Monoacylglycerol Lipase/2-Arachidonoylglycerol Modulation

Intravenous self-administration of 2-AG has been shown both in nonhuman primates and rats (De Luca et al., 2014; Justinova, Yasar, Redhi, & Goldberg, 2011). In rats, 2-AG increases extracellular dopamine levels mainly in the NAc shell (De Luca et al., 2014), which has also been observed after the intravenous injection of THC and WIN 55,212-2 in rats (Tanda et al., 1997), and also during intravenous self-administration of cannabinoids and other drugs including heroin and alcohol (Caille et al., 2007; Fadda et al., 2006; Lecca, Cacciapaglia, Valentini, & Di Chiara, 2006). Intravenous self-administration of 2-AG in squirrel monkeys occurred in animals previously trained to self-administer either anandamide or nicotine (Justinova et al., 2011). However, the study by Justinova et al. revealed that intravenous 2-AG was a very effective reinforcer, maintaining high self-administration responding rates across a wide range of doses. This study also revealed that 2-AG self-administration was blocked by rimonabant without affecting responding maintained by food under similar conditions (Justinova et al., 2011), suggesting the possible mediation of CB_1 receptors in 2-AG self-administration. Similarly to the clinical studies investigating the effects of modulating anandamide on THC withdrawal, referred in the preferred section, future studies might also address the impact of modulating 2-AG on THC-reinforcing effects or in mitigating THC withdrawal.

CB_2 Modulation

There is a current lack of studies looking at the involvement of CB_2 receptors in the rewarding properties of cannabinoids. This is particularly important to study as we know that THC is a partial agonist to CB_1 and to CB_2.

CONCLUSIONS

Although here we have covered only the preclinical literature, it is important to note that some of these findings have been explored/validated in humans. It is now feasible to explore the status of some components of the ECS, such as CB_1 expression or FAAH levels using positron emission tomography (PET). A study using PET measured the availability of cerebral CB_1 receptors in both healthy controls and subjects with alcohol dependence (Ceccarini et al., 2014). The results of the study showed that, in healthy controls, the acute intravenous administration of ethanol increased the availability of CB_1 receptors (+15%). On the other hand, subjects with alcohol dependence showed reduced availability of CB_1 receptors not only following chronic heavy drinking (−16%) but also 1 month after monitored abstinence (−17%) (Ceccarini et al., 2014). In a similar study using PET, CB_1 receptor binding in subjects with alcohol dependence was reduced by 20%–30% as compared with controls (Hirvonen et al., 2013). Remarkably, in both studies, the changes in CB_1 receptor in subjects with alcohol dependence were still present after 2–4 weeks of monitored abstinence (Ceccarini et al., 2014; Hirvonen et al., 2013), suggesting the existence of long-lasting adaptations in the ECS in response to chronic use of alcohol in humans. There is a PET tracer available to study FAAH in humans (Wilson et al., 2011). Reduction of FAAH levels have been found in subjects with cannabis dependence and also with alcohol use disorders (Boileau et al., 2015, 2016). Pharmacological approaches have also been used in humans. The largest amount of work in the area has been done with rimonabant, a CB_1 antagonists/inverse agonist. In humans, rimonabant was effective in increasing abstinence rates in smokers trying to quit smoking (Le Foll, Forget, Aubin, & Goldberg, 2008). Rimonabant seems to have some positive effects as a possible treatment for alcohol use disorder, but the results of a first clinical trial were not conclusive, possibly due to a high rate of response in the placebo arm and an overall limited sample (Soyka et al., 2008). In addition, a laboratory study showed no changes in alcohol consumption or endocrine measures in non treatment-seeking heavy alcohol drinkers (George et al., 2010). Further clinical studies are clearly needed in this area. Rimonabant could also decrease some psychoactive effects induced by THC (Huestis et al., 2007, 2001). However, due to the increased incidence of neuropsychiatric side effects, this drug was removed from the market. New evidence suggests that the use of neutral CB_1 antagonists would be a better alternative for retaining the therapeutic efficacy, but potentially devoid of the undesirable side effects. However, this strategy remained to be tested in humans.

Another research strategy to study the role of the ECS in addiction consists of altering the endogenous cannabinoid levels. In this regard, preclinical models have shown that modifying endogenous cannabinoid levels might serve to alleviate withdrawal symptoms and prevent relapse but could also have other interesting beneficial pharmacological effects such as analgesia [e.g., FAAH inhibitors have analgesic and antiinflammatory effects in animal models (Roques, Fournie-Zaluski, & Wurm, 2012)]. However, new events underlie the necessity of being cautious in our expectations based only on preclinical research. FAAH inhibitors were no effective in reducing pain in humans (Huggins, Smart, Langman, Taylor, & Young, 2012). It should be mentioned here that a short while ago a specific FAAH inhibitor resulted in a fatal outcome in an early clinical trial (Kerbrat et al., 2016; Moore, 2016). Nevertheless, the fact that several other FAAH inhibitors have been previously used safely (Huggins et al., 2012) seem to indicate that this adverse event was isolated and not a class effect. So this single negative event should not stop the clinical research on the utility of FAAH inhibitors in humans.

The CB_2 receptors have attracted a lot of attention lately. However, their role still remains controversial and further studies are required to determine if CB_2 receptors might still represent an interesting alternative and complement the traditional focus on pharmacotherapies targeting CB_1 receptors.

REFERENCES

Adamczyk, P., McCreary, A. C., Przegalinski, E., Mierzejewski, P., Bienkowski, P., & Filip, M. (2009). The effects of fatty acid amide hydrolase inhibitors on maintenance of cocaine and food self-administration and on reinstatement of cocaine-seeking and food-taking behavior in rats. *Journal of Physiology and Pharmacology, 60*(3), 119–125.

Adamczyk, P., Miszkiel, J., McCreary, A. C., Filip, M., Papp, M., & Przegalinski, E. (2012). The effects of cannabinoid CB1, CB2 and vanilloid TRPV1 receptor antagonists on cocaine addictive behavior in rats. *Brain Research, 1444*, 45–54. http://dx.doi.org/10.1016/j.brainres.2012.01.030.

Al Mansouri, S., Ojha, S., Al Maamari, E., Al Ameri, M., Nurulain, S. M., & Bahi, A. (2014). The cannabinoid receptor 2 agonist, beta-caryophyllene, reduced voluntary alcohol intake and attenuated ethanol-induced place preference and sensitivity in mice. *Pharmacology, Biochemistry, and Behavior, 124*, 260–268. http://dx.doi.org/10.1016/j.pbb.2014.06.025.

Alvarez-Jaimes, L., Stouffer, D. G., & Parsons, L. H. (2009). Chronic ethanol treatment potentiates ethanol-induced increases in interstitial nucleus accumbens endocannabinoid levels in rats. *Journal of Neurochemistry, 111*(1), 37–48. http://dx.doi.org/10.1111/j.1471-4159.2009.06301.x.

Aracil-Fernandez, A., Trigo, J. M., Garcia-Gutierrez, M. S., Ortega-Alvaro, A., Ternianov, A., Navarro, D., … Manzanares, J. (2012). Decreased cocaine motor sensitization and self-administration in mice overexpressing cannabinoid CB(2) receptors. *Neuropsychopharmacology*, *37*(7), 1749–1763. http://dx.doi.org/10.1038/npp.2012.22.

Arnone, M., Maruani, J., Chaperon, F., Thiebot, M. H., Poncelet, M., Soubrie, P., … Le Fur, G. (1997). Selective inhibition of sucrose and ethanol intake by SR 141716, an antagonist of central cannabinoid (CB1) receptors. *Psychopharmacology (Berlin)*, *132*(1), 104–106.

Atwood, B. K., & Mackie, K. (2010). CB2: A cannabinoid receptor with an identity crisis. *British Journal of Pharmacology*, *160*(3), 467–479. http://dx.doi.org/10.1111/j.1476-5381.2010.00729.x.

Basavarajappa, B. S., Cooper, T. B., & Hungund, B. L. (1998). Chronic ethanol administration down-regulates cannabinoid receptors in mouse brain synaptic plasma membrane. *Brain Research*, *793*(1–2), 212–218.

Basavarajappa, B. S., & Hungund, B. L. (1999a). Chronic ethanol increases the cannabinoid receptor agonist anandamide and its precursor N-arachidonoylphosphatidylethanolamine in SK-N-SH cells. *Journal of Neurochemistry*, *72*(2), 522–528.

Basavarajappa, B. S., & Hungund, B. L. (1999b). Down-regulation of cannabinoid receptor agonist-stimulated [35S]GTP gamma S binding in synaptic plasma membrane from chronic ethanol exposed mouse. *Brain Research*, *815*(1), 89–97.

Basavarajappa, B. S., Ninan, I., & Arancio, O. (2008). Acute ethanol suppresses glutamatergic neurotransmission through endocannabinoids in hippocampal neurons. *Journal of Neurochemistry*, *107*(4), 1001–1013. http://dx.doi.org/10.1111/j.1471-4159.2008.05685.x.

Basavarajappa, B. S., Saito, M., Cooper, T. B., & Hungund, B. L. (2000). Stimulation of cannabinoid receptor agonist 2-arachidonylglycerol by chronic ethanol and its modulation by specific neuromodulators in cerebellar granule neurons. *Biochimica et Biophysica Acta*, *1535*(1), 78–86.

Basavarajappa, B. S., Saito, M., Cooper, T. B., & Hungund, B. L. (2003). Chronic ethanol inhibits the anandamide transport and increases extracellular anandamide levels in cerebellar granule neurons. *European Journal of Pharmacology*, *466*(1–2), 73–83.

Basavarajappa, B. S., Yalamanchili, R., Cravatt, B. F., Cooper, T. B., & Hungund, B. L. (2006). Increased ethanol consumption and preference and decreased ethanol sensitivity in female FAAH knockout mice. *Neuropharmacology*, *50*(7), 834–844. http://dx.doi.org/10.1016/j.neuropharm.2005.12.005.

Befort, K. (2015). Interactions of the opioid and cannabinoid systems in reward: Insights from knockout studies. *Frontiers in Pharmacology*, *6*, 6. http://dx.doi.org/10.3389/fphar.2015.00006.

Bisogno, T., Howell, F., Williams, G., Minassi, A., Cascio, M. G., Ligresti, A., … Doherty, P. (2003). Cloning of the first sn1-DAG lipases points to the spatial and temporal regulation of endocannabinoid signaling in the brain. *The Journal of Cell Biology*, *163*(3), 463–468. http://dx.doi.org/10.1083/jcb.200305129.

Blanco-Calvo, E., Rivera, P., Arrabal, S., Vargas, A., Pavon, F. J., Serrano, A., … Rodriguez de Fonseca, F. (2014). Pharmacological blockade of either cannabinoid CB1 or CB2 receptors prevents both cocaine-induced conditioned locomotion and cocaine-induced reduction of cell proliferation in the hippocampus of adult male rat. *Frontiers in Integrative Neuroscience*, *7*, 106. http://dx.doi.org/10.3389/fnint.2013.00106.

Blednov, Y. A., Cravatt, B. F., Boehm, S. L., 2nd, Walker, D., & Harris, R. A. (2007). Role of endocannabinoids in alcohol consumption and intoxication: Studies of mice lacking fatty acid amide hydrolase. *Neuropsychopharmacology*, *32*(7), 1570–1582. http://dx.doi.org/10.1038/sj.npp.1301274.

Blum, K., Hudson, K. C., Friedman, R. N., & Wallace, J. E. (1975). Tetrahydrocannabinol: Inhibition of alcohol-induced withdrawal symptoms in mice. In J. L. Singh (Ed.), *Drug addiction, neurobiology and influences on behavior* (Vol. 3) (pp. 39–53). New York: Stratton Intercon Med. Book Corp.

Boileau, I., Mansouri, E., Williams, B., Kish, S. J., Payer, D., DeLuca, V., ... Tong, J. (2015). *Investigating endocannabinoid metabolism in chronic cannabis and alcohol users: Neuroimaging studies with the Novel FAAH Probe, [11C]-CURB.*

Boileau, I., Mansouri, E., Williams, B., Le Foll, B., Rusjan, P., Mizrahi, R., ... Tong, J. (2016). Fatty acid amide hydrolase binding in brain of cannabis users: Imaging with the novel radiotracer [11C]CURB. *Biological Psychiatry, 80*(9), 691–701. http://dx.doi.org/10.1016/j.biopsych.2016.04.012.

Booth, M. (2003). *Cannabis: A history: Transworld.*

Braida, D., Iosue, S., Pegorini, S., & Sala, M. (2005). 3,4 Methylenedioxymethamphetamine-induced conditioned place preference (CPP) is mediated by endocannabinoid system. *Pharmacological Research, 51*(2), 177–182. http://dx.doi.org/10.1016/j.phrs.2004.07.009.

Braida, D., Pozzi, M., Cavallini, R., & Sala, M. (2001). Conditioned place preference induced by the cannabinoid agonist CP 55,940: Interaction with the opioid system. *Neuroscience, 104*(4), 923–926.

Braida, D., Pozzi, M., Parolaro, D., & Sala, M. (2001). Intracerebral self-administration of the cannabinoid receptor agonist CP 55,940 in the rat: Interaction with the opioid system. *European Journal of Pharmacology, 413*(2–3), 227–234.

Bystrowska, B., Smaga, I., Frankowska, M., & Filip, M. (2014). Changes in endocannabinoid and N-acylethanolamine levels in rat brain structures following cocaine self-administration and extinction training. *Progress in Neuro-Psychopharmacology & Biological Psychiatry, 50*, 1–10. http://dx.doi.org/10.1016/j.pnpbp.2013.12.002.

Caille, S., Alvarez-Jaimes, L., Polis, I., Stouffer, D. G., & Parsons, L. H. (2007). Specific alterations of extracellular endocannabinoid levels in the nucleus accumbens by ethanol, heroin, and cocaine self-administration. *The Journal of Neuroscience, 27*(14), 3695–3702. http://dx.doi.org/10.1523/JNEUROSCI.4403-06.2007.

Caille, S., & Parsons, L. H. (2003). SR141716A reduces the reinforcing properties of heroin but not heroin-induced increases in nucleus accumbens dopamine in rats. *The European Journal of Neuroscience, 18*(11), 3145–3149.

Caille, S., & Parsons, L. H. (2006). Cannabinoid modulation of opiate reinforcement through the ventral striatopallidal pathway. *Neuropsychopharmacology, 31*(4), 804–813. http://dx.doi.org/10.1038/sj.npp.1300848.

Castane, A., Robledo, P., Matifas, A., Kieffer, B. L., & Maldonado, R. (2003). Cannabinoid withdrawal syndrome is reduced in double mu and delta opioid receptor knockout mice. *The European Journal of Neuroscience, 17*(1), 155–159.

Castane, A., Valjent, E., Ledent, C., Parmentier, M., Maldonado, R., & Valverde, O. (2002). Lack of CB1 cannabinoid receptors modifies nicotine behavioural responses, but not nicotine abstinence. *Neuropharmacology, 43*(5), 857–867.

Ceccarini, J., Casteels, C., Koole, M., Bormans, G., & Van Laere, K. (2013). Transient changes in the endocannabinoid system after acute and chronic ethanol exposure and abstinence in the rat: A combined PET and microdialysis study. *European Journal of Nuclear Medicine and Molecular Imaging, 40*(10), 1582–1594. http://dx.doi.org/10.1007/s00259-013-2456-1.

Ceccarini, J., Hompes, T., Verhaeghen, A., Casteels, C., Peuskens, H., Bormans, G., ... Van Laere, K. (2014). Changes in cerebral CB1 receptor availability after acute and chronic alcohol abuse and monitored abstinence. *The Journal of Neuroscience, 34*(8), 2822–2831. http://dx.doi.org/10.1523/JNEUROSCI.0849-13.2014.

Chaperon, F., Soubrie, P., Puech, A. J., & Thiebot, M. H. (1998). Involvement of central cannabinoid (CB1) receptors in the establishment of place conditioning in rats. *Psychopharmacology (Berlin), 135*(4), 324–332.

Cheer, J. F., Wassum, K. M., Heien, M. L., Phillips, P. E., & Wightman, R. M. (2004). Cannabinoids enhance subsecond dopamine release in the nucleus accumbens of awake rats. *The Journal of Neuroscience, 24*(18), 4393–4400. http://dx.doi.org/10.1523/JNEUROSCI.0529-04.2004.

Cheer, J. F., Wassum, K. M., Sombers, L. A., Heien, M. L., Ariansen, J. L., Aragona, B. J., … Wightman, R. M. (2007). Phasic dopamine release evoked by abused substances requires cannabinoid receptor activation. *The Journal of Neuroscience*, *27*(4), 791–795. http://dx.doi.org/10.1523/JNEUROSCI.4152-06.2007.

Chen, J. P., Paredes, W., Li, J., Smith, D., Lowinson, J., & Gardner, E. L. (1990). Delta 9-tetrahydrocannabinol produces naloxone-blockable enhancement of presynaptic basal dopamine efflux in nucleus accumbens of conscious, freely-moving rats as measured by intracerebral microdialysis. *Psychopharmacology (Berlin)*, *102*(2), 156–162.

Chicca, A., Marazzi, J., Nicolussi, S., & Gertsch, J. (2012). Evidence for bidirectional endocannabinoid transport across cell membranes. *The Journal of Biological Chemistry*, *287*(41), 34660–34682. http://dx.doi.org/10.1074/jbc.M112.373241.

Cichewicz, D. L., & McCarthy, E. A. (2003). Antinociceptive synergy between delta(9)-tetrahydrocannabinol and opioids after oral administration. *The Journal of Pharmacology and Experimental Therapeutics*, *304*(3), 1010–1015. http://dx.doi.org/10.1124/jpet.102. 045575.

Cippitelli, A., Cannella, N., Braconi, S., Duranti, A., Tontini, A., Bilbao, A., … Ciccocioppo, R. (2008). Increase of brain endocannabinoid anandamide levels by FAAH inhibition and alcohol abuse behaviours in the rat. *Psychopharmacology (Berlin)*, *198*(4), 449–460. http://dx.doi.org/10.1007/s00213-008-1104-0.

Clarke, R. B., & Adermark, L. (2010). Acute ethanol treatment prevents endocannabinoid-mediated long-lasting disinhibition of striatal output. *Neuropharmacology*, *58*(4–5), 799–805. http://dx.doi.org/10.1016/j.neuropharm.2009.12.006.

Cohen, C., Perrault, G., Griebel, G., & Soubrie, P. (2005). Nicotine-associated cues maintain nicotine-seeking behavior in rats several weeks after nicotine withdrawal: Reversal by the cannabinoid (CB(1)) receptor antagonist, rimonabant (SR141716). *Neuropsychopharmacology*, *30*(1), 145–155.

Cohen, C., Perrault, G., Voltz, C., Steinberg, R., & Soubrie, P. (2002). SR141716, a central cannabinoid (CB(1)) receptor antagonist, blocks the motivational and dopamine-releasing effects of nicotine in rats. *Behavioural Pharmacology*, *13*(5–6), 451–463.

Colombo, G., Agabio, R., Fa, M., Guano, L., Lobina, C., Loche, A., … Gessa, G. L. (1998). Reduction of voluntary ethanol intake in ethanol-preferring sP rats by the cannabinoid antagonist SR-141716. *Alcohol and Alcoholism*, *33*(2), 126–130.

Colombo, G., Serra, S., Brunetti, G., Gomez, R., Melis, S., Vacca, G., … Gessa, L. (2002). Stimulation of voluntary ethanol intake by cannabinoid receptor agonists in ethanol-preferring sP rats. *Psychopharmacology (Berlin)*, *159*(2), 181–187. http://dx.doi.org/ 10.1007/s002130100887.

Corchero, J., Avila, M. A., Fuentes, J. A., & Manzanares, J. (1997). Delta-9-Tetrahydrocannabinol increases prodynorphin and proenkephalin gene expression in the spinal cord of the rat. *Life Sciences*, *61*(4), PL 39–43.

Corchero, J., Fuentes, J. A., & Manzanares, J. (1997). Delta 9-Tetrahydrocannabinol increases proopiomelanocortin gene expression in the arcuate nucleus of the rat hypothalamus. *European Journal of Pharmacology*, *323*(2–3), 193–195.

Corchero, J., Fuentes, J. A., & Manzanares, J. (1999). Chronic treatment with CP-55,940 regulates corticotropin releasing factor and proopiomelanocortin gene expression in the hypothalamus and pituitary gland of the rat. *Life Sciences*, *64*(11), 905–911.

Corchero, J., Manzanares, J., & Fuentes, J. A. (2004). Cannabinoid/opioid crosstalk in the central nervous system. *Critical Reviews in Neurobiology*, *16*(1–2), 159–172.

Corchero, J., Oliva, J. M., Garcia-Lecumberri, C., Martin, S., Ambrosio, E., & Manzanares, J. (2004). Repeated administration with Delta9-tetrahydrocannabinol regulates mu-opioid receptor density in the rat brain. *Journal of Psychopharmacology*, *18*(1), 54–58. http://dx.doi.org/10.1177/0269881104040237.

Corchero, J., Romero, J., Berrendero, F., Fernandez-Ruiz, J., Ramos, J. A., Fuentes, J. A., & Manzanares, J. (1999). Time-dependent differences of repeated administration with Delta9-tetrahydrocannabinol in proenkephalin and cannabinoid receptor gene expression and G-protein activation by mu-opioid and CB1-cannabinoid receptors in the caudate-putamen. *Brain Research. Molecular Brain Research*, *67*(1), 148–157.

Cossu, G., Ledent, C., Fattore, L., Imperato, A., Bohme, G. A., Parmentier, M., ... Fratta, W. (2001). Cannabinoid CB1 receptor knockout mice fail to self-administer morphine but not other drugs of abuse. *Behavioural Brain Research*, *118*(1), 61–65.

De Luca, M. A., Valentini, V., Bimpisidis, Z., Cacciapaglia, F., Caboni, P., & Di Chiara, G. (2014). Endocannabinoid 2-arachidonoylglycerol self-administration by Sprague-Dawley rats and stimulation of in vivo dopamine transmission in the nucleus accumbens shell. *Frontiers in Psychiatry*, *5*, 140. http://dx.doi.org/10.3389/fpsyt.2014.00140.

De Vries, T. J., Homberg, J. R., Binnekade, R., Raaso, H., & Schoffelmeer, A. N. (2003). Cannabinoid modulation of the reinforcing and motivational properties of heroin and heroin-associated cues in rats. *Psychopharmacology (Berlin)*, *168*(1–2), 164–169. http://dx.doi.org/10.1007/s00213-003-1422-1.

De Vries, T. J., Shaham, Y., Homberg, J. R., Crombag, H., Schuurman, K., Dieben, J., ... Schoffelmeer, A. N. (2001). A cannabinoid mechanism in relapse to cocaine seeking. *Nature Medicine*, *7*(10), 1151–1154. http://dx.doi.org/10.1038/nm1001-1151.

Deiana, S., Fattore, L., Spano, M. S., Cossu, G., Porcu, E., Fadda, P., ... Fratta, W. (2007). Strain and schedule-dependent differences in the acquisition, maintenance and extinction of intravenous cannabinoid self-administration in rats. *Neuropharmacology*, *52*(2), 646–654. http://dx.doi.org/10.1016/j.neuropharm.2006.09.007.

Del Arco, I., Navarro, M., Bilbao, A., Ferrer, B., Piomelli, D., & Rodriguez De Fonseca, F. (2002). Attenuation of spontaneous opiate withdrawal in mice by the anandamide transport inhibitor AM404. *European Journal of Pharmacology*, *454*(1), 103–104.

Devane, W. A., Hanus, L., Breuer, A., Pertwee, R. G., Stevenson, L. A., Griffin, G., ... Mechoulam, R. (1992). Isolation and structure of a brain constituent that binds to the cannabinoid receptor. *Science*, *258*(5090), 1946–1949.

Di Chiara, G., & Imperato, A. (1988). Drugs abused by humans preferentially increase synaptic dopamine concentrations in the mesolimbic system of freely moving rats. *Proceedings of the National Academy of Sciences of the United States of America*, *85*(14), 5274–5278.

Diergaarde, L., de Vries, W., Raaso, H., Schoffelmeer, A. N., & De Vries, T. J. (2008). Contextual renewal of nicotine seeking in rats and its suppression by the cannabinoid-1 receptor antagonist Rimonabant (SR141716A). *Neuropharmacology*, *55*(5), 712–716. http://dx.doi.org/10.1016/j.neuropharm.2008.06.003.

Fadda, P., Scherma, M., Spano, M. S., Salis, P., Melis, V., Fattore, L., ... Fratta, W. (2006). Cannabinoid self-administration increases dopamine release in the nucleus accumbens. *Neuroreport*, *17*(15), 1629–1632. http://dx.doi.org/10.1097/01.wnr.0000236853.40221.8e.

Fattore, L., Cossu, G., Martellotta, C. M., & Fratta, W. (2001). Intravenous self-administration of the cannabinoid CB1 receptor agonist WIN 55,212-2 in rats. *Psychopharmacology (Berlin)*, *156*(4), 410–416.

Fattore, L., Spano, M. S., Cossu, G., Deiana, S., & Fratta, W. (2003). Cannabinoid mechanism in reinstatement of heroin-seeking after a long period of abstinence in rats. *The European Journal of Neuroscience*, *17*(8), 1723–1726. http://dx.doi.org/10.1046/j.1460-9568.2003.02607.x.

Fattore, L., Spano, M., Melis, V., Fadda, P., & Fratta, W. (2011). Differential effect of opioid and cannabinoid receptor blockade on heroin-seeking reinstatement and cannabinoid substitution in heroin-abstinent rats. *British Journal of Pharmacology*, *163*(7), 1550–1562. http://dx.doi.org/10.1111/j.1476-5381.2011.01459.x.

Femenia, T., Garcia-Gutierrez, M. S., & Manzanares, J. (2010). CB1 receptor blockade decreases ethanol intake and associated neurochemical changes in fawn-hooded rats. *Alcoholism, Clinical and Experimental Research*, *34*(1), 131–141. http://dx.doi.org/10.1111/j.1530-0277.2009.01074.x.

Filip, M., Golda, A., Zaniewska, M., McCreary, A. C., Nowak, E., Kolasiewicz, W., … Przegalinski, E. (2006). Involvement of cannabinoid CB1 receptors in drug addiction: Effects of rimonabant on behavioral responses induced by cocaine. *Pharmacological Reports*, *58*(6), 806–819.

Forget, B., Coen, K. M., & Le Foll, B. (2009). Inhibition of fatty acid amide hydrolase reduces reinstatement of nicotine seeking but not break point for nicotine self-administration–comparison with CB(1) receptor blockade. *Psychopharmacology (Berlin)*, *205*(4), 613–624. http://dx.doi.org/10.1007/s00213-009-1569-5.

Forget, B., Guranda, M., Gamaleddin, I., Goldberg, S. R., & Le Foll, B. (2016). Attenuation of cue-induced reinstatement of nicotine seeking by URB597 through cannabinoid CB1 receptor in rats. *Psychopharmacology (Berlin)*, *233*(10), 1823–1828. http://dx.doi.org/10.1007/s00213-016-4232-y.

Forget, B., Hamon, M., & Thiebot, M. H. (2005). Cannabinoid CB1 receptors are involved in motivational effects of nicotine in rats. *Psychopharmacology (Berlin)*, *181*(4), 722–734.

Gamage, T. (2013). *Differential effects of endocannabinoid catabolic inhibitors on opioid withdrawal in mice*. Virginia Commonwealth University. Retrieved from http://scholarscompass.vcu.edu/etd/3293/.

Gamage, T. F., Ignatowska-Jankowska, B. M., Muldoon, P. P., Cravatt, B. F., Damaj, M. I., & Lichtman, A. H. (2015). Differential effects of endocannabinoid catabolic inhibitors on morphine withdrawal in mice. *Drug and Alcohol Dependence*, *146*, 7–16. http://dx.doi.org/10.1016/j.drugalcdep.2014.11.015.

Gamaleddin, I., Guranda, M., Goldberg, S. R., & Le Foll, B. (2011). The selective anandamide transport inhibitor VDM11 attenuates reinstatement of nicotine seeking behaviour, but does not affect nicotine intake. *British Journal of Pharmacology*, *164*(6), 1652–1660. http://dx.doi.org/10.1111/j.1476-5381.2011.01440.x.

Gamaleddin, I., Guranda, M., Scherma, M., Fratta, W., Makriyannis, A., Vadivel, S. K., … Le Foll, B. (2013). AM404 attenuates reinstatement of nicotine seeking induced by nicotine-associated cues and nicotine priming but does not affect nicotine- and food-taking. *Journal of Psychopharmacology*, *27*(6), 564–571. http://dx.doi.org/10.1177/0269881113477710.

Gamaleddin, I., Wertheim, C., Zhu, A. Z., Coen, K. M., Vemuri, K., Makryannis, A., … Le Foll, B. (2012). Cannabinoid receptor stimulation increases motivation for nicotine and nicotine seeking. *Addiction Biology*, *17*(1), 47–61. http://dx.doi.org/10.1111/j.1369-1600.2011.00314.x.

Gamaleddin, I., Zvonok, A., Makriyannis, A., Goldberg, S. R., & Le Foll, B. (2012). Effects of a selective cannabinoid CB2 agonist and antagonist on intravenous nicotine self administration and reinstatement of nicotine seeking. *PLoS One*, *7*(1), e29900. http://dx.doi.org/10.1371/journal.pone.0029900.

George, D. T., Herion, D. W., Jones, C. L., Phillips, M. J., Hersh, J., Hill, D., … Kunos, G. (2010). Rimonabant (SR141716) has no effect on alcohol self-administration or endocrine measures in nontreatment-seeking heavy alcohol drinkers. *Psychopharmacology (Berlin)*, *208*(1), 37–44. http://dx.doi.org/10.1007/s00213-009-1704-3.

Gessa, G. L., Serra, S., Vacca, G., Carai, M. A., & Colombo, G. (2005). Suppressing effect of the cannabinoid CB1 receptor antagonist, SR147778, on alcohol intake and motivational properties of alcohol in alcohol-preferring sP rats. *Alcohol and Alcoholism*, *40*(1), 46–53. http://dx.doi.org/10.1093/alcalc/agh114.

Global-Drug-Survey. (2016). *Global drug survey 2016*.

Goldberg, S. R., Spealman, R. D., & Goldberg, D. M. (1981). Persistent behavior at high rates maintained by intravenous self-administration of nicotine. *Science*, *214*(4520), 573–575.

Gonzalez, S., Cascio, M. G., Fernandez-Ruiz, J., Fezza, F., Di Marzo, V., & Ramos, J. A. (2002). Changes in endocannabinoid contents in the brain of rats chronically exposed to nicotine, ethanol or cocaine. *Brain Research*, *954*(1), 73–81.

Grim, T. W., Wiebelhaus, J. M., Morales, A. J., Negus, S. S., & Lichtman, A. H. (2015). Effects of acute and repeated dosing of the synthetic cannabinoid CP55,940 on intracranial self-stimulation in mice. *Drug and Alcohol Dependence, 150*, 31–37. http://dx.doi.org/10.1016/j.drugalcdep.2015.01.022.

Grobin, A. C., Matthews, D. B., Devaud, L. L., & Morrow, A. L. (1998). The role of GABA(A) receptors in the acute and chronic effects of ethanol. *Psychopharmacology (Berlin), 139*(1–2), 2–19.

Gueye, A. B., Pryslawsky, Y., Trigo, J. M., Poulia, N., Delis, F., Antoniou, K., … Le Foll, B. (2016). The CB1 neutral antagonist AM4113 retains the therapeutic efficacy of the inverse agonist rimonabant for nicotine dependence and weight loss with better psychiatric tolerability. *International Journal of Neuropsychopharmacology.* http://dx.doi.org/10.1093/ijnp/pyw068.

Gutierrez-Lopez, M. D., Llopis, N., Feng, S., Barrett, D. A., O'Shea, E., & Colado, M. I. (2010). Involvement of 2-arachidonoyl glycerol in the increased consumption of and preference for ethanol of mice treated with neurotoxic doses of methamphetamine. *British Journal of Pharmacology, 160*(3), 772–783. http://dx.doi.org/10.1111/j.1476-5381.2010.00720.x.

Hashemizadeh, S., Sardari, M., & Rezayof, A. (2014). Basolateral amygdala CB1 cannabinoid receptors mediate nicotine-induced place preference. *Progress in Neuro-Psychopharmacology & Biological Psychiatry, 51*, 65–71. http://dx.doi.org/10.1016/j.pnpbp.2014.01.010.

Henderson-Redmond, A. N., Guindon, J., & Morgan, D. J. (2016). Roles for the endocannabinoid system in ethanol-motivated behavior. *Progress in Neuro-Psychopharmacology & Biological Psychiatry, 65*, 330–339. http://dx.doi.org/10.1016/j.pnpbp.2015.06.011.

Herkenham, M., Lynn, A. B., Johnson, M. R., Melvin, L. S., de Costa, B. R., & Rice, K. C. (1991). Characterization and localization of cannabinoid receptors in rat brain: A quantitative in vitro autoradiographic study. *The Journal of Neuroscience, 11*(2), 563–583.

Hirvonen, J., Zanotti-Fregonara, P., Umhau, J. C., George, D. T., Rallis-Frutos, D., Lyoo, C. H., … Heilig, M. (2013). Reduced cannabinoid CB1 receptor binding in alcohol dependence measured with positron emission tomography. *Molecular Psychiatry, 18*(8), 916–921. http://dx.doi.org/10.1038/mp.2012.100.

Houchi, H., Babovic, D., Pierrefiche, O., Ledent, C., Daoust, M., & Naassila, M. (2005). CB1 receptor knockout mice display reduced ethanol-induced conditioned place preference and increased striatal dopamine D2 receptors. *Neuropsychopharmacology, 30*(2), 339–349. http://dx.doi.org/10.1038/sj.npp.1300568.

Huestis, M. A., Boyd, S. J., Heishman, S. J., Preston, K. L., Bonnet, D., Le Fur, G., … Gorelick, D. A. (2007). Single and multiple doses of rimonabant antagonize acute effects of smoked cannabis in male cannabis users. *Psychopharmacology (Berlin), 194*(4), 505–515. http://dx.doi.org/10.1007/s00213-007-0861-5.

Huestis, M. A., Gorelick, D. A., Heishman, S. J., Preston, K. L., Nelson, R. A., Moolchan, E. T., … Frank, R. A. (2001). Blockade of effects of smoked marijuana by the CB1-selective cannabinoid receptor antagonist SR141716. *Archives of General Psychiatry, 58*(4), 322–328.

Huggins, J. P., Smart, T. S., Langman, S., Taylor, L., & Young, T. (2012). An efficient randomised, placebo-controlled clinical trial with the irreversible fatty acid amide hydrolase-1 inhibitor PF-04457845, which modulates endocannabinoids but fails to induce effective analgesia in patients with pain due to osteoarthritis of the knee. *Pain, 153*(9), 1837–1846. http://dx.doi.org/10.1016/j.pain.2012.04.020.

Hungund, B. L., Goldstein, D. B., Villegas, F., & Cooper, T. B. (1988). Formation of fatty acid ethyl esters during chronic ethanol treatment in mice. *Biochemical Pharmacology, 37*(15), 3001–3004.

Hungund, B. L., Szakall, I., Adam, A., Basavarajappa, B. S., & Vadasz, C. (2003). Cannabinoid CB1 receptor knockout mice exhibit markedly reduced voluntary alcohol consumption and lack alcohol-induced dopamine release in the nucleus accumbens. *Journal of Neurochemistry, 84*(4), 698–704.

Ignatowska-Jankowska, B. M., Muldoon, P. P., Lichtman, A. H., & Damaj, M. I. (2013). The cannabinoid CB2 receptor is necessary for nicotine-conditioned place preference, but not other behavioral effects of nicotine in mice. *Psychopharmacology (Berlin)*, *229*(4), 591–601. http://dx.doi.org/10.1007/s00213-013-3117-6.

Ishiguro, H., Iwasaki, S., Teasenfitz, L., Higuchi, S., Horiuchi, Y., Saito, T., … Onaivi, E. S. (2007). Involvement of cannabinoid CB2 receptor in alcohol preference in mice and alcoholism in humans. *The Pharmacogenomics Journal*, *7*(6), 380–385. http://dx.doi.org/10.1038/sj.tpj.6500431.

Jones, R. T., & Stone, G. C. (1970). Psychological studies of marijuana and alcohol in man. *Psychopharmacologia*, *18*(1), 108–117.

Justinova, Z., Goldberg, S. R., Heishman, S. J., & Tanda, G. (2005). Self-administration of cannabinoids by experimental animals and human marijuana smokers. *Pharmacology, Biochemistry, and Behavior*, *81*(2), 285–299. http://dx.doi.org/10.1016/j.pbb.2005.01.026.

Justinova, Z., Mangieri, R. A., Bortolato, M., Chefer, S. I., Mukhin, A. G., Clapper, J. R., … Goldberg, S. R. (2008). Fatty acid amide hydrolase inhibition heightens anandamide signaling without producing reinforcing effects in primates. *Biological Psychiatry*, *64*(11), 930–937. http://dx.doi.org/10.1016/j.biopsych.2008.08.008.

Justinova, Z., Munzar, P., Panlilio, L. V., Yasar, S., Redhi, G. H., Tanda, G., … Goldberg, S. R. (2008). Blockade of THC-seeking behavior and relapse in monkeys by the cannabinoid CB(1)-receptor antagonist rimonabant. *Neuropsychopharmacology*, *33*(12), 2870–2877. http://dx.doi.org/10.1038/npp.2008.21.

Justinova, Z., Panlilio, L. V., Moreno-Sanz, G., Redhi, G. H., Auber, A., Secci, M. E., … Goldberg, S. R. (2015). Effects of fatty acid amide hydrolase (FAAH) inhibitors in non-human primate models of nicotine reward and relapse. *Neuropsychopharmacology*, *40*(9), 2185–2197. http://dx.doi.org/10.1038/npp.2015.62.

Justinova, Z., Solinas, M., Tanda, G., Redhi, G. H., & Goldberg, S. R. (2005). The endogenous cannabinoid anandamide and its synthetic analog R(+)-methanandamide are intravenously self-administered by squirrel monkeys. *The Journal of Neuroscience*, *25*(23), 5645–5650. http://dx.doi.org/10.1523/JNEUROSCI.0951-05.2005.

Justinova, Z., Tanda, G., Munzar, P., & Goldberg, S. R. (2004). The opioid antagonist naltrexone reduces the reinforcing effects of Delta 9 tetrahydrocannabinol (THC) in squirrel monkeys. *Psychopharmacology (Berlin)*, *173*(1–2), 186–194. http://dx.doi.org/10.1007/s00213-003-1693-6.

Justinova, Z., Tanda, G., Redhi, G. H., & Goldberg, S. R. (2003). Self-administration of delta9-tetrahydrocannabinol (THC) by drug naive squirrel monkeys. *Psychopharmacology (Berlin)*, *169*(2), 135–140.

Justinova, Z., Yasar, S., Redhi, G. H., & Goldberg, S. R. (2011). The endogenous cannabinoid 2-arachidonoylglycerol is intravenously self-administered by squirrel monkeys. *The Journal of Neuroscience*, *31*(19), 7043–7048. http://dx.doi.org/10.1523/JNEUROSCI.6058-10.2011.

Kerbrat, A., Ferre, J. C., Fillatre, P., Ronziere, T., Vannier, S., Carsin-Nicol, B., … Edan, G. (2016). Acute neurologic disorder from an inhibitor of fatty acid amide hydrolase. *The New England Journal of Medicine*, *375*(18), 1717–1725. http://dx.doi.org/10.1056/NEJMoa1604221.

Koob, G. F., & Le Moal, M. (2008). Review. Neurobiological mechanisms for opponent motivational processes in addiction. *Philosophical Transactions of the Royal Society B: Biological Sciences*, *363*(1507), 3113–3123. http://dx.doi.org/10.1098/rstb.2008.0094.

Kupferschmidt, D. A., Klas, P. G., & Erb, S. (2012). Cannabinoid CB1 receptors mediate the effects of corticotropin-releasing factor on the reinstatement of cocaine seeking and expression of cocaine-induced behavioural sensitization. *British Journal of Pharmacology*, *167*(1), 196–206. http://dx.doi.org/10.1111/j.1476-5381.2012.01983.x.

Kupferschmidt, D. A., Newman, A. E., Boonstra, R., & Erb, S. (2012). Antagonism of cannabinoid 1 receptors reverses the anxiety-like behavior induced by central injections of corticotropin-releasing factor and cocaine withdrawal. *Neuroscience*, *204*, 125–133. http://dx.doi.org/10.1016/j.neuroscience.2011.07.022.

Lallemand, F., & De Witte, P. (2006). SR147778, a CB1 cannabinoid receptor antagonist, suppresses ethanol preference in chronically alcoholized Wistar rats. *Alcohol, 39*(3), 125–134. http://dx.doi.org/10.1016/j.alcohol.2006.08.001.

Laposata, E. A., & Lange, L. G. (1986). Presence of nonoxidative ethanol metabolism in human organs commonly damaged by ethanol abuse. *Science, 231*(4737), 497–499.

Le Foll, B., Forget, B., Aubin, H. J., & Goldberg, S. R. (2008). Blocking cannabinoid CB1 receptors for the treatment of nicotine dependence: Insights from pre-clinical and clinical studies. *Addiction Biology, 13*(2), 239–252. http://dx.doi.org/10.1111/j.1369-1600.2008.00113.x.

Le Foll, B., & Goldberg, S. R. (2004). Rimonabant, a CB_1 antagonist, blocks nicotine-conditioned place preferences. *Neuroreport, 15*(13), 2139–2143.

Lecca, D., Cacciapaglia, F., Valentini, V., & Di Chiara, G. (2006). Monitoring extracellular dopamine in the rat nucleus accumbens shell and core during acquisition and maintenance of intravenous WIN 55,212-2 self-administration. *Psychopharmacology (Berlin), 188*(1), 63–74. http://dx.doi.org/10.1007/s00213-006-0475-3.

Ledent, C., Valverde, O., Cossu, G., Petitet, F., Aubert, J. F., Beslot, F., ... Parmentier, M. (1999). Unresponsiveness to cannabinoids and reduced addictive effects of opiates in CB1 receptor knockout mice. *Science, 283*(5400), 401–404.

Lefever, T. W., Marusich, J. A., Antonazzo, K. R., & Wiley, J. L. (2014). Evaluation of WIN 55,212-2 self-administration in rats as a potential cannabinoid abuse liability model. *Pharmacology, Biochemistry, and Behavior, 118*, 30–35. http://dx.doi.org/10.1016/j.pbb.2014.01.002.

Lesscher, H. M., Hoogveld, E., Burbach, J. P., van Ree, J. M., & Gerrits, M. A. (2005). Endogenous cannabinoids are not involved in cocaine reinforcement and development of cocaine-induced behavioural sensitization. *European Neuropsychopharmacology, 15*(1), 31–37. http://dx.doi.org/10.1016/j.euroneuro.2004.04.003.

Linsenbardt, D. N., & Boehm, S. L., 2nd (2009). Agonism of the endocannabinoid system modulates binge-like alcohol intake in male C57BL/6J mice: Involvement of the posterior ventral tegmental area. *Neuroscience, 164*(2), 424–434. http://dx.doi.org/10.1016/j.neuroscience.2009.08.007.

Little, P. J., Compton, D. R., Johnson, M. R., Melvin, L. S., & Martin, B. R. (1988). Pharmacology and stereoselectivity of structurally novel cannabinoids in mice. *The Journal of Pharmacology and Experimental Therapeutics, 247*(3), 1046–1051.

Lu, H. C., & Mackie, K. (2016). An introduction to the endogenous cannabinoid system. *Biological Psychiatry, 79*(7), 516–525. http://dx.doi.org/10.1016/j.biopsych.2015.07.028.

Luchicchi, A., Lecca, S., Carta, S., Pillolla, G., Muntoni, A. L., Yasar, S., ... Pistis, M. (2010). Effects of fatty acid amide hydrolase inhibition on neuronal responses to nicotine, cocaine and morphine in the nucleus accumbens shell and ventral tegmental area: Involvement of PPAR-alpha nuclear receptors. *Addiction Biology, 15*(3), 277–288. http://dx.doi.org/10.1111/j.1369-1600.2010.00222.x.

Maccioni, P., Colombo, G., & Carai, M. A. (2010). Blockade of the cannabinoid CB1 receptor and alcohol dependence: Preclinical evidence and preliminary clinical data. *CNS & Neurological Disorders Drug Targets, 9*(1), 55–59.

Maldonado, R., & Valverde, O. (2003). Participation of the opioid system in cannabinoid-induced antinociception and emotional-like responses. *European Neuropsychopharmacology, 13*(6), 401–410.

Manzanares, J., Corchero, J., Romero, J., Fernandez-Ruiz, J. J., Ramos, J. A., & Fuentes, J. A. (1998). Chronic administration of cannabinoids regulates proenkephalin mRNA levels in selected regions of the rat brain. *Brain Research. Molecular Brain Research, 55*(1), 126–132.

Manzanares, J., Corchero, J., Romero, J., Fernandez-Ruiz, J. J., Ramos, J. A., & Fuentes, J. A. (1999). Pharmacological and biochemical interactions between opioids and cannabinoids. *Trends in Pharmacological Sciences, 20*(7), 287–294.

Marinho, E. A., Oliveira-Lima, A. J., Santos, R., Hollais, A. W., Baldaia, M. A., Wuo-Silva, R., … Frussa-Filho, R. (2015). Effects of rimonabant on the development of single dose-induced behavioral sensitization to ethanol, morphine and cocaine in mice. *Progress in Neuro-Psychopharmacology & Biological Psychiatry*, *58*, 22–31. http://dx.doi.org/10.1016/j.pnpbp.2014.11.010.

Martellotta, M. C., Cossu, G., Fattore, L., Gessa, G. L., & Fratta, W. (1998). Self-administration of the cannabinoid receptor agonist WIN 55,212-2 in drug-naive mice. *Neuroscience*, *85*(2), 327–330.

Martin, M., Ledent, C., Parmentier, M., Maldonado, R., & Valverde, O. (2000). Cocaine, but not morphine, induces conditioned place preference and sensitization to locomotor responses in CB1 knockout mice. *The European Journal of Neuroscience*, *12*(11), 4038–4046.

Mas-Nieto, M., Pommier, B., Tzavara, E. T., Caneparo, A., Da Nascimento, S., Le Fur, G., … Noble, F. (2001). Reduction of opioid dependence by the CB(1) antagonist SR141716A in mice: Evaluation of the interest in pharmacotherapy of opioid addiction. *British Journal of Pharmacology*, *132*(8), 1809–1816. http://dx.doi.org/10.1038/sj.bjp.0703990.

Massi, P., Vaccani, A., Romorini, S., & Parolaro, D. (2001). Comparative characterization in the rat of the interaction between cannabinoids and opiates for their immunosuppressive and analgesic effects. *Journal of Neuroimmunology*, *117*(1–2), 116–124.

Matsuda, L. A., Lolait, S. J., Brownstein, M. J., Young, A. C., & Bonner, T. I. (1990). Structure of a cannabinoid receptor and functional expression of the cloned cDNA. *Nature*, *346*(6284), 561–564. http://dx.doi.org/10.1038/346561a0.

Mechoulm, R., Ben-Shabat, S., Hanus, L., Ligumsky, M., Kaminski, N.E., Schatz, A.R., … Zvi, V. (1995). Identification of an endogenous 2-monoglyceride, present in canine gut, that binds to cannabinoid receptors. *Biochemical Pharmacology*, *50*(1), 83–90.

Mechoulam, R., & Gaoni, Y. (1964). Isolation, structure, and partial synthesis of an active constituent of Hashish. *Journal of the American Chemical Society*, *86*(8), 1646–1647.

Melis, M., Sagheddu, C., De Felice, M., Casti, A., Madeddu, C., Spiga, S., … Pistis, M. (2014). Enhanced endocannabinoid-mediated modulation of rostromedial tegmental nucleus drive onto dopamine neurons in Sardinian alcohol-preferring rats. *The Journal of Neuroscience*, *34*(38), 12716–12724. http://dx.doi.org/10.1523/JNEUROSCI.1844-14.2014.

Mendizabal, V., Zimmer, A., & Maldonado, R. (2006). Involvement of kappa/dynorphin system in WIN 55,212-2 self-administration in mice. *Neuropsychopharmacology*, *31*(9), 1957–1966. http://dx.doi.org/10.1038/sj.npp.1300957.

Merritt, L. L., Martin, B. R., Walters, C., Lichtman, A. H., & Damaj, M. I. (2008). The endogenous cannabinoid system modulates nicotine reward and dependence. *The Journal of Pharmacology and Experimental Therapeutics*, *326*(2), 483–492. http://dx.doi.org/10.1124/jpet.108.138321.

Miller, L. L., Ward, S. J., & Dykstra, L. A. (2008). Chronic unpredictable stress enhances cocaine-conditioned place preference in type 1 cannabinoid receptor knockout mice. *Behavioural Pharmacology*, *19*(5–6), 575–581. http://dx.doi.org/10.1097/FBP.0b013e32830ded11.

Mitrirattanakul, S., Lopez-Valdes, H. E., Liang, J., Matsuka, Y., Mackie, K., Faull, K. F., … Spigelman, I. (2007). Bidirectional alterations of hippocampal cannabinoid 1 receptors and their endogenous ligands in a rat model of alcohol withdrawal and dependence. *Alcoholism, Clinical and Experimental Research*, *31*(5), 855–867. http://dx.doi.org/10.1111/j.1530-0277.2007.00366.x.

Moore, N. (2016). Lessons from the fatal French study BIA-10-2474. *British Medical Journal*, *353*, i2727. http://dx.doi.org/10.1136/bmj.i2727.

Moranta, D., Esteban, S., & Garcia-Sevilla, J. A. (2006). Ethanol desensitizes cannabinoid CB1 receptors modulating monoamine synthesis in the rat brain in vivo. *Neuroscience Letters*, *392*(1–2), 58–61. http://dx.doi.org/10.1016/j.neulet.2005.08.061.

Morrow, A. L., Suzdak, P. D., Karanian, J. W., & Paul, S. M. (1988). Chronic ethanol adminis-
tration alters gamma-aminobutyric acid, pentobarbital and ethanol-mediated 36Cl-
uptake in cerebral cortical synaptoneurosomes. *The Journal of Pharmacology and
Experimental Therapeutics, 246*(1), 158–164.
Most, D., Ferguson, L., & Harris, R. A. (2014). Molecular basis of alcoholism. *Handbook of
Clinical Neurology, 125,* 89–111. http://dx.doi.org/10.1016/B978-0-444-62619-
6.00006-9.
Muldoon, P. P., Chen, J., Harenza, J. L., Abdullah, R. A., Sim-Selley, L. J., Cravatt, B. F., …
Damaj, M. I. (2015). Inhibition of monoacylglycerol lipase reduces nicotine withdrawal.
British Journal of Pharmacology, 172(3), 869–882. http://dx.doi.org/10.1111/bph.12948.
Munro, S., Thomas, K. L., & Abu-Shaar, M. (1993). Molecular characterization of a periph-
eral receptor for cannabinoids. *Nature, 365*(6441), 61–65. http://dx.doi.org/
10.1038/365061a0.
Nashmi, R., & Lester, H. A. (2006). CNS localization of neuronal nicotinic receptors. *Journal
of Molecular Neuroscience, 30*(1–2), 181–184. http://dx.doi.org/10.1385/JMN:30:1:181.
Navarrete, F., Rodriguez-Arias, M., Martin-Garcia, E., Navarro, D., Garcia-Gutierrez, M. S.,
Aguilar, M. A., … Manzanares, J. (2013). Role of CB2 cannabinoid receptors in the
rewarding, reinforcing, and physical effects of nicotine. *Neuropsychopharmacology, 38*(12),
2515–2524. http://dx.doi.org/10.1038/npp.2013.157.
Navarro, M., Carrera, M. R., Fratta, W., Valverde, O., Cossu, G., Fattore, L., … Rodriguez
de Fonseca, F. (2001). Functional interaction between opioid and cannabinoid recep-
tors in drug self-administration. *The Journal of Neuroscience, 21*(14), 5344–5350.
Newman, L. M., Lutz, M. P., Gould, M. H., & Domino, E. F. (1972). 9 -Tetrahydrocannabinol
and ethyl alcohol: Evidence for cross-tolerance in the rat. *Science, 175*(4025),
1022–1023.
Nicolussi, S., & Gertsch, J. (2015). Endocannabinoid transport revisited. *Vitamins and
Hormones, 98,* 441–485. http://dx.doi.org/10.1016/bs.vh.2014.12.011.
Oliere, S., Joliette-Riopel, A., Potvin, S., & Jutras-Aswad, D. (2013). Modulation of the endocannabinoid system: Vulnerability factor and new treatment target for stimulant addiction. *Frontiers in Psychiatry, 4,* 109. http://dx.doi.org/10.3389/fpsyt.2013.00109.
Onaivi, E. S., Carpio, O., Ishiguro, H., Schanz, N., Uhl, G. R., & Benno, R. (2008). Behavioral
effects of CB2 cannabinoid receptor activation and its influence on food and alcohol consumption. *Annals of the New York Academy of Sciences, 1139,* 426–433. http://dx.doi.org/
10.1196/annals.1432.035.
Onaivi, E. S., Ishiguro, H., Gong, J. P., Patel, S., Meozzi, P. A., Myers, L., … Uhl, G. R.
(2008). Brain neuronal CB2 cannabinoid receptors in drug abuse and depression:
From mice to human subjects. *PLoS One, 3*(2), e1640. http://dx.doi.org/10.1371/
journal.pone.0001640.
Ortega-Alvaro, A., Ternianov, A., Aracil-Fernandez, A., Navarrete, F., Garcia-Gutierrez, M. S.,
& Manzanares, J. (2015). Role of cannabinoid CB2 receptor in the reinforcing actions of
ethanol. *Addiction Biology, 20*(1), 43–55. http://dx.doi.org/10.1111/adb.12076.
Ortiz, S., Oliva, J. M., Perez-Rial, S., Palomo, T., & Manzanares, J. (2004a). Chronic ethanol
consumption regulates cannabinoid CB1 receptor gene expression in selected regions of
rat brain. *Alcohol and Alcoholism, 39*(2), 88–92.
Ortiz, S., Oliva, J. M., Perez-Rial, S., Palomo, T., & Manzanares, J. (2004b). Differences in
basal cannabinoid CB1 receptor function in selective brain areas and vulnerability to
voluntary alcohol consumption in Fawn Hooded and Wistar rats. *Alcohol and Alcoholism,
39*(4), 297–302. http://dx.doi.org/10.1093/alcalc/agh063.
Paldy, E., Bereczki, E., Santha, M., Wenger, T., Borsodi, A., Zimmer, A., … Benyhe, S. (2008).
CB(2) cannabinoid receptor antagonist SR144528 decreases mu-opioid receptor
expression and activation in mouse brainstem: Role of CB(2) receptor in pain.
Neurochemistry International, 53(6–8), 309–316. http://dx.doi.org/10.1016/j.neuint.
2008.08.005.

Paldyova, E., Bereczki, E., Santha, M., Wenger, T., Borsodi, A., & Benyhe, S. (2008). Noladin ether, a putative endocannabinoid, inhibits mu-opioid receptor activation via CB2 cannabinoid receptors. *Neurochemistry International, 52*(1–2), 321–328. http://dx.doi.org/10.1016/j.neuint.2007.06.033.

Palomino, A., Pavon, F. J., Blanco-Calvo, E., Serrano, A., Arrabal, S., Rivera, P., ... Suarez, J. (2014). Effects of acute versus repeated cocaine exposure on the expression of endocannabinoid signaling-related proteins in the mouse cerebellum. *Frontiers in Integrative Neuroscience, 8*, 22. http://dx.doi.org/10.3389/fnint.2014.00022.

Panlilio, L. V., & Goldberg, S. R. (2007). Self-administration of drugs in animals and humans as a model and an investigative tool. *Addiction, 102*(12), 1863–1870. http://dx.doi.org/10.1111/j.1360-0443.2007.02011.x.

Parolaro, D., Rubino, T., Vigano, D., Massi, P., Guidali, C., & Realini, N. (2010). Cellular mechanisms underlying the interaction between cannabinoid and opioid system. *Current Drug Targets, 11*(4), 393–405.

Parsons, L. H., & Hurd, Y. L. (2015). Endocannabinoid signalling in reward and addiction. *Nature Reviews. Neuroscience, 16*(10), 579–594. http://dx.doi.org/10.1038/nrn4004.

Pava, M. J., Blake, E. M., Green, S. T., Mizroch, B. J., Mulholland, P. J., & Woodward, J. J. (2012). Tolerance to cannabinoid-induced behaviors in mice treated chronically with ethanol. *Psychopharmacology (Berlin), 219*(1), 137–147. http://dx.doi.org/10.1007/s00213-011-2387-0.

Pava, M. J., & Woodward, J. J. (2012). A review of the interactions between alcohol and the endocannabinoid system: Implications for alcohol dependence and future directions for research. *Alcohol, 46*(3), 185–204. http://dx.doi.org/10.1016/j.alcohol.2012.01.002.

Perra, S., Pillolla, G., Luchicchi, A., & Pistis, M. (2008). Alcohol inhibits spontaneous activity of basolateral amygdala projection neurons in the rat: Involvement of the endocannabinoid system. *Alcoholism, Clinical and Experimental Research, 32*(3), 443–449. http://dx.doi.org/10.1111/j.1530-0277.2007.00588.x.

Perra, S., Pillolla, G., Melis, M., Muntoni, A. L., Gessa, G. L., & Pistis, M. (2005). Involvement of the endogenous cannabinoid system in the effects of alcohol in the mesolimbic reward circuit: Electrophysiological evidence in vivo. *Psychopharmacology (Berlin), 183*(3), 368–377. http://dx.doi.org/10.1007/s00213-005-0195-0.

Pickel, V. M., Chan, J., Kash, T. L., Rodriguez, J. J., & MacKie, K. (2004). Compartment-specific localization of cannabinoid 1 (CB1) and mu-opioid receptors in rat nucleus accumbens. *Neuroscience, 127*(1), 101–112. http://dx.doi.org/10.1016/j.neuroscience.2004.05.015.

Ramesh, D., Gamage, T. F., Vanuytsel, T., Owens, R. A., Abdullah, R. A., Niphakis, M. J., ... Lichtman, A. H. (2013). Dual inhibition of endocannabinoid catabolic enzymes produces enhanced antiwithdrawal effects in morphine-dependent mice. *Neuropsychopharmacology, 38*(6), 1039–1049. http://dx.doi.org/10.1038/npp.2012.269.

Ramesh, D., Ross, G. R., Schlosburg, J. E., Owens, R. A., Abdullah, R. A., Kinsey, S. G., ... Lichtman, A. H. (2011). Blockade of endocannabinoid hydrolytic enzymes attenuates precipitated opioid withdrawal symptoms in mice. *The Journal of Pharmacology and Experimental Therapeutics, 339*(1), 173–185. http://dx.doi.org/10.1124/jpet.111.181370.

Rau, M., Nicolussi, S., Chicca, A., & Gertsch, J. (2016). Assay of endocannabinoid uptake. *Methods in Molecular Biology, 1412*, 191–203. http://dx.doi.org/10.1007/978-1-4939-3539-0_20.

Reddy, I. A., Pino, J. A., Weikop, P., Osses, N., Sorensen, G., Bering, T., ... Galli, A. (2016). Glucagon-like peptide 1 receptor activation regulates cocaine actions and dopamine homeostasis in the lateral septum by decreasing arachidonic acid levels. *Translational Psychiatry, 6*, e809. http://dx.doi.org/10.1038/tp.2016.86.

Rice, O. V., Gordon, N., & Gifford, A. N. (2002). Conditioned place preference to morphine in cannabinoid CB1 receptor knockout mice. *Brain Research, 945*(1), 135–138.

Riegel, A. C., & Lupica, C. R. (2004). Independent presynaptic and postsynaptic mechanisms regulate endocannabinoid signaling at multiple synapses in the ventral tegmental area. *The Journal of Neuroscience, 24*(49), 11070–11078. http://dx.doi.org/10.1523/JNEUROSCI.3695-04.2004.

Rios, C., Gomes, I., & Devi, L. A. (2006). Mu opioid and CB1 cannabinoid receptor interactions: Reciprocal inhibition of receptor signaling and neuritogenesis. *British Journal of Pharmacology, 148*(4), 387–395. http://dx.doi.org/10.1038/sj.bjp.0706757.

Rivera, P., Miguens, M., Coria, S. M., Rubio, L., Higuera-Matas, A., Bermudez-Silva, F. J., … Ambrosio, E. (2013). Cocaine self-administration differentially modulates the expression of endogenous cannabinoid system-related proteins in the hippocampus of Lewis vs. Fischer 344 rats. *Journal of Neuropsychopharmacology, 16*(6), 1277–1293. http://dx.doi.org/10.1017/S1461145712001186.

Rodriguez-Arias, M., Roger-Sanchez, C., Vilanova, I., Revert, N., Manzanedo, C., Minarro, J., … Aguilar, M. A. (2016). Effects of cannabinoid exposure during adolescence on the conditioned rewarding effects of WIN 55212-2 and cocaine in mice: Influence of the novelty-seeking trait. *Neural Plasticity, 2016*, 6481862. http://dx.doi.org/10.1155/2016/6481862.

Rodriguez-Arias, M., Valverde, O., Daza-Losada, M., Blanco-Gandia, M. C., Aguilar, M. A., & Minarro, J. (2013). Assessment of the abuse potential of MDMA in the conditioned place preference paradigm: Role of CB1 receptors. *Progress in Neuro-Psychopharmacology & Biological Psychiatry, 47*, 77–84. http://dx.doi.org/10.1016/j.pnpbp.2013.07.013.

Rodriguez, J. J., Mackie, K., & Pickel, V. M. (2001). Ultrastructural localization of the CB1 cannabinoid receptor in mu-opioid receptor patches of the rat Caudate putamen nucleus. *The Journal of Neuroscience, 21*(3), 823–833.

Ronan, P. J., Wongngamnit, N., & Beresford, T. P. (2016). Molecular mechanisms of cannabis signaling in the brain. *Progress in Molecular Biology and Translational Science, 137*, 123–147. http://dx.doi.org/10.1016/bs.pmbts.2015.10.002.

Roques, B. P., Fournie-Zaluski, M. C., & Wurm, M. (2012). Inhibiting the breakdown of endogenous opioids and cannabinoids to alleviate pain. *Nature Reviews. Drug Discovery, 11*(4), 292–310. http://dx.doi.org/10.1038/nrd3673.

Salio, C., Fischer, J., Franzoni, M. F., Mackie, K., Kaneko, T., & Conrath, M. (2001). CB1-cannabinoid and mu-opioid receptor co-localization on postsynaptic target in the rat dorsal horn. *Neuroreport, 12*(17), 3689–3692.

Scherma, M., Justinova, Z., Zanettini, C., Panlilio, L. V., Mascia, P., Fadda, P., … Goldberg, S. R. (2012). The anandamide transport inhibitor AM404 reduces the rewarding effects of nicotine and nicotine-induced dopamine elevations in the nucleus accumbens shell in rats. *British Journal of Pharmacology, 165*(8), 2539–2548. http://dx.doi.org/10.1111/j.1476-5381.2011.01467.x.

Scherma, M., Panlilio, L. V., Fadda, P., Fattore, L., Gamaleddin, I., Le Foll, B., … Goldberg, S. R. (2008). Inhibition of anandamide hydrolysis by URB597 reverses abuse-related behavioral and neurochemical effects of nicotine in rats. *The Journal of Pharmacology and Experimental Therapeutics, 327*(2), 482–490. http://dx.doi.org/10.1124/jpet.108.142224.

Schindler, C. W., Redhi, G. H., Vemuri, K., Makriyannis, A., Le Foll, B., Bergman, J., … Justinova, Z. (2016). Blockade of nicotine and cannabinoid reinforcement and relapse by a cannabinoid CB1-receptor neutral antagonist AM4113 and inverse agonist rimonabant in squirrel monkeys. *Neuropsychopharmacology, 41*(9), 2283–2293. http://dx.doi.org/10.1038/npp.2016.27.

Schindler, C. W., Scherma, M., Redhi, G. H., Vadivel, S. K., Makriyannis, A., Goldberg, S. R., … Justinova, Z. (2016). Self-administration of the anandamide transport inhibitor AM404 by squirrel monkeys. *Psychopharmacology (Berlin), 233*(10), 1867–1877. http://dx.doi.org/10.1007/s00213-016-4211-3.

Serra, S., Brunetti, G., Pani, M., Vacca, G., Carai, M. A., Gessa, G. L., … Colombo, G. (2002). Blockade by the cannabinoid CB(1) receptor antagonist, SR 141716, of alcohol deprivation effect in alcohol-preferring rats. *European Journal of Pharmacology, 443*(1–3), 95–97.

Serrano, A., Rivera, P., Pavon, F. J., Decara, J., Suarez, J., Rodriguez de Fonseca, F., … Parsons, L. H. (2012). Differential effects of single versus repeated alcohol withdrawal on the expression of endocannabinoid system-related genes in the rat amygdala. *Alcoholism, Clinical and Experimental Research, 36*(6), 984–994. http://dx.doi.org/10.1111/j.1530-0277.2011.01686.x.

Shahidi, S., & Hasanein, P. (2011). Behavioral effects of fatty acid amide hydrolase inhibition on morphine withdrawal symptoms. *Brain Research Bulletin, 86*(1–2), 118–122. http://dx.doi.org/10.1016/j.brainresbull.2011.06.019.

Siemens, A. J., & Doyle, O. L. (1979). Cross-tolerance between delta9-tetrahydrocannabinol and ethanol: The role of drug disposition. *Pharmacology, Biochemistry, and Behavior, 10*(1), 49–55.

Singh, M. E., Verty, A. N., McGregor, I. S., & Mallet, P. E. (2004). A cannabinoid receptor antagonist attenuates conditioned place preference but not behavioural sensitization to morphine. *Brain Research, 1026*(2), 244–253. http://dx.doi.org/10.1016/j.brainres.2004.08.027.

Solinas, M., Panlilio, L. V., Antoniou, K., Pappas, L. A., & Goldberg, S. R. (2003). The cannabinoid CB1 antagonist N-piperidinyl-5-(4-chlorophenyl)-1-(2,4-dichlorophenyl)-4-methylpyrazole-3-carboxamide (SR-141716A) differentially alters the reinforcing effects of heroin under continuous reinforcement, fixed ratio, and progressive ratio schedules of drug self-administration in rats. *The Journal of Pharmacology and Experimental Therapeutics, 306*(1), 93–102. http://dx.doi.org/10.1124/jpet.102.047928.

Solinas, M., Panlilio, L. V., & Goldberg, S. R. (2004). Exposure to delta-9-tetrahydrocannabinol (THC) increases subsequent heroin taking but not heroin's reinforcing efficacy: A self-administration study in rats. *Neuropsychopharmacology, 29*(7), 1301–1311. http://dx.doi.org/10.1038/sj.npp.13004311300431.

Solinas, M., Panlilio, L. V., Tanda, G., Makriyannis, A., Matthews, S. A., & Goldberg, S. R. (2005). Cannabinoid agonists but not inhibitors of endogenous cannabinoid transport or metabolism enhance the reinforcing efficacy of heroin in rats. *Neuropsychopharmacology, 30*(11), 2046–2057. http://dx.doi.org/10.1038/sj.npp.1300754.

Soria, G., Mendizabal, V., Tourino, C., Robledo, P., Ledent, C., Parmentier, M., … Valverde, O. (2005). Lack of CB1 cannabinoid receptor impairs cocaine self-administration. *Neuropsychopharmacology, 30*(9), 1670–1680. . http://dx.doi.org/10.1038/sj.npp.1300707.

Soyka, M., Koller, G., Schmidt, P., Lesch, O. M., Leweke, M., Fehr, C., …(2008). ACTOL Study Investigators. Cannabinoid receptor 1 blocker rimonabant (SR 141716) for treatment of alcohol dependence: results from a placebo-controlled, double-blind trial. *Journal of Clinical Psychopharmacology, 28*(3), 317–324.

Sprague, G. L., & Craigmill, A. L. (1976). Ethanol and delta-9-tetrahydrocannabinol: Mechanism for cross-tolerance in mice. *Pharmacology, Biochemistry, and Behavior, 5*(4), 409–415.

Sprague, G. L., & Craigmill, A. L. (1978). Effects of two cannabinoids upon abstinence signs in ethanol-dependent mice. *Pharmacology, Biochemistry, and Behavior, 9*(1), 11–15.

Sugiura, T., Kondo, S., Sukagawa, A., Nakane, S., Shinoda, A., Itoh, K., … Waku, K. (1995). 2-Arachidonoylglycerol: A possible endogenous cannabinoid receptor ligand in brain. *Biochemical and Biophysical Research Communications, 215*(1), 89–97.

Talani, G., & Lovinger, D. M. (2015). Interactions between ethanol and the endocannabinoid system at GABAergic synapses on basolateral amygdala principal neurons. *Alcohol, 49*(8), 781–794. http://dx.doi.org/10.1016/j.alcohol.2015.08.006.

Tanda, G., Munzar, P., & Goldberg, S. R. (2000). Self-administration behavior is maintained by the psychoactive ingredient of marijuana in squirrel monkeys. *Nature Neuroscience, 3*(11), 1073–1074. http://dx.doi.org/10.1038/80577.

Tanda, G., Pontieri, F. E., & Di Chiara, G. (1997). Cannabinoid and heroin activation of mesolimbic dopamine transmission by a common mu1 opioid receptor mechanism. *Science, 276*(5321), 2048–2050.

Thanos, P. K., Dimitrakakis, E. S., Rice, O., Gifford, A., & Volkow, N. D. (2005). Ethanol self-administration and ethanol conditioned place preference are reduced in mice lacking cannabinoid CB1 receptors. *Behavioural Brain Research, 164*(2), 206–213. . http://dx.doi.org/10.1016/j.bbr.2005.06.021.

Tourino, C., Ledent, C., Maldonado, R., & Valverde, O. (2008). CB1 cannabinoid receptor modulates 3,4-methylenedioxymethamphetamine acute responses and reinforcement. *Biological Psychiatry, 63*(11), 1030–1038. http://dx.doi.org/10.1016/j.biopsych.2007.09.003.

Trigo, J. M., & Le Foll, B. (2016). Inhibition of monoacylglycerol lipase (MAGL) enhances cue-induced reinstatement of nicotine-seeking behavior in mice. *Psychopharmacology (Berlin), 233*(10), 1815–1822. http://dx.doi.org/10.1007/s00213-015-4117-5.

Tsou, K., Brown, S., Sanudo-Pena, M. C., Mackie, K., & Walker, J. M. (1998). Immunohistochemical distribution of cannabinoid CB1 receptors in the rat central nervous system. *Neuroscience, 83*(2), 393–411.

Tung, L. W., Lu, G. L., Lee, Y. H., Yu, L., Lee, H. J., Leishman, E., ... Chiou, L. C. (2016). Orexins contribute to restraint stress-induced cocaine relapse by endocannabinoid-mediated disinhibition of dopaminergic neurons. *Nature Communications, 7*, 12199. http://dx.doi.org/10.1038/ncomms12199.

Valjent, E., & Maldonado, R. (2000). A behavioural model to reveal place preference to delta 9-tetrahydrocannabinol in mice. *Psychopharmacology (Berlin), 147*(4), 436–438.

Valjent, E., Mitchell, J. M., Besson, M. J., Caboche, J., & Maldonado, R. (2002). Behavioural and biochemical evidence for interactions between Delta 9-tetrahydrocannabinol and nicotine. *British Journal of Pharmacology, 135*(2), 564–578. http://dx.doi.org/10.1038/sj.bjp.0704479.

Valverde, O., Noble, F., Beslot, F., Dauge, V., Fournie-Zaluski, M. C., & Roques, B. P. (2001). Delta9-tetrahydrocannabinol releases and facilitates the effects of endogenous enkephalins: Reduction in morphine withdrawal syndrome without change in rewarding effect. *The European Journal of Neuroscience, 13*(9), 1816–1824.

Vaughn, L. K., Mantsch, J. R., Vranjkovic, O., Stroh, G., Lacourt, M., Kreutter, M., ... Hillard, C. J. (2012). Cannabinoid receptor involvement in stress-induced cocaine reinstatement: Potential interaction with noradrenergic pathways. *Neuroscience, 204*, 117–124. http://dx.doi.org/10.1016/j.neuroscience.2011.08.021.

Vigano, D., Rubino, T., & Parolaro, D. (2005). Molecular and cellular basis of cannabinoid and opioid interactions. *Pharmacology, Biochemistry, and Behavior, 81*(2), 360–368. http://dx.doi.org/10.1016/j.pbb.2005.01.021.

Vigano, D., Valenti, M., Cascio, M. G., Di Marzo, V., Parolaro, D., & Rubino, T. (2004). Changes in endocannabinoid levels in a rat model of behavioural sensitization to morphine. *The European Journal of Neuroscience, 20*(7), 1849–1857. http://dx.doi.org/10.1111/j.1460-9568.2004.03645.x.

Vinklerova, J., Novakova, J., & Sulcova, A. (2002). Inhibition of methamphetamine self-administration in rats by cannabinoid receptor antagonist AM 251. *Journal of Psychopharmacology, 16*(2), 139–143.

Vinod, K. Y., Sanguino, E., Yalamanchili, R., Manzanares, J., & Hungund, B. L. (2008). Manipulation of fatty acid amide hydrolase functional activity alters sensitivity and dependence to ethanol. *Journal of Neurochemistry, 104*(1), 233–243. http://dx.doi.org/10.1111/j.1471-4159.2007.04956.x.

Vinod, K. Y., Yalamanchili, R., Xie, S., Cooper, T. B., & Hungund, B. L. (2006). Effect of chronic ethanol exposure and its withdrawal on the endocannabinoid system. *Neurochemistry International, 49*(6), 619–625. http://dx.doi.org/10.1016/j.neuint.2006.05.002.

Wang, H., Treadway, T., Covey, D. P., Cheer, J. F., & Lupica, C. R. (2015). Cocaine-induced endocannabinoid mobilization in the ventral tegmental area. *Cell Reports, 12*(12), 1997–2008. http://dx.doi.org/10.1016/j.celrep.2015.08.041.

Wang, L., Zou, F., Zhai, T., Lei, Y., Tan, S., Jin, X., ... Yang, Z. (2016). Abnormal gray matter volume and resting-state functional connectivity in former heroin-dependent individuals abstinent for multiple years. *Addiction Biology, 21*(3), 646–656. http://dx.doi.org/10.1111/adb.12228.

Ward, S. J., Rosenberg, M., Dykstra, L. A., & Walker, E. A. (2009). The CB1 antagonist rimonabant (SR141716) blocks cue-induced reinstatement of cocaine seeking and other context and extinction phenomena predictive of relapse. *Drug and Alcohol Dependence, 105*(3), 248–255. http://dx.doi.org/10.1016/j.drugalcdep.2009.07.002.

Warnault, V., Houchi, H., Barbier, E., Pierrefiche, O., Vilpoux, C., Ledent, C., ... Naassila, M. (2007). The lack of CB1 receptors prevents neuroadaptations of both NMDA and GABA(A) receptors after chronic ethanol exposure. *Journal of Neurochemistry, 102*(3), 741–752. http://dx.doi.org/10.1111/j.1471-4159.2007.04577.x.

Wilkerson, J. L., Niphakis, M. J., Grim, T. W., Mustafa, M. A., Abdullah, R. A., Poklis, J. L., ... Lichtman, A. H. (2016). The selective monoacylglycerol lipase inhibitor MJN110 produces opioid-sparing effects in a mouse neuropathic pain model. *The Journal of Pharmacology and Experimental Therapeutics, 357*(1), 145–156. http://dx.doi.org/10.1124/jpet.115.229971.

Wills, K. L., Vemuri, K., Kalmar, A., Lee, A., Limebeer, C. L., Makriyannis, A., ... Parker, L. A. (2014). CB1 antagonism: Interference with affective properties of acute naloxone-precipitated morphine withdrawal in rats. *Psychopharmacology (Berlin), 231*(22), 4291–4300. http://dx.doi.org/10.1007/s00213-014-3575-5.

Wilson, A. A., Garcia, A., Parkes, J., Houle, S., Tong, J., & Vasdev, N. (2011). [11C]CURB: Evaluation of a novel radiotracer for imaging fatty acid amide hydrolase by positron emission tomography. *Nuclear Medicine and Biology, 38*(2), 247–253. http://dx.doi.org/10.1016/j.nucmedbio.2010.08.001.

Wiskerke, J., Pattij, T., Schoffelmeer, A. N., & De Vries, T. J. (2008). The role of CB1 receptors in psychostimulant addiction. *Addiction Biology, 13*(2), 225–238. http://dx.doi.org/10.1111/j.1369-1600.2008.00109.x.

Xi, Z. X., Gilbert, J. G., Peng, X. Q., Pak, A. C., Li, X., & Gardner, E. L. (2006). Cannabinoid CB1 receptor antagonist AM251 inhibits cocaine-primed relapse in rats: Role of glutamate in the nucleus accumbens. *The Journal of Neuroscience, 26*(33), 8531–8536. http://dx.doi.org/10.1523/JNEUROSCI.0726-06.2006.

Xi, Z. X., Peng, X. Q., Li, X., Song, R., Zhang, H. Y., Liu, Q. R., ... Gardner, E. L. (2011). Brain cannabinoid CB(2) receptors modulate cocaine's actions in mice. *Nature Neuroscience, 14*(9), 1160–1166. http://dx.doi.org/10.1038/nn.2874.

Xi, Z. X., Spiller, K., Pak, A. C., Gilbert, J., Dillon, C., Li, X., ... Gardner, E. L. (2008). Cannabinoid CB1 receptor antagonists attenuate cocaine's rewarding effects: Experiments with self-administration and brain-stimulation reward in rats. *Neuropsychopharmacology, 33*(7), 1735–1745. http://dx.doi.org/10.1038/sj.npp.1301552.

Yamaguchi, T., Hagiwara, Y., Tanaka, H., Sugiura, T., Waku, K., Shoyama, Y., ... Yamamoto, T. (2001). Endogenous cannabinoid, 2-arachidonoylglycerol, attenuates naloxone-precipitated withdrawal signs in morphine-dependent mice. *Brain Research, 909*(1–2), 121–126.

Yu, L. L., Zhou, S. J., Wang, X. Y., Liu, J. F., Xue, Y. X., Jiang, W., ... Lu, L. (2011). Effects of cannabinoid CB(1) receptor antagonist rimonabant on acquisition and reinstatement of psychostimulant reward memory in mice. *Behavioural Brain Research, 217*(1), 111–116. http://dx.doi.org/10.1016/j.bbr.2010.10.008.

Zhang, H. Y., Bi, G. H., Li, X., Li, J., Qu, H., Zhang, S. J., ... Liu, Q. R. (2015). Species differences in cannabinoid receptor 2 and receptor responses to cocaine self-administration in mice and rats. *Neuropsychopharmacology, 40*(4), 1037–1051. http://dx.doi.org/10.1038/npp.2014.297.

Zhang, H.Y., Gao, M., Liu, Q. R., Bi, G. H., Li, X., Yang, H. J., ... Xi, Z. X. (2014). Cannabinoid CB2 receptors modulate midbrain dopamine neuronal activity and dopamine-related behavior in mice. *Proceedings of the National Academy of Sciences of the United States of America, 111*(46), E5007–E5015. http://dx.doi.org/10.1073/pnas.1413210111.

Zhang, H. Y., Gao, M., Shen, H., Bi, G. H., Yang, H. J., Liu, Q. R., ... Xi, Z. X. (2016). Expression of functional cannabinoid CB2 receptor in VTA dopamine neurons in rats. *Addiction Biology.* http://dx.doi.org/10.1111/adb.12367.

Zheng, L., Wu, X., Dong, X., Ding, X., & Song, C. (2015). Effects of chronic alcohol exposure on the modulation of ischemia-induced glutamate release via cannabinoid receptors in the dorsal hippocampus. *Alcoholism, Clinical and Experimental Research, 39*(10), 1908–1916. http://dx.doi.org/10.1111/acer.12845.

Zou, F., Wu, X., Zhai, T., Lei, Y., Shao, Y., Jin, X., ... Yang, Z. (2015). Abnormal resting-state functional connectivity of the nucleus accumbens in multi-year abstinent heroin addicts. *Journal of Neuroscience Research, 93*(11), 1693–1702. http://dx.doi.org/10.1002/jnr.23608.

INDEX

'Note: Page numbers followed by "f" indicate figures and "t" indicate tables.'

Printed in the United States
By Bookmasters